Ecotourism, fourth edition

Ecotourism continues to be embraced as the antithesis of mass tourism because of its promise of achieving sustainability through conservation mindedness, community development, education and learning, and the promotion of nature-based activities that are sensitive to both ecological and social systems. The extent to which this promise has been realised is open to debate.

Focusing on an array of economic, social and ecological inconsistencies that continue to plague ecotourism in theory and practice, the volume examines ecotourism in reference to other related forms of tourism, impacts, conservation, sustainability, education and interpretation, policy and governance, and the ethical imperative of ecotourism as these apply to the world's greenest form of tourism. Ecotourism is a growing field attracting increasing attention from students and academics. David Fennell provides an authoritative and comprehensive review of the most important issues that continue to both plague ecotourism and make it one of the most dynamic sectors in the tourism industry. It covers a comprehensive range of themes and geographical regions.

Building on the success of prior editions, *Ecotourism* has been revised throughout to incorporate recent research and benefits from the introduction of real-life case studies and summaries of recent literature. An essential reference for those interested in ecotourism, the book is accessible to students but retains the depth required for use by researchers and practitioners in the field. New chapters on the theory and application of animal ethics; community development in sustainable tourism; and education and learning in the field have added further value to an already very comprehensive book. It will be of interest to students across a range of disciplines including geography, economics, business, ethics, biology and environmental studies.

David A. Fennell is a Professor in Tourism Management at Brock University in Ontario, Canada. His previous books include *Tourism and Animal Ethics*, *Tourism Ethics* and *Ecotourism Programme Planning*. David is also the founding Editor in Chief of the *Journal of Ecotourism*.

'Now in its fourth edition, David Fennell's *Ecotourism* has become a guidebook in its own right. No other work offers a more thoughtful or thorough history of the ideas and ideals that have informed the study and practice of ecotourism. Fennell is a pioneer in the field, and he writes with the authority of someone who's followed and documented the many successes and failures of ecotourism that have kept it a topic of debate for so many years. This is an important text for those seeking to understand the potential value and ethics of ecotourism for communities, economies, and ecosystems.'

Dr Amanda Stronza, *Texas A&M University, USA*

'Fennell captures the very essence of ecotourism by emphasising the ethical imperative, natural history, and the deeper integration with sustainable development. His comprehensive approach to the conceptual and philosophical foundations, as well as the thoughtful consideration of the issues associated with ecotourism help the reader to integrate theory and practice. Now in its fourth edition, *Ecotourism* remains a must for students, researchers and practitioners in the field.'

Professor Jarkko Saarinen, *University of Oulu, Finland, and University of Johannesburg, South Africa*

'David Fennell's latest contribution to ecotourism is a masterpiece. He has unravelled the many complexities of the topic and presented them in three easy to understand parts. This book is essential reading for everyone associated with ecotourism; I am confident it will become a classic in the field.'

Professor Ross K. Dowling OAM, *Edith Cowan University, Australia*

Ecotourism

Fourth edition

David A. Fennell

Routledge
Taylor & Francis Group

LONDON AND NEW YORK

First published 2015
by Routledge
2 Park Square, Milton Park, Abingdon, Oxon OX14 4RN

and by Routledge
711 Third Avenue, New York, NY 10017

Routledge is an imprint of the Taylor & Francis Group, an informa business

British Library Cataloguing in Publication Data
A catalogue record for this book is available from the British Library

Library of Congress Cataloging in Publication Data
Fennell, David A., 1963-
Ecotourism / David A. Fennell. – Fourth edition.
    pages cm
ISBN 978-0-415-82964-9 (hardback) – ISBN 978-0-415-82965-6 (paperback)
– ISBN 978-0-203-38211-0 (e-book)   1. Ecotourism.   I. Title.
G156.5.E26F465 2014
338.4'791–dc23

                                                      2014012615

ISBN: 978-0-415-82964-9 (hbk)
ISBN: 978-0-415-82965-6 (pbk)
ISBN: 978-0-203-38211-0 (ebk)

Typeset in Times New Roman
by RefineCatch Limited, Bungay, Suffolk

To my family

Contents

Plates

Figures

Tables

 **Case studies**

 **Preface**

One of the benefits of coming to ecotourism in its earliest days as a field of study is that I have seen first-hand how the concept has evolved over time. During the latter part of the 1980s, ecotourism was defined by a rather restricted range of opportunities in a few charismatic destinations that were essential in defining the nature of the experience. The market, typically birdwatchers and scientists, was much more predictable for these types of experiences: ecotourists were affiliated with conservation organisations; they heavily invested in the gear that would allow them to better capture these travel experiences; they travelled as ecotourists frequently; and they were long-staying, well educated, financially well off and allocentric in their travel desires. Ecotourism ('eco' standing for ecological) also represented a promising platform for conservationists to affect meaningful change in an industry that was increasingly defined by negative socio-cultural and ecological change. The allure of this new type of travel, no doubt stemming from the onset of sustainable development and the media hype generated from its coverage, gave way to an expanding market clamouring to take advantage of new alternative tourism opportunities in places that were virtually *terra incognita*. Concurrent to this growth in demand were opportunities for service providers to develop their own small-scale ecotourism packages, in their own little corner of the world, without being tied to large tourism organisations. Not surprisingly, the growth in ecotourism supply and demand over such a relatively short period of time has been accompanied by some very serious philosophical and practical inconsistencies that continue to plague ecotourism in study and practice. For example, in a sobering account of her travel experience in the Peruvian rainforest, Arlen (1995) writes that ecotourism has reached a critical juncture in its evolution. She speaks graphically of instances where tourists endured swimming in water with human waste; guides capturing sloths and caiman for tourists to photograph; raw sewage openly dumped into the ocean; mother cheetahs killing their cubs to avoid the harassment of cheetah-chasing tourists; and an ecotourism industry under-regulated with little hope for enforcement. Other writers have recorded similar experiences. Farquharson (1992) argues that ecotourism is a dream that has been severely diluted. She writes that whereas birding once prevailed, ecotourism has fallen into the clutches of many of the mega-resorts like Cancún: the word [ecotourism] changes colour like a chameleon. What began as a concept designed by ecologists to actively prevent the destruction of the environment has become a marketing term for tourism developers who want to publicise clean beaches, fish-filled seas and a bit of culture for when the sunburn begins to hurt (Farquharson 1992: 8). It comes as no surprise therefore that mass tourism industry developers have capitalised on the concept in implementing their own version of ecotourism: one that is defined by a larger and softer market that is perhaps less ecologically knowledgeable and sensitive than their more traditional counterpart.

A second stimulus for undertaking this book was to attempt to represent the vast literature that has emerged on ecotourism. This continues to be an important motive in the fourth edition, where over 200 new references have been incorporated to bring the discussion up to date. Most of these sources come from the *Journal of Ecotourism* – the field's only international periodical – which continues to publish timely articles that stretch our thinking on ecotourism in new and different ways. These references are coupled with scholarly work from outside tourism, and ecotourism more specifically. Interdisciplinary research is vital to the advancement of ecotourism both in theory and how such translates into practice.

In addition to the many new sources included in this edition is the inclusion of a new conceptual framework. This change serves a dual purpose. First, it has helped me to better organise the material throughout the book. As such, some of the discussion in previous editions has been moved to reflect this new direction. Second, and most importantly, I believe the framework will allow students to benefit by virtue of a clearer path, which is focused on (1) identifying the essence of ecotourism, (2) a section on the core criteria used to define ecotourism, and (3) a third section dedicated to many of the most important issues in ecotourism theory and practice.

Acknowledgements

I would like to acknowledge the work of several people at Routledge, including Andrew Mould, Faye Leerink, and especially Carrie Bell and Sarah Gilkes for coordinating the project. Mandy Gentle got me back on track through the copyediting stage. All of the team were very professional, on so many levels, in my efforts to pull together all the various strands required to improve upon previous editions. Thanks are also extended to Mike Fennell for providing an excellent photograph for the cover. I wish him well on his journey.

Every effort has been made to contact copyright holders and we apologise for any inadvertent omissions. If any acknowledgement is missing it would be appreciated if contact could be made care of the publishers so that this can be rectified in any future edition.

The essence of ecotourism

This book is divided into three main parts (Figure P1). The first part discusses the nature of ecotourism and how it differs from other types of tourism, along with a more detailed investigation of the ecotourist. Part II focuses on the core criteria used to define ecotourism, and Part III focuses on many of the main topics and issues that are important to ecotourism both in theory and practice.

Part I deals with the *essence* of ecotourism, or the nature of ecotourism. I have characterised this as travel with a primary interest in the natural history of a destination. In exploring this dimension recently I found assistance in looking more specifically at the literature on biology and natural history (Fennell 2012d). Natural history is that branch of science premised on observational rather than experimental practices. According to Bartholomew (1986: 326), 'A student of natural history, or a naturalist, studies the world by observing plants and animals directly . . .' From this perspective, Wilcove and Eisner (2000) add that natural history is broadly the observation of different organisms, including their evolution and behaviour as well as how they interact with other species.

Explained as such, ecotourists may be regarded as students of natural history. They are motivated to pay close attention through observation of organisms, their role and function within the environment, and those of a more dedicated kind (hard-path ecotourists) do this through patient observation. Furthermore, it would seem logical to characterise ecotourists as naturalists rather than ecologists, because of the observational tendencies of the former and the experimental practices of the latter. (See Schmidley 2005 for a discussion on how natural history needs to become more scientific and theory-based.) Natural history,

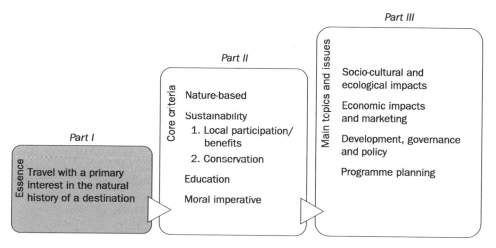

Figure P1 The structure of ecotourism

and those who practise it, is taken to encompass the following areas in the broadest sense: botany, general biology, geology, palaeontology and zoology. It follows that the ecotourist, according to this manner of viewing ecotourism, would be interested in these types of attractions, and not just wildlife as in the case of wildlife tourism. For those theorists looking for a line of demarcation between ecotourism and wildlife tourism, the foregoing may be of some use.

Part I contains two chapters. The first chapter focuses on the nature of ecotourism and, in particular, on investigating ecotourism as a distinct form of tourism as compared to other types. This includes an in-depth look at the roots of ecotourism as well as a series of different definitions that have been used in the past. Mass and alternative forms of tourism are also examined. The second chapter places emphasis on understanding the ecotourist, especially in regard to how this particular type of traveller differs from other types of tourists along the lines of motivations, expectations and behaviours.

There are questions about whether ecotourism is in fact a distinct market or not, and this discussion is examined at length. This discussion provides the background for a more specific look at the core criteria of ecotourism including: (1) the nature-based foundation of ecotourism; (2) the sustainability dimension of ecotourism from the perspective of conservation; (3) the human dimension of sustainability in the form of local participation and benefits; (4) learning and education as part of the ecotourism experience; and (5) the ethical imperative.

1 The nature of ecotourism

In this chapter the structure of the tourism industry is discussed, with more of a focus on attractions as fundamental components of the tourist experience. Both mass tourism and alternative tourism (AT) paradigms are introduced for the purpose of exploring the range of different approaches to tourism planning, development and management. As a form of AT, ecotourism is introduced and defined, and emphasis is placed on exploring ecotourism's roots; that is, how ecotourism has evolved over time. Initial steps are taken towards differentiating ecotourism from other forms of tourism through a discussion on the introduction of key defining criteria. This discussion provides the necessary backdrop from which to examine more closely the ecotourist, in Chapter 2.

Tourism

As one of the world's largest industries, tourism is associated with many of the prime sectors of the world's economy. According to Yeoman *et al.* (2006) tourism has had an average annual increase of 6.6 per cent over the last half century, with international travel rising from 25 million in 1950 to over 700 million by 2002. In 2012, the number of tourists crossing international borders reached 1.035 billion, up from 995 million in 2011 (UNWTO 2013). More specifically, and of interest to our discussion of ecotourism to follow, is the fact that in 1950 the top five travel destination (in Europe and the Americas) held 71 per cent of the travel market, but by 2002 they held only 35 per cent. Yeoman *et al.* ascribe this to an increasing desire to visit new places, which in turn has been stimulated by an emergence of newly accessible destinations in Asia, Africa, the Middle East and the Pacific.

Because of this magnitude, tourism has proved difficult to define because of its reliance on primary, secondary and tertiary levels of production and service, and the fact that it is so intricately interwoven into the fabric of life economically, socio-culturally and environmentally. This difficulty is mirrored in a 1991 issue of *The Economist*:

> There is no accepted definition of what constitutes the [tourism] industry; any definition runs the risk of either overestimating or underestimating economic activity. At its simplest, the industry is one that gets people from their home to somewhere else (and back), and which provides lodging and food for them while they are away. But that does not get you far. For example, if all the sales of restaurants were counted as travel and tourism, the figure would be artificially inflated by sales to locals. But to exclude all restaurant sales would be just as misleading.

It is this complex integration within our socio-economic system, according to Clawson and Knetsch (1966) and Mitchell (1984) that complicates efforts to define tourism.

Tourism studies are often placed poles apart in terms of philosophical approach, methodological orientation or intent of the investigation. A variety of tourism definitions, each with disciplinary attributes, reflect research initiatives corresponding to various fields. For example, tourism shares strong fundamental characteristics and theoretical foundations with the recreation and leisure studies field. According to Jansen-Verbeke and Dietvorst (1987) the terms 'leisure', 'recreation' and 'tourism' represent a type of loose, harmonious unity which focuses on the experiential and activity-based features that typify these terms. On the other hand, economic and technical/statistical definitions generally ignore the human experiential elements of the concept in favour of an approach based on the movement of people over political borders and the amount of money generated from this movement.

It is this relationship with other disciplines; for example, psychology, sociology, anthropology, geography and economics, which seems to have defined the complexion of tourism. However, despite its strong reliance on such disciplines, some, including Leiper (1981), have advocated a move away in favour of a distinct tourism discipline. To Leiper the way in which we need to approach the tourism field should be built around the structure of the industry, which he considers as an open system of five elements interacting with broader environments: (1) a dynamic human element; (2) a generating region; (3) a transit region; (4) a destination region; and (5) the tourist industry. This definition is similar to one established by Mathieson and Wall (1982), who see tourism as comprising three basic elements: (1) a dynamic element, which involves travel to a selected destination; (2) a static element, which involves a stay at the destination; and (3) a consequential element, resulting from the above two, which is concerned with the effects on the economic, social and physical subsystems with which the tourist is directly or indirectly in contact. Others, including Mill and Morrison, define tourism as a system of interrelated parts. The system is 'like a spider's web – touch one part of it and reverberations will be felt throughout' (Mill and Morrison 1985: xix). Included in their tourism system are four component parts, including Market (reaching the marketplace), Travel (the purchase of travel products), Destination (the shape of travel demand) and Marketing (the selling of travel).

In recognition of the difficulty in defining tourism, Smith (1990a) feels that it is more realistic to accept the existence of a number of definitions, each designed to serve different purposes. This may in fact prove to be the most practical of approaches to follow. In this book, tourism is defined as the interrelated system that includes tourists and the associated services that are provided and utilised (facilities, attractions, transportation and accommodation) to aid in their movement, while a tourist, as established by the World Tourism Organization, is defined as a person travelling for pleasure for a period of at least one night, but not more than one year for international tourists and six months for persons travelling in their own countries, with the main purpose of the visit being other than to engage in activities for remuneration in the place(s) visited.

Tourism attractions

The tourism industry includes a number of key elements that tourists rely on to achieve their general and specific goals and needs within a destination. Broadly categorised, they include facilities, accommodation, transportation and attractions, as noted above. Although an in-depth discussion of each is beyond the scope of this book, there is value in elaborating upon the importance of tourism attractions as a fundamental element of the tourist experience. These may be loosely categorised as cultural (e.g. historical sites, museums), natural (e.g. parks, flora and fauna), events (e.g. festivals, religious events), recreation (e.g. golf, hiking) and entertainment (e.g. theme parks, cinemas), according to Goeldner *et al.* (2000). Past tourism research has tended to rely more on the understanding of attractions, and how they affect tourists, than of other components of the industry. As Gunn has suggested, 'they [attractions] represent the most important reasons for travel to destinations' (1972: 24).

MacCannell described tourism attractions as, 'empirical relationships between a tourist, a site and a marker' (1989: 41). The tourist represents the human component, the site includes the actual destination or physical entity, and the marker represents some form of information that the tourist uses to identify and give meaning to a particular attraction. Lew (1987), however, took a different view, arguing that under the conditions of tourist-site-marker, virtually anything could become an attraction, including services and facilities. Lew chose to emphasise the objective and subjective characteristics of attractions by suggesting that researchers ought to be concerned with three main areas of the attraction:

1   *Ideographic.* Describes the concrete uniqueness of a site. Sites are individually identified by name and usually associated with small regions. This is the most frequent form of attraction studied in tourism research.
2   *Organisational.* The focus is not on the attractions themselves, but rather on their spatial capacity and temporal nature. Scale continua are based on the size of the area which the attraction encompasses.
3   *Cognitive.* A place that fosters the feeling of being a tourist, attractions are places that elicit feelings related to what Relph (1976) termed 'insider' 'outsider', and the authenticity of MacCannell's (1989) front and back regions.

Leiper (1990:381) further added to the debate by adapting MacCannell's model into a systems definition. He wrote that:

> A tourist attraction is a systematic arrangement of three elements: a person with touristic needs, a nucleus (any feature or characteristic of a place they might visit) and at least one marker (information about the nucleus).

The type of approach established by Leiper is also reflected in the efforts of Gunn (1972), who has written at length on the importance of attractions in tourism research. Gunn produced a model of tourist attractions that contained three separate zones, including: (1) the nuclei, or core of the attraction; (2) the inviolate belt, which is the space needed to set the nuclei in a context; and (3) the zone of closure, which includes desirable tourism infrastructure such as toilets and information. Gunn argued that an attraction missing one of these zones will be incomplete and difficult to manage.

Some authors, including Pearce (1982), Gunn (1988) and Leiper (1990), have made reference to the fact that attractions occur on various hierarchies of scale, from very specific and small objects within a site to entire countries and continents. This scale variability further complicates the analysis of attractions as both sites and regions. Consequently, there exists a series of attraction cores and attraction peripheries, within different regions, between regions, and from the perspective of the types of tourists who visit them. Spatially, and with the influence of time, the number and type of attractions visited by tourists and tourist groups may create a niche; a role certain types of tourists occupy within a vacation destination. Through an analysis of space, time and other behavioural factors, tourists can be fitted into a typology based on their utilisation and travel between selected attractions. One could make the assumption that tourist groups differ on the basis of the type of attractions they choose to visit, and according to how much time they spend at them (see Fennell 1996). The implications for the tourism industry are that often it must provide a broad range of experiences for tourists interested in different aspects of a region. A specific destination region, for example, may recognise the importance of providing a mix of touristic opportunities, from the very specific, to more general interest experiences for the tourists in search of cultural and natural experiences, in urban, rural and back-country settings.

Attractions have also been referred to as sedentary, physical entities of a cultural or natural form (Gunn 1988). In their natural form, such attractions form the basis for distinctive types of tourism which are based predominantly on aspects of the natural world, such as wildlife tourism (see Reynolds and Braithwaite 2001) and ecotourism (see Page and Dowling 2002). For example, to a birdwatcher individual species become attractions of the most specific and most sought-after kind. A case in point is the annual return of a single albatross at the Hermaness National Nature Reserve in Unst, Shetland, Scotland. The albatross has become a major attraction for birder-tourists, while Hermaness, in a broader context, acts as a medium (attraction cluster) by which to present the attraction (bird). Natural attractions can be transitory in space and time, and this time may be measured for particular species in seconds, hours, days, weeks, months, seasons or years. For tourists who travel with the prime reason to experience these transitory attractions, their movement is a source of both challenge and frustration.

Mass tourism and AT: competing paradigms

Tourism has been both lauded and denounced for its capacity to transform regions physically. In the former case, tourism is the provider of long-term development opportunities; in the latter the ecological and sociological disturbance to transformed regions can be overwhelming. While there are many cases describing impacts in the more developed countries, most of the documented cases of the negative impacts of tourism are in the developing world. Young (1983), for example, documented the transformation of a small fishing farming community in Malta by graphically illustrating the extent to which tourism development – through an increasingly complex system of transportation, resort development and social behaviour – overwhelms such areas over time.

These days we are more prone to vilify or characterise conventional mass tourism as a beast, a monstrosity which has few redeeming qualities for the destination region, their people and the natural resource base. Consequently, mass tourism has been criticised for the fact that it dominates tourism within a region owing to its non-local orientation, and the fact that very little money spent within the destination actually stays and generates more income. It is quite often the hotel or mega-resort that is the symbol of mass tourism's domination of a region, which are built using non-local products, have little requirement for local food products, and are owned by metropolitan interests. Hotel marketing occurs on the basis of high volume, attracting as many people as possible, often over seasonal periods of time. The implications of this seasonality are such that local people are at times moved in and out of paid positions that are based solely on this volume of touristic traffic. Development exists as a means by which to concentrate people in very high densities, displacing local people from traditional subsistence-style livelihoods (as outlined by Young 1983) to ones that are subservience-based. Finally, the attractions that lie in and around these massive developments are created and transformed to meet the expectations and demands of visitors. Emphasis is often on commercialisation of natural and cultural resources, and the result is a contrived and inauthentic representation of, for example, a cultural theme or event that has been eroded into a distant memory.

The picture of mass tourism painted above is outlined to illustrate the point that the tourism industry has not always operated with the interests of local people and the resource

Plate 1.1 Tourist development at Cancún, Mexico

base in mind. This has been reinforced through much of the tourism research that emerged in the 1980s, which argued for a new, more socially and ecologically benign alternative to mass tourism development. According to Krippendorf (1982), the philosophy behind AT – forms of tourism that advocate an approach opposite to mass conventional tourism – was to ensure that tourism policies should no longer concentrate on economic and technical necessities alone, but rather emphasise the demand for an unspoiled environment and consideration of the needs of local people. This 'softer' approach places the natural and cultural resources at the forefront of planning and development, instead of as an afterthought. Also, as an inherent function, alternative forms of tourism provide the means for countries to eliminate outside influences, and to sanction projects themselves and to participate in their development – in essence, to win back the decision-making power in essential matters rather than conceding to outside people and institutions.

AT is a generic term that encompasses a whole range of tourism strategies (e.g. 'appropriate', 'eco-', 'soft', 'responsible', 'people to people', 'controlled', 'small-scale', 'cottage' and 'green' tourism), all of which purport to offer a more benign alternative to conventional mass tourism (Conference Report 1990, cited in Weaver 1991). Dernoi (1981) illustrates that the advantages of AT will be felt in five ways:

1   There will be benefits for the individual or family: accommodation based in local homes will channel revenue directly to families. Also families will acquire managerial skills.
2   The local community will benefit: AT will generate direct revenue for community members, in addition to upgrading housing standards while avoiding huge public infrastructure expenses.
3   For the host country, AT will help avoid the leakage of tourism revenue outside the country. AT will also help prevent social tensions and may preserve local traditions.

4   For those in the industrialised generating country, AT is ideal for cost-conscious travellers or for people who prefer close contacts with locals.
5   There will be benefits for international relations: AT may promote international, interregional and intercultural understanding.

More specifically, Weaver (1993) has analysed the potential benefits of an AT design from the perspective of accommodation, attractions, market, economic impact and regulation (Table 1.1). This more sensitive approach to tourism development strives to satisfy the needs of local people, tourists and the resource base in a complementary rather than competitive manner. The importance, as well as the challenge, of AT as a softer and more responsible form of tourism is demonstrated by the fact that in Europe, tourism is supposed to double over the course of the next 25 years, with most of this coming in the form of AT (European Commission 2004).

Some researchers, however, are quick to point out that as an option to mass tourism, fully-fledged AT cannot replace conventional tourism simply because of mass tourism's varied and many-sided associated phenomena (Cohen 1987). Instead, it is more realistic to concentrate efforts in attempts to reform the worst prevailing situations, not the development

Table 1.1 *Potential benefits derived from an AT strategy*

Accommodation

- Does not overwhelm the community.
- Benefits (jobs, expenditures) are more evenly distributed.
- Less competition with homes and businesses for the use of infrastructure.
- A larger percentage of revenues accrue to local areas.
- Greater opportunity for local entrepreneurs to participate in the tourism sector.

Attractions

- Authenticity and uniqueness of community is promoted and enhanced.
- Attractions are educational and promote self-fulfilment.
- Locals can benefit from existence of the attractions even if tourists are not present.

Market

- Tourists do not overwhelm locals in numbers; stress is avoided.
- 'Drought/deluge' cycles are avoided, and equilibrium is fostered.
- A more desirable visitor type.
- Less vulnerability to disruption within a single major market.

Economic impact

- Economic diversity is promoted to avoid single-sector dependence.
- Sectors interact and reinforce each other.
- Net revenues are proportionally higher; money circulates within the community.
- More jobs and economic activity are generated.

Regulation

- Community makes the critical development/strategy decisions.
- Planning to meet ecological, social and economic carrying capacities.
- Holistic approach stresses integration and well-being of community interests.
- Long-term approach takes into account the welfare of future generations.
- Integrity of foundation assets is protected.
- Possibility of irreversibilities is reduced.

Source: Weaver (1993)

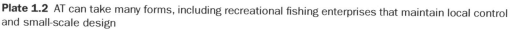

Plate 1.2 AT can take many forms, including recreational fishing enterprises that maintain local control and small-scale design

of alternatives. Butler (1990) feels that mass tourism has not been rejected outright for two main reasons. The first is economic, in that it provides a significant amount of foreign exchange for countries; the second is socio-psychological and relates to the fact that

> many people seem to enjoy being a mass tourist. They actually like not having to make their own travel arrangements, not having to find accommodation when they arrive at a destination, being able to obtain goods and services without learning a foreign language, being able to stay in reasonable, in some cases considerable comfort, being able to eat reasonably familiar food, and not having to spend vast amounts of money or time to achieve these goals.
>
> (Butler 1990: 40)

Ecotourism's roots

Until recently, there has been some confusion surrounding the etymology or origin of the term 'ecotourism', as evident in the tremendous volume of literature on the topic. For example, Orams (1995) and Hvenegaard (1994) write that the term can be traced back only to the late 1980s, while others (Higgins 1996) suggest that it can be traced to the late 1970s through the work of Miller (1989) on ecodevelopment. One of the consistent themes emergent in the literature supports the fact that Ceballos-Lascuráin was the first to coin the phrase in the early 1980s (see Thompson 1995). He defined it as, 'traveling to relatively undisturbed or uncontaminated natural areas with the specific objective of studying, admiring, and enjoying the scenery and its wild plants and animals, as well as any existing cultural manifestations (both past and present) found in these areas' (Boo 1990: xiv). Ceballos-Lascuráin himself states that his initial reference to the phrase occurred in 1983, while he was in the process of developing PRONATURA, a non-governmental organisation (NGO) in Mexico (van der Merwe 1996).

Apparently, however, the term has been traced further back to the work of Hetzer (1965), who used it to explain the intricate relationship between tourists and the environments and cultures in which they interact. Hetzer identified four fundamental pillars that need to be followed for a more responsible form of tourism. These include: (1) minimum environmental impact; (2) minimum impact on – and maximum respect for – host cultures; (3) maximum economic benefits to the host country's grassroots; and (4) maximum 'recreational' satisfaction to participating tourists. The development of the concept of ecotourism grew, according to Hetzer (personal communication, October 1997), as a culmination of dissatisfaction with negative approaches to development, especially from an ecological point of view. Nelson (1994) also adopts this particular stand in illustrating that the idea of ecotourism is in fact an old one, which manifested itself during the late 1960s and early 1970s when researchers became concerned over inappropriate use of natural resources.

Even before this time, however, Lothar Machura's (1954) paper, 'Nature protection and tourism: with particular reference to Austria', was perhaps the first academic work to plant the seeds of the relationship between tourism and conservation. He discussed how tourism could cooperate with nature protection or how it would be incompatible. Tourism, as Machura wrote, could be an agent to arouse or express a love of nature. Of interest is that fact that a Google Scholar search of Machura's paper as of October 2013 yielded no citations.

In other related research, Fennell (1998) found evidence of Canadian government 'ecotours' which were operational during the mid-1970s. These ecotours centred around the Trans-Canada Highway and were developed on the basis of different ecological zones found along the course of the highway – the first of which was developed in 1976. This Canadian version of ecotourism is felt to be rather progressive for the time despite the lack of an explicit look at low impact, sustainability, community development and the moral philosophy labels that are attached to ecotourism in the present day. The ecotours were developed at a time when the Canadian government felt it important to allow Canadian and foreign travellers to appreciate the human–land relationship in Canada, through the interpretation of the natural environment. Although a set definition of ecotourism was not provided, each of the ecotour guides contains the following foreword:

> Ecotours are prepared by the Canadian Forestry Service to help you, as a traveller, understand the features of the landscape you see as you cross the country. Both natural and human history are described and interpreted. The route covered by the Ecotours is divided into major landscape types, or Ecozones, and a map of each Ecozone shows the location of interesting features (identified by code numbers). While most features can be seen from your car, stops are suggested for some of them. Distances between points of interest are given in kilometres. Where side trips are described, distances are given to the turnoff from the highway. You will derive the maximum value from this Ecotour if you keep a record of the distance travelled and read the information on each point of interest before reaching it.
>
> (Fennell 1998: 32)

This prompted Fennell to suggest that ecotourism most likely has a convergent evolution, 'where many places and people independently responded to the need for more nature travel opportunities in line with society's efforts to become more ecologically minded' (Fennell 1998: 234), as also suggested by Nelson (see above). This evidence comes at a time when researchers have been struggling to find common ground between ecotourism and its relationship to other forms of tourism. (For other early references on ecotourism see Mathieson and Wall 1982; Romeril 1985.)

There seems to be acceptance of the fact that ecotourism was viable long before the 1980s, however, in practice, if not in name. For example, Blangy and Nielson (1993) illustrate that the travel department of the American Museum of Natural History has conducted natural history tours since 1953. Probably the finest examples of the evolution of ecotourism can be found in the African wildlife-based examples of tourism developed in the early twentieth century and, to some, the nature tourism enterprises of the mid-nineteenth century (Wilson 1992). Machura's paper, above, may have been a reaction to these natural history-based tours. Furthermore, there are numerous references to the fact that human beings, at least since the Romantic period, have travelled to the wilderness for intrinsic reasons. Nash writes that during the nineteenth century many people travelled both in Europe and North America for the primary purpose of enjoying the outdoors, as illustrated in the following passage:

> Alexis de Tocqueville resolved to see wilderness during his 1831 trip to the United States, and in Michigan Territory in July the young Frenchman found himself at last on the fringe of civilization. But when he informed the frontiersmen of his desire to travel for pleasure into the primitive forest, they thought him mad. The Americans required considerable persuasion from Tocqueville to convince them that his interests lay in matters other than lumbering or land speculation.
>
> (Nash 1982: 23)

Tocqueville was after something that we consider as an essential psychological factor in travel: novelty. Nash (1982) credits the intellectual revolution in the eighteenth and nineteenth centuries as the push needed to inspire the belief that unmodified nature could act as a deep spiritual and psychological tonic. It required the emergence of a group of affluent and cultured persons who largely resided in urban environments to garnish this appreciation (e.g. Jean-Jacques Rousseau and John Ruskin). For these people, Nash (1982: 347) writes, 'wilderness could become an intriguing novelty and even a deep spiritual and psychological need'. In the USA, the sentiment at the time was not as strong as it was in Europe, where, 'as late as the 1870s almost all nature tourists on the American frontier continued to be foreigners' (1982: 348).

When Americans did start travelling to the wild parts of their country it was the privileged classes that held the exclusive rights. A trip to Yellowstone in the 1880s, according to O'Gara (1996), was about three times as expensive as travel to Europe at the time. There was no question that those from the city were especially taken with Yellowstone's majesty, but their mannerisms left much to be desired, as is evident in an account of such tourists by Rudyard Kipling (1996: 56):

> It is not the ghastly vulgarity, the oozing, rampant Bessemer steel self-sufficiency and ignorance of the men that revolts me, so much as the display of these same qualities in the womenfolk. . . . All the young ladies . . . remarked that [Old Faithful] was 'elegant' and betook themselves to writing their names in the bottoms of the shallow pools. Nature fixes the insult indelibly, and the after-years will learn that 'Hattie,' 'Sadie,' 'Mamie,' 'Sophie,' and so forth, have taken out their hairpins and scrawled in the face of Old Faithful.

Defining ecotourism

Given the ambiguity associated with the historical origins of ecotourism, the purpose of the present section is to identify the key principles of the term, especially the link between nature tourism (or nature-oriented tourism) and ecotourism. For example, Laarman and Durst, in their early reference to ecotourism, defined it as a nature tourism in which the 'traveler is drawn to a destination because of his or her interest in one or more features of that destination's natural history. The visit combines education, recreation, and often adventure' (Laarman and Durst 1987: 5). In addition, these authors were perhaps the first to make reference to nature tourism's hard and soft dimensions, based on the physical rigour of the experience and also the level of interest in natural history (Figure 1.1). Laarman and Durst suggested that scientists would in most likelihood be more dedicated than casual in their pursuit of ecotourism, and that some types of ecotourists would be more willing to endure hardships than others in order to secure their experiences. The letter 'B' in Figure 1.1 identifies a harder ecotourism experience based on a more difficult or rigorous experience, and also based on the dedication shown by the ecotourist relative to the interest in the activity. The hard and soft path characteristics have been theoretically positioned in work by Acott *et al.* (1998) on deep and shallow ecotourism. Deep ecotourism is characterised according to intrinsic value, small-scale development, community identity, community participation and the notion that materialism for its own sake is wrong. Conversely, shallow ecotourism is characterised as a business-as-usual attitude to the natural world, nature is seen as a resource to be exploited in maximising human benefits, management decisions are based on utilitarian reasoning and sustainability is viewed from a weak or very weak perspective. The void between deep and shallow ecotourism, acknowledging that each are dichotomous positions on a continuum, prompted the authors to observe that shallow ecotourism verges on mass ecotourism. The only difference, they note, is in the way each is promoted, where shallow ecotourism would make ecotourism claims in its advertising (e.g. wildlife viewing of one sort or another), with profit taking precedence over social and ecological considerations. (See Weaver 2001a and Fennell 2002a for a more in-depth discussion of the hard and soft dimensions of ecotourism.)

A subsequent definition by Laarman and Durst (1993) identifies a conceptual difference between ecotourism and nature tourism. In recognising the difficulties in defining

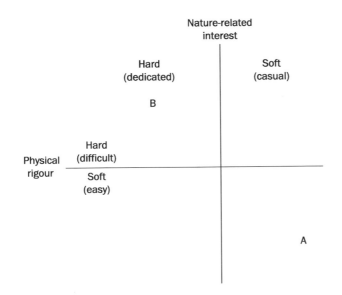

Figure 1.1 Hard and soft dimensions of ecotourism

Source: Laarman and Durst (1987)

nature tourism, they establish both a narrow and broad scope to its definition. Narrowly, they say, it refers to operators running nature-oriented tours; however, broadly it applies to tourism's use of natural resources including beaches and country landscapes. They define nature tourism as 'tourism focused principally on natural resources such as relatively undisturbed parks and natural areas, wetlands, wildlife reserves, and other areas of protected flora, fauna, and habitats' (1993: 2). Given this perspective, there appears to be consensus in the literature that describes ecotourism as one part of a broader nature-based tourism (NBT). This becomes evident in the discussion by Goodwin (1996: 287), who wrote that *nature tourism*:

> encompasses all forms of tourism – mass tourism, adventure tourism, low-impact tourism, ecotourism – which use natural resources in a wild or undeveloped form – including species, habitat, landscape, scenery and salt and fresh-water features. Nature tourism is travel for the purpose of enjoying undeveloped natural areas or wildlife.

And conversely, that *ecotourism* is:

> low impact nature tourism which contributes to the maintenance of species and habitats either directly through a contribution to conservation and/or indirectly by providing revenue to the local community sufficient for local people to value, and therefore protect, their wildlife heritage area as a source of income.
>
> (Goodwin 1996: 288)

The emergence of a basic foundation clarifying the relationship between (NBT) and ecotourism has not, however, precluded the development of numerous definitions of ecotourism, each seeking to find the right mix of terms. Beyond the early definitions discussed above, Ziffer (1989) discussed NBT and ecotourism by first considering a variety of terms, such as 'nature travel', 'adventure travel' and 'cultural travel', which are largely activity-based; and also the value-laden terms, such as 'responsible', 'alternative' and 'ethical' tourism, which underscore the need to consider impacts and the consequences of travel. Ziffer feels that nature tourism, while not necessarily ecologically sound in principle, concentrates more on the motivation and the behaviour of the individual tourist. Conversely, ecotourism is much more difficult to attain owing to its overall comprehensiveness (the need for planning and the achievement of societal goals). She defines ecotourism as follows:

> a form of tourism inspired primarily by the natural history of an area, including its indigenous cultures. The ecotourist visits relatively undeveloped areas in the spirit of appreciation, participation and sensitivity. The ecotourist practises a non-consumptive use of wildlife and natural resources and contributes to the visited area through labor or financial means aimed at directly benefiting the conservation of the site and the economic well-being of the local residents. The visit should strengthen the ecotourist's appreciation and dedication to conservation issues in general, and to the specific needs of the locale. Ecotourism also implies a managed approach by the host country or region which commits itself to establishing and maintaining the sites with the participation of local residents, marketing them appropriately, enforcing regulations, and using the proceeds of the enterprise to fund the area's land management as well as community development.
>
> (Ziffer 1989: 6)

Like Ziffer's, the following definition by Wallace and Pierce (1996: 848) is also comprehensive, acknowledging the importance of a broad number of variables. To these authors, ecotourism is:

travel to relatively undisturbed natural areas for study, enjoyment, or volunteer assistance. It is travel that concerns itself with the flora, fauna, geology, and ecosystems of an area, as well as the people (caretakers) who live nearby, their needs, their culture, and their relationship to the land. it [*sic*] views natural areas both as 'home to all of us' in a global sense ('eco' meaning home) but 'home to nearby residents' specifically. It is envisioned as a tool for both conservation and sustainable development – especially in areas where local people are asked to forgo the consumptive use of resources for others.

Wallace and Pierce (1996; see also Honey 2008 in regard to her seven principles of authentic ecotourism) suggest that tourism may be ecotourism if it addresses six key principles, including:

1   a type of use that minimises negative impacts to the environment and to local people;
2   the awareness and understanding of an area's natural and cultural systems and the subsequent involvement of visitors in issues affecting those systems;
3   the conservation and management of legally protected and other natural areas;
4   the early and long-term participation of local people in the decision-making process that determines the kind and amount of tourism that should occur;
5   directing economic and other benefits to local people that complement rather than overwhelm or replace traditional practices (farming, fishing, social systems, etc.);
6   the provision of special opportunities for local people and nature tourism employees to utilise and visit natural areas and learn more about the wonders that other visitors come to see.

Donohoe and Needham (2006) embarked on an in-depth content analysis of ecotourism definitions and came up with similar results to an earlier and similar study by Fennell (2001). The themes that occurred most consistently in the ecotourism definition literature included: (1) nature-based; (2) preservation; (3) education; (4) sustainability; (5) distribution of benefits; and (6) ethics/responsibility. The absence of many of these core components of a definition of ecotourism has contributed to greenwashing, environmental opportunism and eco-exploitation according to the authors.

In subsequent work, these authors identified a continuum of ecotourism according to the operational congruency with ecotourism tenets (see also Honey 2003). Figure 1.2 shows that genuine ecotourism is that which abides by all of the tenets of ecotourism, while pseudo ecotourism can be characterised in two ways. The first, ecotourism lite, includes those operators or products that apply guidelines some of the time, with a focus on NBT. By contrast, greenwashing includes those products that rarely use the tenets of ecotourism, and where the focus is on marketing as a form of opportunism – taking the opportunity to market oneself as an ecotourism operator but without the intent of living up to the lofty goals of ecotourism.

Donohoe and Needham, Honey, Ziffer, Wallace and Pierce, and others recognise that for ecotourism to succeed it must strive to reach lofty goals. By comparison, however, the Ecotourism Society (now The International Ecotourism Society) advocated a much more general definition of the term; one that advocates a 'middle-of-the-road' or passive position (see Orams 1995), and one that is more easily articulated. This organisation defined ecotourism as, 'responsible travel to natural areas which conserves the environment and improves the welfare of local people' (Western 1993: 8). Preece *et al.* (1995) used the Australian National Ecotourism Strategy definition of ecotourism in their overview of biodiversity and ecotourism, which is also one that is quite general in nature. The strategy defines ecotourism as NBT that involves education and interpretation of the natural environment and is managed to be ecologically sustainable.

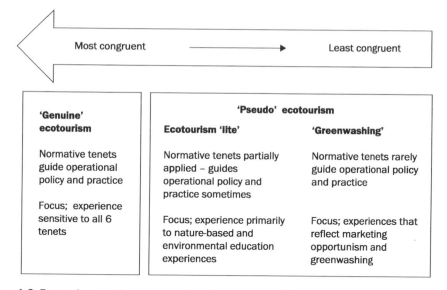

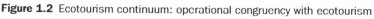

Figure 1.2 Ecotourism continuum: operational congruency with ecotourism

Source: Donohoe and Needham (2008)

These definitions are representative of what appears to be an emerging set of core principles that serve to delineate ecotourism. This core includes sustainability, education, a nature base, and a conservation mandate or orientation (see Blamey 1995; Diamantis 1999; Buckley 1994; Kutay 1989; Wight 1993a; Hawkes and Williams 1993; Wallace and Pierce 1996; Weaver and Lawton 2007). The Quebec Declaration (UNEP/WTO 2002), the penultimate meeting of the International Year of Ecotourism in 2002, suggested that five distinct criteria should be used to define ecotourism, namely: nature-based product, minimal impact management, environmental education, contribution to conservation and contribution to community.

The foregoing, however, also serves to illustrate that there has been no quick or easy formula to define ecotourism, despite the emergence of core criteria. While basic definitions of the term leave much to the interpretation of the reader (see Table 1.2 for an overview of definitions over time); comprehensive definitions risk placing too many constraints on service providers such that the term becomes impossible to implement. If we examine closely some of the weaker definitions of ecotourism we can see how these may be subject to misrepresentation. The following two examples serve to illustrate this point:

Responsible travel that conserves the environment and sustains the well-being of local people.

Responsible travel to natural areas that conserves the environment and improves the welfare of local people.

The first definition was used as a long-standing definition of ecotourism by the International Ecotourism Society of the USA. The second is a definition used by the South Carolina Nature-Based Tourism Association, also of the USA, which defines a type of tourism inclusive of backpacking, boat tours, cycling, farm tours, fishing, hunting and ecotourism. Although defined in a similar context, in reality the two terms are fundamentally different.

Table 1.2 *Comparison of selected ecotourism and nature tourism definitions*

Main principles of definition[a]	Definitions														
	1	2	3	4	5	6	7	8	9	10	11	12	13	14	15
Interest in nature	✓	✓			✓	✓	✓	✓		✓	✓				✓
Contributes to conservation				✓	✓	✓	✓	✓	✓	✓			✓	✓	✓
Reliance on parks and protected areas	✓		✓		✓	✓		✓	✓				✓	✓	✓
Benefits local people/long-term benefits				✓	✓	✓	✓		✓				✓	✓	✓
Education and study	✓	✓	✓			✓					✓			✓	✓
Low impact/non-consumptive						✓						✓	✓	✓	✓
Ethics/responsibility				✓					✓	✓				✓	✓
Management						✓		✓		✓					✓
Sustainable								✓		✓					✓
Enjoyment/appreciation	✓					✓							✓		
Culture	✓					✓							✓		
Adventure			✓										✓		
Small scale											✓				✓

Sources: 1 Ceballos-Lascuráin (1987); 2 Laarman and Durst (1987)[b]; 3 Halbertsma (1988)[b]; 4 Kutay (1989); 5 Ziffer (1989); 6 Fennell and Eagles (1990); 7 CEAC (1992); 8 Valentine (1993); 9 The Ecotourism Society (1993) in Goodwin (1996); 10 Western (1993) in Goodwin (1996); 11 Australian National Ecotourism Strategy (1993) in Goodwin (1996); 12 Brandon (1996); 13 Goodwin (1996); 14 Wallace and Pierce (1996); 15 The present study.

Notes:
[a] Variables ranked by frequency of response
[b] Nature tourism definitions

In the case of the latter, it is a definition, which describes a number of types of tourism that rely on the natural environment. In the words of Weaver (2001a: 350), NBT is, 'any type of tourism that relies mainly on attractions directly related to the natural environment, Ecotourism and 3S tourism are both types of nature-based tourism'. As noted above, ecotourism is then only one of many forms of NBT that rely on the open-air environment. This corroborates what others, have said about the relationship between ecotourism and NBT.

Defining both terms in a similar capacity or treating them both as synonymous, however, has many implications. For example, the province of Saskatchewan in Canada uses a similar definition to the ones described above (Ecotourism is 'responsible travel to areas which conserves the environment and improved the welfare of local people') (Ecotourism Society of Saskatchewan 2000). Using the example of fishing, which is a popular activity in Saskatchewan, such an activity can be responsible, in the implementation of catch limits; it can conserve the environment, in the way hatcheries contribute fish stocks to the lakes and rivers; and it can contribute to the welfare of local people, through the use of an Aboriginal fishing guide. The point is that fishing is certainly acceptable as a form of NBT, but it is questionable as a form of ecotourism because of the practical and philosophical issues surrounding the pursuit and capture of game. In failing to effectively conceptualise ecotourism as a distinct form of NBT, industry stakeholders have misinterpreted and mismarketed ecotourism and in the process created a much bigger – but not necessarily better – industry (more on this in Chapter 7).

But why all the fuss over attempts to arrive at the right definition of ecotourism? Bottrill and Pearce (1995) observe that definitional variables are important because they are often used to observe, measure and evaluate what is and what is not ecotourism (see also Wallace and Pierce 1996, and their evaluation of ecotourism principles in Amazonas). In an analysis of 22 ecotourism ventures, Bottrill and Pearce found that only five were classified as ecotourism using the following criteria: motivation (physical activity, education, participation), sensitive management and protected area status. The authors submit that more work should follow to further define and modify the points and criteria raised. Their paper quite nicely addressed the need to move beyond definition to a position where ecotourism operators should be open to ethical and operational scrutiny by the public and other concerned stakeholders. This resonates with the work of Miller and Kaye (1993: 37), who suggest that 'the merits or deficiencies of ecotourism . . . are not to be found in any label *per se*, but in the quality and intensity of specific environmental and social impacts of human activity in an ecological system'.

In view of the preceding, I offer the following definition of ecotourism, which has emerged from a review of the abundant literature on the topic (see Fennell 2001), as well as personal experience. The definition is thought to be comprehensive enough to avoid being misapplied, but not so wide-ranging to be overly restrictive. Ecotourism, therefore, is:

> Travel with a primary interest in the natural history of a destination. It is a form of nature-based tourism that places about nature first-hand emphasis on learning, sustainability (conservation and local participation/benefits), and ethical planning, development and management.

This definition is structured in recognising that having been identified as a separate form of tourism, ecotourism must be classified and defined as such in order to maintain an element of distinctiveness, even though most ecotourists demand a softer, easily accessible, front-country type of experience (Kearsley 1997 in Weaver 1998), and that the 'popular' form of ecotourism demands mechanised transport, easy accessibility and a high level of services (Queensland Draft Ecotourism Strategy in Weaver 1998). The relationship between this very soft form of ecotourism and other types of tourism is a topic that is discussed later in this book. That said, it is important to consider that much more than simply demand must go into the understanding of ecotourism and ecotourists. In an attempt to stay clear of this fine line, at least at this stage, a harder stance on ecotourism is adopted. Furthermore, a stricter definition of ecotourism begs for the employment of measurable indicators in determining what is and is not ecotourism. (See Orams 1995 for a description of the hard–soft path ecotourism continuum.) An early example of this type of thinking can be seen through the efforts of Shores (1992), who identified the need for higher standards in the ecotourism industry through the implementation of a scale to measure the level of achievement according to the principles of ecotourism. The scale ranges from 0 (travellers made aware of the fragility of the environment in a general capacity) to 5 (a trip where the entire system was operating in an environmental way).

The reader will most certainly recognise the absence of culture as a fundamental principle of ecotourism in the aforementioned definition. This definition views culture only inasmuch as the benefits from ecotourism accrue to local people, recognising that culture, whether exotic or not, is part of any tourism experience. If culture was a primary theme of ecotourism then it would be cultural tourism – not ecotourism. There is no doubt that culture can be part of the ecotourism experience; the point is, however, that it is more likely to be a secondary motivation to the overall experience, not primary as in the case of nature and natural resources. For example, in a study completed by Fennell (1990), it was found that there was no statistically significant difference between the average Canadian traveller and ecotourists as regards many cultural attractions, including museums and art galleries, local festivals and events, and local crafts. Furthermore, in a study on whale watching brochures Kur and Hevenegaard (2012) found that these promotional devices did not emphasise cultural or historical attractions, but tended to favour more the education, natural environment and sustainability of the whale watching experience. The authors concluded that the marketing of culture has limits according to the interests of ecotourists.

Sustainable development plays a vital role in ecotourism because it forces us to consider not only the needs of local people – people who need to have an opportunity to participate in decision-making and who must benefit economically and socially from these decisions – but also the need to conserve or indeed preserve the natural world for now and for future generations. These are values that transcend the interests of the corporation or other stakeholders that would take a more self-interested approach to ecotourism development (Chapters 4 and 5).

Learning about the environment through highly structured environmental education programmes is an essential aspect of the ecotourism experience and past research (see Bachert 1990) has sought to examine how it relates to the need to gain knowledge on-site through interpretation and the information provided by guides and other facilitators. In whale watching, for example, many people just want to see a whale (novelty or curiosity), while conversely others want a more comprehensive learning-based whale watching experience (the focus of Chapter 6). Knowledge can be thought of as information one applies to a situation, whereas learning is something that results from participation. And it is important to view learning in terms of a primary motivation of the ecotourist. Hultman and Cederholm (2006: 78) argue that despite the fact that ecotourism is about learning or acquiring knowledge about nature, ecotourists 'cannot bodily engage with nature in any invasive way; nature must remain pristine'. This discussion on being non-invasive is consistent with the perspective adopted in this book. Even though ecotourists experience nature first-hand, it is a type of interaction that places the interests of the natural world, including individual entities, first over the interests of the ecotourism industry (see also Butler 1992). Experiencing nature first-hand relates to Wilson's (1984: 214) concept of biophilia defined as 'The innate tendency to be attracted by other life forms and to affiliate with natural living systems'. Ecotourism is thus an outlet for this mindset, whereby the feelings of such users may dictate the forms of recreation participated in and the potential to negate effects of such activities on the natural world (see also Kellert 1985 who notes people and the activities they choose can be correlated with these values towards wildlife).

Focusing on the right values is important for an industry that purports to be ethical, which, along with the thoughts of an increasing number of scholars, is seen to be an integral aspect of the definition in theory and practice (the focus of Chapter 7). This means appropriate planning, development and management of ecotourism. This would include ethical marketing, the ethical treatment of animals and other aspects of the natural world, which in the past has translated into discussions on low impact and non-consumptiveness. Acott et al. (1998: 239) may have got it right when they argued that 'There are many problems in trying to define ecotourism without proper attention being paid to underlying philosophical and ethical principles'. One's geographic location (e.g. within a protected area or other space deemed important as an ecotourism region) alone, for example, was not enough to determine the environmental values of that individual. Furthermore, Powell and Ham (2008) bring many of these strands together when they argue that if ecotourism is going to be sustainable it will need to consider 4Es: environmental conservation, education, economic benefits and equity. Finally, Cater (1994) sheds more light on the need for ethical insight when she argues that there must be the willingness on the part of the external forces to give way to the needs and priorities of people who have otherwise been

Plate 1.3 Mayan ruins: major attractions in the peripheral regions of the Yucatan Peninsula, Mexico

marginalised by tourism. These, she says, are moral issues in ecotourism. These latter perspectives are viewed as being essential to ecotourism, and are simply made more explicit in the pages that follow.

Conclusion

The attractiveness of ecotourism to tourism scholars is a theme investigated by McKercher (2010), who argues that academics have played a major role in nurturing an interest in ecotourism far beyond other types of tourism. The proliferation of definitions of ecotourism is one manifestation of this perspective – a fetish or fascination over what ecotourism ought to be. The magnitude of the ecotourism industry has been addressed by the International Ecotourism Society (TIES 2006), which claims that since 1990 ecotourism has experienced annual growth rates of 20–34 per cent; a number which has easily outpaced the tourism industry on the whole. By the year 2024, ecotourism holidays will have grown three times faster than conventional trips, representing 5 per cent of the global market (Starmer-Smith 2004). Much of this sustained growth will have implications for the regions, developed and lesser developed, that play host to the escalating numbers and demands of ecotourists.

As the discussion advances on to ecotourists and other core criteria of ecotourism, it is perhaps useful to identify some of the iconic ecotourism destinations in the developed and lesser developed countries. These include the Monteverde Cloud Forest Reserve, Costa Rica, Galapagos Islands, Ecuador, Iguaçu Falls, the Amazon basin, Patagonia, Antarctica, the Great Barrier Reef in Australia, the Serengeti Plain, Kruger National Park in South Africa and polar bear watching in Churchill, Canada. Students should also know that ecotourism thrives in all environments including rainforests, mountain regions, polar environments, islands and coasts, deserts and grasslands, and marine regions, and includes attractions from the blue whale to lichens and mosses. Many of these places, like the Galapagos Islands, represent a once-in-a-lifetime experience. And while many do not return, their loyalty to the destination is represented in their willingness to recommend a place like Galapagos to others by word-of-mouth (Rivera and Croes 2010).

Summary questions

1   Why is tourism so hard to define?
2   What are some of the characteristics that have been used to define attractions?
3   What are the basic differences between mass tourism and AT?
4   Why is a discussion of values and ethics so important in the understanding of sustainable tourism?

2 The ecotourist

While the preceding chapter placed emphasis on describing the characteristics of eco-tourism, the present chapter focuses on the characteristics of the tourists who choose to travel as such. The chapter begins with a brief discussion and some examples of tourist typologies, before venturing into a more specific treatment of ecotourist typologies. The literature on ecotourist typologies is voluminous, so I have elected to partition this research into sub-sections: general ecotourism typologies, groups based on socio-demographics, psychographic information and finally specialisation. The chapter concludes with a discussion on whether or not there really is a distinct ecotourism segment that is as different than other types of tourists, including mass tourists.

Tourist typologies and profiles

The general literature

Given the demand for new and different travel experiences, the tourism industry has kept pace through the development of a rich array of tourism types. Tourism research has also kept pace through studies that concentrate on the various individual traits, characteristics, motivations, needs and so on of travellers that render them similar or different to other types. This has enabled both researchers and practitioners to better understand tourists on the basis of the types of experiences they seek, as individuals and groups. The following section briefly examines a number of pertinent motivational/behavioural and social/cultural typologies, as a basis for understanding ecotourists.

Christaller effectively grasped the notion that, over time, a destination will play host to a rich variety of different types of travellers. He wrote:

> The typical course of development has the following pattern. Painters search out untouched unusual places to paint. Step by step the place develops as a so-called artist colony. Soon a cluster of poets follows, kindred to the painters; then cinema people, gourmets, and the *jeunesse dorée*. The place becomes fashionable and the entrepreneur takes note. The fisherman's cottage, the shelter huts become converted into boarding houses and hotels come on the scene . . . Only the painters with a commercial inclination who like to do well in business remain; they capitalize on the good name of this former painter's corner and on the gullibility of tourists. More and more townsmen choose this place, now *en vogue* and advertised in the newspapers . . . At last the tourist agencies come with their package rate travelling parties; now, the indulged public avoids such places. At the same time, in other places the cycle occurs again; more and more places come into fashion, change their type, turn into everybody's tourist haunt.
>
> (1963: 103)

While tourist destinations were transformed under the pressure and influence of tourism, the type of tourists also changed. Cohen (1972: 172) commented, echoing Christaller, that 'attractions and facilities which were previously frequented by the local population are gradually abandoned. As Greenwich Village became a tourist attraction, many of the original bohemians moved to the east Village'. Cohen used this as an analogy to demonstrate that travellers were inherently different on the basis of their relationship to both the tourist business establishment and host country. Accordingly, he grouped tourists into four categories, including organised mass tourists, individual mass tourists, explorers and drifters. This typology reflected a continuum such that the organised mass tourist is seen as the least adventuresome with little motivation to leave the confines of his or her home environmental bubble. The drifter, in contrast, shunned the tourist establishment, searching for the most authentic travel experiences available (see Wickens 2002 for a breakdown of five micro-types of Cohen's Individual Mass Tourism type in Greece; see also Smith 1989, and Plog 1973).

MacCannell (1989) also considered the fundamental differences between traveller types in examining the social structure of tourist space based on 'front' and 'back' regions. Front regions are those readily experienced by tourists and places where hosts and guests regularly interact. Conversely, back regions are the preserve of the host and are essentially non-tourism-oriented in their function. Tourists in search of authenticity penetrate back regions in the hope of acquiring real day-to-day mannerisms of residents. To what degree tourists are willing or able to penetrate back regions – in identified access zones of tourism areas – may be important in achieving their overall purpose. Although these 'places' are actual locations in the conventional economic/geographical context, in the mind of the tourist or tourist group they hold special value in defining the travel experience.

Motivation, or the drive to satisfy inner physiological and psychological needs, has been fundamental to tourism researchers interested in the 'why' people travel. In general motivation theory posits that the emergence of a need, motive or drive creates a state of disequilibrium in an individual, which is reducible through participation in an activity that leads to a desired goal. Positive feedback results through satisfying experiences while negative feedback leads to activity cessation or modification (Mannell and Kleiber 1997; see Gnoth 1997). Figure 2.1 provides a graphic illustration of the relationship between these various factors.

Other researchers contend that travel motivation is purely psychological and not sociological in nature. Iso-Ahola (1982) argued that people travel for basically two reasons: (1) to seek intrinsic rewards (novelty), and (2) to escape their everyday environments. These two motivations may be personal (personal troubles or failures) or interpersonal (related to co-workers, family or neighbours). The amalgam of these elements is a

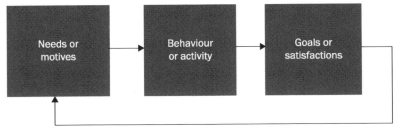

Feedback

Figure 2.1 Model of motivation

Source: Mannell and Kleiber (1997)

four-cell matrix, where a single tourist could theoretically go through one or all aspects during the course of one trip (see also Pearce 1982).

But even more deeply, tourism researchers need to better understand more than just the outward manifestations of travel motivation. More emphasis needs to be placed on understanding the link between physiology and psychology when it comes to motivation. In his exploration into the nature of pleasure in pleasure travel, Fennell (2009) argues that anticipation and novelty are key factors in the onset of pleasure. The neurotransmitter dopamine is responsible for turning on appetitive (goal-directed behaviour) motivation, helping to commit to memory rewards that create incentive value for previously neutral stimuli (Wise 2004). As such, dopamine is the transmitter of choice for the anticipation of rewards like thrill seeking, desire and seeking to engage with the world (Berridge 2003). It has also been shown that the anticipation of an expected reward may be just as pleasurable as the reward itself (Bressan and Crippa 2005). In addition, we often grow tired of the same stimulus after some level of learning and habituation (Phillips 2003), a state referred to as sensory-specific satiety. For those hedonists who thought that pleasure was unending – that there was no natural limit – this appears not to be the case. Activities that delight us initially wear thin over time. This is perhaps why we choose to travel to different destinations, seek different attractions and environments, and enjoy a range of different travel experiences like ecotourism. An explanation of why we have a spectrum of different travel types – and even a spectrum of different ecotourism types – is perhaps wrapped up in this deeper understanding of the nature of pleasure. More research is required in this area.

The ecotourist

General descriptions and socio-demographic research

As research on ecotourism gathered momentum in the late 1980s and early 1990s, theorists grew curious about what it was that compelled ecotourists to travel as such. The work that emerged was both descriptive and empirical, and is characterised by a range of different frameworks and approaches.

An early example of this research comes from Kusler (1991) who argued that ecotourists are affiliated with one of three general types. *Do-it-yourself ecotourists* comprise the largest percentage of all ecotourists; they stay in a variety of different types of accommodation and have the mobility to visit any number of settings. Their experience is marked by a high degree of flexibility. *Ecotourists on tours* expect a high degree of organisation within their tour, and travel to exotic destinations (e.g. Antarctica). The final group, *School groups or scientific groups*, is characterised by the readiness of ecotourists to become involved directly in scientific research by an organisation or individual, stay in the same region for extensive periods of time, and be willing to endure harsher site conditions than other ecotourists (more on this topic in Chapter 6).

In other research, Lindberg (1991: 3) emphasised the importance of dedication and time as a function of defining different types of ecotourists, including what tourists wish to experience from ecotourism, where they wish to travel and how they wish to travel. Lindberg identified four basic types (ecotourism was synonymous with nature tourism at the time):

1   *Hard-core nature tourists:* scientific researchers or members of tours specifically designed for education, removal of litter, or similar purposes;

2   *Dedicated nature tourists:* people who take trips specifically to see protected areas and who want to understand local natural and cultural history;

3   *Mainstream nature tourists:* people who visit the Amazon, the Rwandan gorilla park, or other destinations primarily to take an unusual trip; and

4   *Casual nature tourists:* people who experience nature incidentally as part of a broader trip.

The concept of time has been included in other studies on the ecotourist. Ballantine and Eagles (1994) suggested that ecotourists could be described on the basis of an intention to learn about nature, an intention to visit undisturbed areas and a commitment of at least 33 per cent of their time to the first two criteria. However, Blamey (1995) sees some inherent problems with the implementation of such a measure. The time factor may be applicable in the safari ecotourism settings of Africa, but is thought to be more problematic in less-structured ecotourism settings and situations. For instance, Blamey asks if a ten-minute guided nature tour qualifies as a day ecotourism visit. Yet one cannot help but grapple for acceptable definitions of ecotourism, as Ballantine and Eagles and others have done; the same has occurred with the meaning of 'tourism' in the past. While the tourism literature has seen fit to define 'tourism' under many different circumstances – time, space, economics, whole systems models – the same will likely occur for ecotourism. It depends on who is operationalising the concept, and for what purpose.

In other work, Mowforth (1993) developed a threefold classification of ecotourists on the basis of age, type of travel, organisation of the event, budget and type of tourism. His three types include: (1) the rough ecotourist, who is young to middle-aged, travels individually or in small groups, is independent, on a low budget and seeks sport and adventure; (2) the smooth ecotourist, who is middle-aged to old, travels in groups, depends on a tour operator, has a high budget and seeks nature and safari type experiences; and (3) the specialist ecotourist, who is young or old, travels individually, is independent and relies on specialist tours, has a mid to high budget and ranges from scientific interests to hobby pursuits.

Some of these early studies focused on developing a profile of ecotourists on the basis of age, education and income, as well as other socio-demographic characteristics. Many of these studies pointed to the fact that the ecotourists were predominantly male, well educated, wealthy and longer staying than other types of tourists. For example, Wilson (1987: 21) reported the following in researching 62 tourists visiting Ecuador:

> The male/female ratio was 52 per cent to 48 per cent, and the mean average age was 42. . . Twenty-seven percent earned a family income between US $30,000 to $60,000, before taxes annually. Approximately one-quarter earned more than $90,000 per year. About 30% had bachelors degrees, and a little over 10% had doctoral degrees.

Both Fennell and Smale (1992) and Reingold (1993) report similar results in their work on Canadian ecotourists. On average, the Canadian ecotourists in the Fennell and Smale study were 54 years of age, with the majority in the 60–69 age cohort. The sample was predominantly male (55 per cent), earned on average about CDN$60,000, with almost one-third and two-thirds combined having undergraduate and graduate degrees, respectively. According to these authors, this education is well above the national average of 19 per cent and 4 per cent of Canadians having a bachelor's degree and graduate training, respectively. In Reingold's work, 24 per cent of Canadian ecotourists were 55 to 64 years of age, 36 per cent had annual incomes over $70,000, 65 per cent had a university degree; however, 64 per cent of the respondents of this study were female. This latter statistic goes against much of the previous research suggesting that males are slightly more representative of this group. Furthermore, polar bear viewers in Churchill, Manitoba, Canada, were found to be middle aged to older aged, well educated and financially successful (see MacKay and McIlraith 1997). In a more recent study of this region, Lemelin *et al.* (2002) found that these tourists numbered slightly more as males than females, 44.4 per cent were aged 45–64, 40 per cent earned in excess of $100,000 as a household income, and 88 per cent had a post-secondary degree or diploma.

The trend towards more female representation in ecotourism, as observed by Reingold above, has been discussed by Weaver (2001a) who refers to this change as the 'feminisation' of the ecotourism sector. Diamantis (1999), for example, found that 57 per cent of occasional ecotourists were female, while a sample of ecotourists in Australia by Weaver and Lawton (2002) was 62 per cent female. More dramatically, a study conducted by Fennell (2002a) in Costa Rica found that 70 per cent of Canadians visiting Costa Rica were female. This trend has also been observed in research by Nowaczek and Smale (2005) in a study of 228 ecotourists visiting a community owned and operated site (Posada Amazonas) adjacent to the Tambopata National Reserve in southeastern Peru. These researchers found that 59.2 per cent of their sample was female, 25 per cent had annual household incomes of US$130,000 or above and 47.1 per cent had graduate level of education.

Studies involving ecotourists have been found to mirror related research on birdwatchers and wildlife tourists. For example, Applegate and Clark (1987) report that more men than women birdwatch and that they have strikingly high levels of affluence and education with more than 50 per cent of the respondents of their study completing four years of college. The findings of Kellert (1985) indicate that committed birders were 73 per cent male with an average age of 42. Committed birders were also far better educated (nearly two-thirds had college and/or graduate school education) and had higher incomes than respondents who did not birdwatch. The general outdoor recreation literature supports the fact that most participants – two-thirds or even three-quarters – are of the male gender (Hendee *et al.* 1990).

Psychographic research

Psychographic research, not to be confused with demographic research, involves the study of values, attitudes, lifestyles and various interests of specific segments of society. An example of one of the earliest studies involving psychographics and ecotourist was completed by Fennell (1990) and Fennell and Smale (1992). Fennell used the results of a 1983 Canadian study (Tourism Attitude and Motivation Study (CTAMS)) on general travel attitudes and benefits of the Canadian population, to compare against a sample of Canadian ecotourists. The questions used in the CTAMS study were duplicated and applied to a sample of Canadian ecotourists who had recently returned from a Costa Rican ecotourism trip between 1988 and 1989. Of 98 surveys mailed to the ecotourists (tourists were contacted by obtaining mailing lists of Canadian ecotour operators offering programmes in Costa Rica), 77 were returned and usable. The results of this study are found in Table 2.1.

In general, the findings illustrate that the benefits sought by ecotourists may be found in new, active and adventuresome activities and involvements, while the Canadian population benefits were more strongly related to sedentary activity and family-related endeavours. Ecotourists were found to pursue attractions related to the outdoors (wilderness areas, parks and protected areas, and rural areas), while attractions related to cities and resorts were more important to the average Canadian traveller. The implications of the research are such that they empirically demonstrate that there is indeed a difference between ecotourists and general travellers in terms of their trip-related needs and focus. Kretchman and Eagles (1990) and Williacy and Eagles (1990) followed the initial survey

Table 2.1 *Relative importance of selected attractions and benefits to Canadian travellers and ecotourists*[a]

Variable	General population		Canadian ecotourists	
	Mean[b]	s.d.	Mean	s.d.
Important benefits to ecotourists[c]				
Experiencing new and different lifestyles	2.67	1.02	1.95	0.78
Trying new foods	2.85	0.99	2.43	0.82
Being physically active	2.36	1.03	1.88	0.92
Visiting historical places	2.71	1.04	2.35	0.74
Seeing as much as possible	2.01	0.99	1.77	0.81
Being daring and adventuresome	2.91	0.98	2.62	1.01
Meeting people with similar interests	2.08	0.94	1.91	0.71
Important benefits to population[c]				
Watching sports	3.16	0.96	3.84	0.37
Visiting friends and relatives	1.83	1.03	3.25	0.82
Doing nothing at all	2.86	1.00	3.72	0.60
Being together as a family	1.66	0.98	2.92	1.04
Reliving past good times	2.46	1.06	3.36	0.79
Visiting places my family came from	2.70	1.16	3.45	0.84
Feeling at home away from home	1.81	0.91	2.64	0.93
Having fun and being entertained	1.95	0.88	2.72	0.94
Important attractions to ecotourists[c]				
Wilderness areas	2.34	1.09	1.06	0.37
National parks and reserves	2.21	1.01	1.14	0.35
Rural areas	2.34	0.94	1.49	0.60
Mountains	2.34	1.07	1.50	0.66
Lakes and streams	2.05	0.99	1.57	0.59
Historic sites and parks	2.37	1.01	2.05	0.81
Cultural activities	2.66	0.97	2.32	0.87
Oceanside	2.15	1.07	1.97	0.78
Important attractions to population[c]				
Indoor sports	2.98	0.94	3.85	0.35
Amusement and theme parks	2.74	0.99	3.80	0.54
Nightlife and entertainment	2.72	1.04	3.70	0.56
Gambling	3.61	0.69	3.96	0.34
Shopping	2.45	1.01	3.14	0.74
Resort areas	2.56	1.01	3.26	0.82
Big cities	2.94	0.93	3.39	0.69
Beaches for swimming/sunning	2.34	1.10	2.79	0.96
Predictable weather	2.11	0.97	2.40	0.81
Live theatre and musicals	2.99	1.01	3.21	0.91

Source: Fennell and Smale (1992)

Notes:

[a] Differences between general population and ecotourists are statistically significant at 0.05 level

[b] Scale: 1 = 'very important'; 4 = 'not at all important'

[c] Attractions/benefits are in descending rank order based on magnitude of difference between groups

design of Fennell (1990) in comparing ecotourists in a variety of different settings. These studies were combined by Eagles (1992) in a comprehensive overview of the travel motivations of Canadian ecotourists. His results generally substantiate the findings of Fennell and Smale (1992), above, in suggesting that ecotourists are fundamentally different in their travel motivations from the general traveller.

Another example comes from Hvenegaard (2002), who noted that there are two broad categories of tourist typologies: (1) interactional, including those where tourists interact with the destination; and (2) cognitive-normative, including those which focus on travel motivations, attitudes and values of tourists (after Murphy 1985; see also Juric *et al.* 2002). Hvenegaard studied trekkers, birders and general interest visitors to Doi Inthanon National Park in Thailand in assessing the congruency of four versions of the aforementioned two basic typologies. These included researcher-based (tourists types defined by the researcher), respondent-based (respondents categorised themselves), activity-based (places visited and activities undertaken as the primary inputs) and motivation-based (focusing on the main reason for visiting the park). Hvenegaard concluded by observing that there are consistent interrelationships between activities, motivations and perceptions of tourist type according to: (1) motivations reflect consistent patterns of activities undertaken and sites visited; (2) researcher-based, respondent-based and activity-based typologies are more congruent than the motivation-based one because tourists are motivated to travel based on a number of motivations; and (3) researcher decisions about tourist types was very close to the respondent decisions suggesting that researchers can in fact relate to the populations they are studying.

Weaver (2002) explored the hard–soft continuum of ecotourists through a sample of 1,180 overnight patrons of two ecolodges in Lamington National Park, Australia. Cluster analysis of ecotourism behaviour on 37 items led Weaver to conclude that ecolodge patrons are not a homogeneous group. Three distinct ecotourism types were discerned from Weaver's work including 'Harder' ecotourists, characterised by a high level of environmental commitment, specialised trips, longer trips, smaller groups, being more physically active and requiring fewer services. A 'Softer' ecotourist cluster was defined by moderate environmental commitment, multi-purpose trips, short trips, larger groups, being physically passive and requiring more services. The third cluster, 'Structured' ecotourists, had a strong pattern of environmental commitment but desired interpretation, escorted tours and services and facilities at a level that was more congruent with mass tourism.

As suggested earlier, an effective way to differentiate types of tourists visiting particular regions is through an examination of various personal characteristics such as primary motives, benefits sought and so on, as above. Turnbull (1981), however, provides an interesting argument against such an approach. He suggests that although on the surface nature tourists travelling to Africa state that it is primarily the animals that bring them to the game parks of East Africa, on the basis of his observations he feels that the reasons for visiting Africa run much deeper. As an anthropologist, Turnbull believes that perhaps latently it is also the 'Africa' and the 'Kilimanjaro' that bring such tourists to this setting. In essence, he believes that people want to tap into their distant past when humans had a much stronger relationship with animals, seeing them as something more than just prey to be hunted, and it is to experience something of this relationship that is really desired by present-day tourists (more on the biology of human nature in Chapter 7). He states that tourists:

expect an indivisible, natural whole made up of both human and animal components. But unless they are unusually lucky this is not what tourists find on the organised safari. All that is observed is man-made: game parks devoid of the herders and hunters who used to live there as an indispensable part of the ecosystem.

(Turnbull 1981: 34)

Plate 2.1 Ecotourists in search of whales

Plate 2.2 Ecotourists learning about the natural history of dry tropical rainforests

Specialisation

The concept of specialisation emerged from work by Bryan (1977) on angling – specifically fly-fishing. Bryan argued that in any activity there are categories or classes of participants, and these classes are organised according to setting, equipment and skill set. Those
of a more advanced or specialised nature require different settings, equipment and possess
advanced skills over other lesser-experienced groups participating in the same activity.
More than this, those more specialised would expect to participate with individuals who
maintained the same dedication and expectations of the activity – they defined a separate
social world or separate culture tied to the activity itself. Bryan's specialisation of anglers'
continuum is as follows:

1   *Occasional fishermen.* Anglers who fish infrequently, are new to the activity and have
    not established it as a regular part of their leisure.
2   *Generalists.* Anglers who have established the sport as a regular leisure activity and
    use a variety of techniques.
3   *Technique specialists.* Anglers who specialise in a particular method, to the exclusion
    of other techniques.
4   *Technique-setting specialists.* Highly committed anglers who specialise in a method
    and prefer specific water types.

Fishermen move into more specialised stages over time, with the most specialised fishermen joining a leisure social world according to shared attitudes, beliefs and behaviours.
As specialisation increases, attitudes/values about the sport change with a focus moving
from consumption to preservation.

Specialisation has become a useful tool to explain differences between bird watchers
and other ecotourist types. For example, Eubanks *et al.* (2004) found that there are many
behavioural and motivational differences between birder populations, which allows for a
degree of specialisation. Specialisation has also been used to study the decline in the
birding visitation at Point Pelee National Park, Canada. Maple *et al.* (2010) discovered
that of three birder groups, beginners, intermediate and experienced, it was the beginners
who were more distinct as a group. The beginners were more likely to be in their first year
as a birder, stayed the least number of nights, spent the least amount of money, participated in more non-birding activities both inside and outside the park, and used more
sources of park information than the other two groups. Experienced and intermediate
level groups were looking for more specialised programmes on bird identification, biology
and bird watching. All of this information is deemed important to help park managers
design programmes specific to the needs of these very different groups (more on
programme planning in Chapter 11).

The need for more specialised programme planning for harder-path ecotourists is a
finding that is not restricted to Point Pelee. In a study of ecotourists visiting Madagascar,
Wollenberg *et al.* (2011) found that those tourists with a relatively high zoological interest
were willing to pay higher prices for specialised biodiversity tours. In related work,
Lindsey *et al.* (2007) investigated the viewing preferences of ecotourists according to the
contention that this group is only interested in viewing charismatic mega-fauna. Based on
their analysis of the preferences of ecotourists in four protected areas in South Africa, the
authors found that while large carnivores and herbivores are indeed important to first-time
and overseas visitors, other more experienced visitors showed different interests. These
latter groups appeared to be more interested in bird and plant diversity along with high-
profile mammals and those animals not easily seen. Lindsey and his colleagues write that
this finding is important because a bias towards just charismatic mega-fauna might direct

Plate 2.3 Bird watching is an activity that attracts both experienced and inexperienced ecotourists

conservation resources to these animals only. Interest in other lower profile animals may provide the incentive for the conservation of these animals and their habitat. In this capacity, Lemelin (2007) has also challenged the view that 'specialised' in ecotourism refers only to a specific interest in birds or large fauna. He argues that there is a significant market of eco-tourists interested in Odonata (the dragonflies and damselflies), and these ecotourists pride themselves just as much on the identification of different types as birders do. In addition, there are increasing numbers of symposia and festivals to support the pursuit of this activity.

A recent addition to the ecotourism literature incorporating Bryan's continuum is work by Lemelin *et al.* (2008) on polar bear viewers at the Churchill Wildlife Management Area. Viewers were examined using a comprehensive index of specialisation and compared according to selected demographic variables and indicators of environmental concern. The results suggest that visitors reflect a wide range of levels of specialisation and that the majority of visitors are novices (56.1 per cent) who do not share the same degree of concern for the environment or the same motives for visiting as their more specialised counterparts.

Is there a distinct ecotourism segment?

Despite all of the research on typologies, Sharpley (2006) argues that the literature remains inconclusive at best in defining *who* the ecotourist is because, in his view, there is just too much variability in socio-demographics and psychographics among the various studies in profiling this group. But more unsettling for ecotourism pundits is the belief that on the basis of tourist motivation, values and consumption, and the consumer culture, the term 'ecotourist' is becoming increasingly meaningless. This is because:

environmental concern is likely to be subordinated to a variety of other values, motivations and desired benefits – in other words, tourists may be demanding new, different products, such as ecotourism, yet the conventional reasons for participating in tourism are unlikely to have changed.

(Sharpley 2006: 19)

If such is the case, we cannot really make any distinction between so-called ecotourists and mass tourists, according to Sharpley, with true ecotourism unlikely to be attained because it is little more than a supply-led market niche. Sharpley argues that:

- Participation in ecotourism is not driven by social and environmental concerns (i.e. the responsibility component built into many definitions of the term).
- Ecotourism development has done very little to reverse power relations within the arena of international tourism (i.e. there is still dependency).
- Tourists are consumers interested in entertainment as a primary motivation (self-reward and self-indulgence).
- It is not known to what extent the values of so-called ecotourists resonate with ecocentric values over instrumental ones (i.e. are they really true ecocitizens?).

(Sharpley 2006: 19)

Sharpley concludes by suggesting that because there is little difference between so-called ecotourists and mass tourists according to values etc. as above, it makes little sense to try to categorise ecotourists. It is not just 'ecotourists' that abide by the many codes of conduct that exist in so many different places, but rather a host of other mainstream tourists that would be likely to do the same.

Beaumont took Sharpley's work one step further in an investigation of domestic and international ecotourists visiting Lamington National Park, Australia. The purpose of the study was to determine the extent to which ecotourists support sustainability in comparison to other tourist types. Ecotourism has been defined, especially in the Australian context, by three key criteria: nature-based, education and sustainability (see Chapter 1). Respondents in the study were classified according to how many of the core criteria they self-reported. Those ticking all three criteria were classified as 'complete ecotourists', those ticking two as 'strong ecotourists', those ticking one as ' peripheral ecotourists' and none 'not an ecotourist'.

Only 15.6 per cent of all respondents were classified as 'complete ecotourists', and only 36.8 per cent of these had strong pro-environmental attitudes, as measured by Olsen *et al.* (1992) Ecological Social Paradigm (ESP) scale (the measure used to determine the level of sustainability concern among the ecotourists). This percentage represented only 5.8 per cent of the total sample of visitors. When comparing the 'complete ecotourist' to the other two ecotourist types and the non-ecotourism group, there was little difference in the results of the ESP scale. This led Beaumont (2011) to conclude that ecotourists are no more interested in sustainability than other types of tourists, confirming the observations of Sharpley (above). As such, Beaumont argues that 'While demand exists for nature and learning experiences, compliance with the sustainability criterion seems to be no more a factor in ecotourist decision-making than for mainstream tourists' (2011: 135).

In contrast to the observations of Sharpley (2006) and Beaumont (2011), Perkins and Grace (2009) contend that there is in fact a distinct market of ecotourists as different from the general market. Based on the results of a self-report survey of 255 tourists, they concluded the following in reference to the differences between mainstream tourists and ecotourists:

In addition, the motivations for ecotourism experiences were qualitatively distinct from those of mainstream experiences. Moreover, and somewhat surprisingly, the motivations for ecotourism-type experiences expressed in this study seem entirely consistent with the key themes expounded in the agreed definitions of ecotourism, namely nature as the focus of the experience, environmental education, and environmental conservation.

(Perkins and Grace 2009: 234)

Perkins and Grace continue by suggesting that this result is interesting given the fact that it is the industry (and not the consumer) that drives the definitions of ecotourism. In fact, the consumer may not even be aware of these three main components of the industry's definition of ecotourism. For Perkins and Grace, there is in fact a distinct ecotourism market segment discernible on the basis of the types of motivations sought.

There is no question that an interface with mass tourism, as noted above, creates a cacophony of overlapping programmes, motivations, values, attitudes and meanings that cloud the basic foundations of the concept. There is value in what Sharpley is saying, but isn't it a generalisation, especially when we compare and contrast the predominant soft end of the ecotourist continuum with the hard? The fact that some ecotour operators consistently have extremely high rates of return customers leads theorists and practitioners (especially) to believe that there is demand out there for what we view as hard-path ecotourism. What occurred to me in reading Sharpley's work, and in view of the previous statement, is that what we might be talking about is a very small and specialised sub-set of the overall travel market made up of ecotourists – a notion supported by many of the earliest studies on ecotourists – who purchase and behave in a consistent manner.

This distinction between hard and the larger soft-path markets in ecotourism is illustrated in Figure 2.2. Starting at the top of the inverted triangle in the figure, the base of interests and attractions represent the spectrum of attractions and activities available to the ecotourist while on vacation. The soft-path ecotourist, the group that is representative of most ecotourists by number, would spend much more time engaged with interests and attractions that fall outside of the natural history attraction realm; that is, many of the interests and attractions sought by the softest part of the soft-path ecotourism segment would be spent on other attractions of the destination, including perhaps adventure, culture, shopping, theme parks and so on. Furthermore, the reliance on built or modified spaces and places would be more important to this part of the soft-path segment, and there would be less specialisation, fewer natural history expectations, and less time spent on these types of attractions.

At the opposite end of the soft-path continuum; that is, just above the hard-path ecotourist part of the triangle, there would be a greater emphasis on specialisation, expectations and time spent, less reliance on modified spaces, and fewer numbers of tourists. The dashed line between the hard-path group of ecotourists, as a very small number of the overall market, and the more dedicated soft-path market is said to be less a matter of kind and more a matter of degree. Even within the hard-path segment there is a slight variation. But in general, the hard-path segment is far more specialised, with higher expectations and a great deal more time spent on attractions within the natural history attraction realm over the course of the vacation. Specialisation means a focus on specific species or groups of species, or it may include other natural history attractions as defined above. If the tour was inadequate in fulfilling the hard-path ecotourists expectations, there would exist a greater measure of dissatisfaction in comparison to the expectations of soft-path ecotourists. As such, hard-path ecotourists are motivated to find ecotour operators that provide the necessary dimensions of a trip to suit their needs, and often stick with these operators on

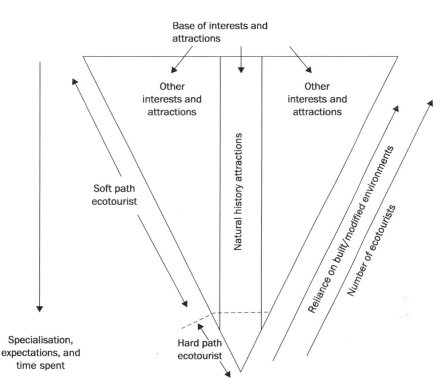

Figure 2.2 Soft- and hard-path dimensions of the ecotourist
Source: Fennell (2002a)

subsequent ecotours (e.g. approximately 67 per cent of Quest Nature Tours participants are repeat visitors).

In returning to the discussion on whether ecotourism is or is not a distinct market, Beaumont further argues that all of the respondents could have been, and perhaps should have been, classified as ecotourists because they were participating in an activity (Lamington) that included the three core criteria. Beaumont continues by suggesting that we cannot assume that just because tourists are participating in an ecotourism activity, they have concerns about sustainability. As a refresher to what was discussed on core criteria in Chapter 1, and as a precursor to a more comprehensive discussion of these criteria to follow, it is important to follow up on Beaumont's line of thought.

Sustainability is not just applicable to ecotourism, but can and should apply to any other type of tourism depending on how the various sub-sectors of the industry like accommodation, transportation, facilities and attractions are planned, developed and managed. Hunting and fishing, for example, can be sustainable in how habitat may be conserved for future fishing and hunting. Furthermore, learning applies to many different types of tourism as much as it does to ecotourism. Examples include museums, art galleries, battlefield tourism and other forms of dark tourism, and the list goes on. So, learning is not particular to ecotourism, but the type of learning that should be part of the ecotourism experience – learning about nature and natural resources – is. And as we also learned in Chapter 1, there are many different types of nature-based tourism (NBT) that share similar environments as ecotourism.

So, while many scholars argue that the combination of these three criteria (sustainability, learning and nature-based) only is enough to define the basis of ecotourism, I believe that the most important criterion is missing. Ecotourism is an attitude and an ethic on how ecotourists and ecotourism practitioners ought to approach the natural world. As such, the three main core criteria will be richly enhanced if they are informed or guided by a moral compass – an ethical underlay.

Ethics can easily be built into environmental education programmes that inform ecotourists, as well as the ways in which ecotourists apply this learning both as tourists and at home. Ethics applied to the nature-based component of ecotourism means how we might control and mitigate the impacts of ecotourism activities. Finally, ethics applied to sustainability means the equity and environmental justice that should be a function of the management of ecotourism. People getting jobs and treated in a way that preserves their dignity – environments conserved and preserved in a way that emphasises the value of other entities besides human beings. In the end, it is all four of these criteria *together* that must form the basis of a legitimate ecotourism industry and experience.

The seeds of this perspective have been planted in earlier works by the author (Fennell and Malloy 1995; Malloy and Fennell 1998a, Malloy and Fennell 1998b). Importantly, many others such as Donohoe and Needham (2006) and Honey (1999), have stressed the need for an ethical approach to the industry. In this capacity, Nowaczek and Smale (2010) have gone so far as to develop an Ecotourist Predisposition Scale, which is a multidimensional scale based on six foundational dimensions including ethics, education, culture, nature, specialisation and contribution. Significantly, the authors found that the scale revealed the highest level of importance for ethics over the other five dimensions.

Plate 2.4 Guides play an important role in the learning process

Nowaczek and Smale suggest that this may mean that ethics is something of an umbrella concept in how it touches on all other aspects of the ecotourism experience.

Conclusion

The volume of research on the ecotourist continues to expand, and this is demonstrated in the range of different studies emerging on the socio-demographic, psychographic and specialisation characteristics of these tourists in many different settings. The purpose of this chapter was to introduce some of the basic characteristics of the ecotourist. Chapters to follow will continue to advance this discussion regarding areas related to differences among nature-based tourists as well as in marketing. The question of a distinct travel segment or market of ecotourists continues to be a compelling one, and presses scholars to look deeper and more comprehensively into ways to better advance our base of knowledge. To be sure, ecotourists are not a homogeneous group. They demonstrate differences on the basis of time spent as ecotourists, their special interests, the places they travel and the attractions they choose to visit. The discussion on hard- and soft-path ecotourists has always been important in emphasising these differences.

Summary questions

1   What are some of the characteristics that have been used to differentiate ecotourists from other types of tourists?
2   List why Sharpley and others suggest that there is not a unique or distinct market of ecotourists.
3   What is the difference between a hard-path ecotourist and a soft-path ecotourist?
4   What is specialisation and how does it apply to ecotourism?
5   How does the average Canadian traveller differ from Canadian ecotourists on the basis of attractions and benefits sought?

Part II

Core criteria used to define ecotourism

In this part, the concepts that are central to defining ecotourism are explored in detail. These include the nature-based focus of ecotourism, sustainability (local benefits) and sustainability (conservation), learning, as well as the ethical imperative.

The first of these, nature-based, is important because ecotourism is a form of tourism that takes place in nature. I have often used the example of a company offering helicopter tours over Hollywood as a form of ecotourism. Set only within the confines of the helicopter, it becomes difficult to qualify this as a form of ecotourism (or even nature-based tourism (NBT)) because there is little if any connection, first-hand, with the natural world. As there are many other types of tourism that take place in nature (fishing, hunting, going to the beach etc.), it becomes essential to define ecotourism using other criteria.

Sustainability – development that meets the needs of the present without compromising the ability of future generations to meet their own needs – is one of these other criteria. I have elected to split sustainability into two different chapters owing to the size of the literature in this area as well as to the importance of emphasising not only conservation, but also local participation and benefits from ecotourism. Theorists argue that for an activity to be viewed as ecotourism, there must be some type of conservation effort extended on behalf of the ecotourist and/or the service provider. This may come in the form of money contributed to parks and protected areas or specific projects like species breeding programmes, rehabilitation of degraded sites, bird banding and so on.

In the case of local participation and local benefits, there are numerous examples of how tourism has marginalised local people through the removal of control and rights. Ecotourism initiatives should be designed to place control back into the hands of local people, who may then be able to govern the pace and scale of development according to their own needs. How community-based projects are organised and implemented, and by whom, is very important in ensuring that opportunities and benefits are realised by a broad spectrum of interests in the community. And as identified above in reference to nature-based, it is important to recognise that just because a type of tourism provides local benefits, this does not suggest it is ecotourism. Ecotourism would be that form of development that contributes to local people while nested in many of the broader criteria discussed here.

Another of the central tenets of ecotourism is learning. A focus on learning about the natural history of a destination is one of the key features that set ecotourists apart from these other forms of tourism such as hunting or fishing. An important but contentious issue surrounding learning is the usefulness of the knowledge gained as an ecotourist. While this information may alter the behaviour of ecotourists on-site, there are questions as to whether it has value over longer periods.

Perhaps the most important of these criteria – one that is often overlooked by theorists – is what I refer to as ecotourism's moral imperative. Zoos and aquaria, for example, are often labelled and marketed as ecotourism because tourists are able to get close to the animals viewed. However, there are important questions to take into consideration in assessing whether these venues are ecotourism or not. Zoos may have the appearance of nature and provide education for visitors, but holding animals captive and removing their ability to live the lives they are meant to live is an ethical issue not often considered. Zoos are often more about satisfying human interests instead of animal interests, when perhaps it should be the other way around.

I argue that all of these criteria together make ecotourism, and a better understanding of these core criteria sets the stage for a better understanding of the many topics and issues that play heavily into the success and challenges of this form of tourism, which is the focus of Part III.

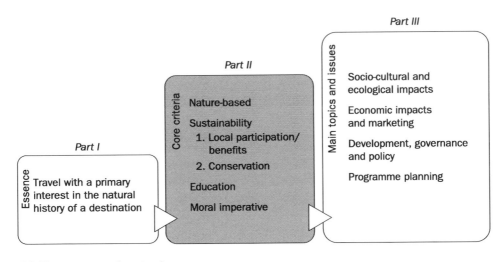

Figure P2 The structure of ecotourism

3 Nature-based

In the previous chapter it was argued that *nature-based* was one of four core criteria used to define ecotourism. However, even though ecotourists gravitate to natural environments, this criterion alone is not enough to differentiate ecotourists from other nature-based tourists who use the same environments. This chapter focuses more specifically on the nature-based dimension, with comparisons and contrasts made between ecotourism and other nature-based tourism (NBT) types like hunting and fishing, before examining related forms of tourism like wildlife tourism, adventure tourism, cultural tourism and adventure, culture and ecotourism (ACE) tourism – especially in relation to the consumptiveness of these types of tourism. The chapter also looks at the prospects for a mass ecotourism market, and the associated issues tied to more people in nature, and how these places have been hardened to accommodate such use.

Natural resources

Zimmerman (1951: 15) wrote that resources 'are not, they become; they are not static but expand and contract in response to human actions'. That is, elements of the resource base such as trees, water, rocks and so on do not become resources until they satisfy human needs. Zimmerman referred to elements of the environment that do not have human utility as 'neutral stuff'. The more humanity uses elements of the natural world, the more this neutral stuff becomes transformed into resources. Culture also has an effect on the use of natural resources. For example, oil becomes a resource if people endeavour to develop the knowledge and ability to extract it, and combine it with the technology required to build the implements we use for work, leisure and survival (cars, furnaces etc.). Resources, therefore, are dynamic both in space and time and very much related to the perception of their worth to a particular person or society, as suggested by Mitchell (1989: 2): 'Natural resources are defined by human perceptions and attitudes, wants, technological skills, legal, financial and institutional arrangements, as well as by political systems . . . Resources, to use Zimmerman's words, are subjective, relative and functional'.

The pursuit of touristic needs occurs along a broad spatial continuum, from those settings that have been substantially modified by humans to pristine environments with very little human intervention. Chubb and Chubb (1981) suggest that the dividing line between what is developed and what is undeveloped is contingent upon one's perception of the meaning of the word 'developed' in relation to the tourism setting. Developed resources include highways, facilities, sewerage, buildings and so on that facilitate the use of a given area. Conversely, undeveloped resources may be found both in urban and wilderness environments but the degree to which they are recognised as such is person dependent as well as situation dependent. These authors outline seven different undeveloped resource types as they apply to outdoor recreation (and tourism). Although independent, these resource types often blend to one degree or another providing a rich diversity of resource conditions that constitute a variety of different tourism experiences.

- *Geographic location.* The characteristics of space that determine the conditions, in association with other variables, for participation (e.g. skiing).
- *Climate and weather.* Determined by latitude and elevation relative to large land-forms, mountains, ocean currents and high-altitude air currents. Along with geology, climate is the prime controller of the physical environment, affecting soils, vegetation, animals and the operation of geomorphological processes such as ice and wind.
- *Topography and landforms.* The general shape of the surface of the earth (topography) and the surface structures that make some geographical areas unique (landforms). A landform region is a section of the earth's surface characterised by a great deal of homogeneity.
- *Surface materials.* The nature of the materials making up the earth's surface, including rocks, sand, fossils, minerals, soil, and so on.
- *Water.* This substance plays a critical role in determining the type and level of outdoor recreational participation in ocean and sea environments as well as freshwater settings (lakes, rivers and wetlands).
- *Vegetation.* Vegetation refers to the total plant life or cover in an area. Recreation quite often is dependent on plant life directly (tourists taking pictures of unique plant species), or indirectly (trees acting as a wind barrier for skiers).
- *Fauna.* Animals have an important role in tourism and outdoor recreation from consumptive and non-consumptive perspectives. Consumptive uses like hunting and fishing recreation are very much different from non-consumptive uses (e.g. bird watching) which have less of an impact on the resource base.

These resources may act either as catalysts in facilitating and drawing people to a tourist region or as constraints to visitation. A case in point is Montserrat. This small island state in the Caribbean is blessed with an abundance of natural features in its 102 km² size, and these usually act as a catalyst to draw people to this destination. Given its relative abundance of natural and climatic features, Montserrat has been referred to as an excellent example of a region where ecotourism may prosper (Reynolds 1992; Weaver 1995). In 1997, however, tourism in Montserrat took a turn for the worse when the island's volcano erupted, leaving the island, and the island's economy, in a critical state. Natural disasters such as these have both immediate and long-term consequences that are felt for years after as Weaver (1995: 601) writes in anticipating the fate of Montserrat's tourism industry:

> The inevitable hurricanes and earthquakes of the future will periodically curtail eco-tourism by damaging not only the physical environment, but the roads, hiking trails, and viewing platforms upon which the sector depends. Furthermore, a prolonged moratorium on tourist activity may be necessary in some areas as the environment recovers from a natural disaster.

Nature-based tourism (NBT)

NBT was defined in Chapter 1 as a form of tourism that encompasses those forms of tourism (e.g. mass tourism, adventure tourism, low-impact tourism, ecotourism) which use natural resources in a wild or undeveloped form. The emerging set of complex social, economic, physical and ecological relationships in so many different settings has elevated

NBT to one of the most important subjects in tourism research today (Hall and Boyd 2005).

The NBT sector is so important in Australia that it represents approximately one-third of the total value of Australia's tourism sector (Buckley and Sommer 2000). In Canada, the sector's value has been measured according to the Survey on the Importance of Nature to Canadians (SINC), which has been used to generate a range of data on the nature-related activities and expenditures of Canadians. Gray *et al.* (2003) examined survey data between the years 1981 and 1996 in finding that more than 80 per cent of Canadians 16 years of age and over engaged in nature-related activities. Furthermore, the frequency of participation in non-consumptive wildlife viewing increased from 56.7 million days in 1981, to 80 million days in 1996. Conversely, hunting declined from 32.3 million days in 1981 to 20.2 million days in 1996; while fishing declined from 88.7 million days to 72.0 million days in 1996 (see Case study 3.1). In looking just at wildlife viewing in 1991 and 1996, the authors reported an increase in female wildlife viewing from 47.6 per cent to 50.7 per cent over this five-year period of time. These figures support newer data on the feminisation of ecotourism, as discussed in the previous chapter. Also, while rural residents represented 27.2 per cent of wildlife viewing (urban: 72.8 per cent) in 1991, the percentage of rural residents dropped to 17.9 per cent by 1996. Overall, nature-related expenditures contributed $11.4 billion to Canada's gross domestic product (GDP) in 1996, with wildlife viewing representing $1.3 billion, fishing $1.9 billion and hunting $815 million. Outdoor activities in natural areas amounted to $7.1 billion. (See Manfredo *et al.* 2003 for a discussion of the factors that have led to a shift in wildlife value orientation in the USA, from the traditional utilitarian perspective to post-materialist values based on higher order psychological needs.) A newer iteration of the SINC, referred to as the Canadian Nature Survey, was administered in 2012, with results due in 2014. It will be interesting to see if these same trends persist in the intervening years.

CASE STUDY 3.1

Nature tourism in Texas

Ecotourism in Texas provides an excellent example of how ecotourism is subsumed by nature tourism. In Texas, nature tourism is defined as 'discretionary travel to natural areas that conserves the environmental, social and cultural values while generating an economic benefit to the local community' (Texas Parks and Wildlife 1996: 2). Although hunting and fishing are reported to be traditional mainstays of nature tourism in Texas, the Task Force on Nature Tourism states that non-consumptive activities such as bird and wildlife watching, nature study and photography, biking, camping, rafting and hiking have experienced the greatest growth over the past few years. In Texas, the task force reports that tourism is the third largest industry in Texas, generating $23 billion annually, with the potential to replace oil and gas, and manufacturing as the highest income earner by the turn of the century.

The Great Texas Coastal Birding Trail is a key ecotourism attraction in the Lone Star State. The goal of this trail is to increase the opportunities for nature tourism in the coastal communities of Texas, in addition to conveying the value of conservation to people living in the region. Although the trail was conceived by the Texas Parks and Wildlife Department, it was made possible by transportation enhancement funds and the Texas Department of Transportation. The trail utilises existing transportation infrastructure (viewing platforms and boardwalks are being added on a continual basis) in creating recreational, economic

continued

and educational opportunities for local people and tourists alike. Upon completion in 1998, it comprised three main sections spanning 600 miles of coastline, incorporating some 300 birding stops in nine wildlife refuges, eleven state parks, one national seashore and several city and county preserves.

CASE STUDY 3.2

Nature tourism in the USA

The Recreation Executive report of the US Forest Service (1994) illustrates that nature-based recreation trends show increases in many non-consumptive outdoor activities. Outdoor photography is up 23 per cent, wildlife watching is up 16 per cent, backpacking is up 34 per cent and day hiking is up 34 per cent. Also, the National Survey on Recreation and the Environment, a study conducted by the US federal government, concluded that of a number of outdoor recreation activities, birding, hiking and backpacking experienced the most significant growth in participation between 1982–83 and 1994–95. Bird watching increased 155 per cent, hiking increased 93 per cent and backpacking increased 73 per cent. Conversely, hunting and fishing experienced negative growth over this period (−12 per cent and −9 per cent, respectively). Further to this, the White House Conference on Tourism in October 1995 indicated the following with respect to the non-consumptive use of birds:

Number of participants:	25 million
Retail sales:	$5 billion
Wages and salaries:	$4 billion
Full- and part-time jobs:	191,000
Tax revenues	
State sales:	$306 million
State income:	$74 million
Federal income:	$516 million
Total economic output:	$16 billion

Just like ecotourism, the literature on NBT has grown tremendously over the last three decades, with much of this research emphasising the diversity of this broad group of tourists. For example, Vespestad and Lindberg (2010) identified four groups of nature-based tourist experiences in their research, based on the belief that the experience itself is the actual product holding value, and not just the outcome or result of the process of consumption. These four groups include nature-based experience as: (1) genuine, where real nature, or back to nature, where the authentic is sought in the creation of a coherent life. The search for a 'holy' natural experience; (2) entertainment, with fun as the ultimate goal and motivation for being in nature; (3) state of being, where the sensation seeker in the search for psychological and physical goals and a new state of being; and (4) socio-cultural community. This last category includes the 'meaning seeker', where nature represents part of the universe for the tribe – seeking recognition of others through participation. The authors argued that the identification of these different types has relevance for marketing and the creation of products that correspond to each type of experience.

Mehmetoglu (2007) found that two variables were central in the investigation of the relationship between trip motive and the importance of nature in decisions of nature-based tourists on choosing a destination: 'novelty and learning' as well as 'everyday life'. Mehmetoglu found that genuine nature-based tourists (those travelling primarily for nature) placed more emphasis on 'a contrast to everyday life' than mixed nature-based tourists (those characterised as travelling for many reasons, including nature). He argues that although environmentalism, changes in consumption and the negative impacts of mass tourism are all reasons for choosing NBT as a travel option, the search for a contrast to everyday life may also be important. Nature provides this contrast (see also Pouta *et al.* 2006 in reference to the spending patterns of nature-based tourists in Finland).

NBT inconsistencies

López-Espinosa (2002) undertook research to determine how closely Mexican ecotour operators in La Paz Bay were fulfilling ecotourism criteria/principles. Based on the responses of 21 sea kayaking, scuba diving, nature cruises and day tour rides, only two did not feel they satisfied the criteria for ecotourism. Further examination found that: (1) many operators disposed of organic wastes in the open ocean even though this practice is not allowed by permit; (2) operators allowed tourists to collect souvenirs such as animal bones and shells despite being prohibited; and (3) companies allowed fishing to take place even though this was not allowed by permit. To complement the study, López-Espinosa interviewed a number of key informants at national, regional and local scales regarding the practice of ecotourism in La Paz. Comments suggested that ecotourism was being sold for short-term profit leading to no conservation benefits; there was little coordination between different levels of government for effective ecotourism management; operators have little knowledge about appropriate carrying capacity levels; and there is confusion over the pros and cons of consumptive versus non-consumptive use of the resource base (for ecotourism). This led the author to question the legitimacy of ecotourism in the area, and to suggest that practices are more akin to NBT.

The sorts of problems encountered by López-Espinosa (2002) above have also been encountered by Naidoo and Adamowicz (2005). These authors prefer not to use the more 'restrictive' term ecotourism because of its inflexibility in describing activities such as hiking and safaris, and prefer NBT because it is 'a non-consumptive activity that should rely on intact natural resources to generate resources' (2005: 160). Part of the problem is that given its name 'nature tourism' or 'NBT', becomes difficult to equate with activities that are more consumptive in their orientation (e.g. hunting). The activities that we now group under nature tourism, including adventure tourism, fishing, hunting, whale watching and ecotourism might best be labelled 'natural resource-based tourism'. This latter categorisation implies an element of use, which we know corresponds to any form of tourism that occurs outside and which relies specifically on the natural resource base. Furthermore, it is more analogous to the continuum of conservation (saving for use) and preservation (saving from use). Ecotourism, although it entails use, should apply more to the preservation end of this continuum, while hunting and fishing relate more to the aspect of conservation. This sentiment relates to the work of Ewert and Shultis (1997), who wrote that 'resource-based' tourism includes a variety of tourism endeavours, including ecotourism, adventure tourism and indigenous tourism, and their various activities.

The aspect of alienation brought on by terminology, above, De la Barre (2005) argues that those involved in wilderness tourism in Canada's Yukon Territory are deliberately not using the term ecotourism based on similar rationale. De la Barre observes that by deliberately *not* labelling the Yukon as ecotourism, constructed space is developed paving the way for divergent local perspectives (consumptive and non-consumptive) to engage in tourism planning, allowing for a diversity of lifestyles and cultural usages of the land that more closely align with NBT. (See Waitt *et al.* (2003) for a discussion of how both the government and the tourism industry have successfully portrayed the Ord River Irrigation Area in the East Kimberly region of Australia as a wild and natural place – for NBT – even if it is artificial.)

Plate 3.1 The kayak continues to be a mainstay of many adventure-related operations in North America and is now used by ecotour operators to take tourists to many different settings

Plate 3.2 Going to the beach is a form of NBT

The consumptive–non-consumptive debate

The foregoing discussion on inconsistencies between NBT and ecotourism often boils down to the issue of consumptiveness. As explained in Fennell (2012a), a use that is consumptive is one that reduces the supply of a resource (e.g. removing water from a source like a river, lake or aquifer without returning an equal amount) (see Mimi 2011). Non-consumptive use may be taken as use of a resource that does not reduce the supply of a target species or feature of the environment; that is, there is no net loss to the environment as a result of our actions. Where NBT and ecotourism diverge, therefore, is usually in consideration of the consumptiveness of the activity. The following few examples serve to emphasise these differences.

Dowsley (2009) argues that hunting can be a form of ecotourism in the high north of Canada, where Inuit are given quotas for the number of polar bears they can kill each year. The total allowable harvest is divided among the Inuit people, and members of the community are allowed to use their quota for the purposes of sport hunting; that is, they can act as guides. This activity is referred to as ecotourism because of its subsistence value, it supports conservation, there is a sustainable harvest, and because it supports local economies. Dowsley refers to this as conservation hunting. In other work, Gunnarsdotter (2006: 178) writes that moose hunting in Sweden can be ecotourism if, 'cultural and social aspects are taken into consideration'. Presumably this means that cultural practices related to hunting are employed (e.g. treatment of the moose after being shot) and that the community is advantaged through economic benefits and cohesion (see also Campbell *et al.* 2011).

The rationale for labelling hunting as ecotourism instead of NBT lies in the relationship to community economic development (Fennell 2003). This is confirmed by Novelli *et al.* (2006), who argue that tourism is seen as one of the most important vehicles for community development for the purpose of meeting basic needs. In over two years of field work in Botswana and Namibia on consumptive forms of tourism (hunting) and non-consumptive wildlife viewing (ecotourism), the authors conclude that, 'where tourism strongly benefits the natural, economic and social environment, whether through consumptive or non-consumptive practices, it fits well within the concept of ecotourism and has the potential to contribute to the ever-sought after sustainable tourism development' (Novelli *et al.* 2006: 77).

In addition to economic reasons (i.e. community economic development) for including activities like fishing and hunting under the ecotourism label, there is the aspect of conservation and sustainability. Novelli *et al.* (2006: 67) argue that, 'hunting has increasingly become part of conservation argument and policies, and is promoted as a low-impact sustainable use approach, adding value to natural resources'. This is an important point, and it is likely that hunting has become part of the conservation firmament because it had to out of necessity. In many parts of the world hunting suffered through a decline in membership (as noted above) because of its social unacceptability. In full view of this, the most effective manner to increase participation is to make the activity socially acceptable through normative measures hinged on sustainability and ethics (i.e. contributions of meat, skins, bones and ivory to local people for commercial reasons). Ecotourism has been a suitable mechanism to accomplish this end, with evidence of this having to do with the power of a name. Based on what we already know about the differences between ecotourism, adventure tourism and NBT, a more fitting label for hunting would be NBT or even adventure tourism – but these terms do not carry the added ethical and sustainable appeal that ecotourism does.

Many of the same arguments have been used for fishing. Zwirn *et al.* (2005) contend that catch-and-release fishing, in association with conservation (from fishing licences) and revenue generation for local communities, makes fishing a legitimate form of ecotourism. Proponents, like Zwirn *et al.* are quick to reference the magnitude of the industry and the various benefits. For example, they argue that Alaska alone generates over $1 billion per year from the recreational fishing industry and it generates 11,000 jobs.

This debate took on a more comprehensive dimension in the example of billfishing (i.e. marlin, swordfish and sailfish) as a form of ecotourism. Holland *et al.* (1998) argued that billfishing could be ecotourism because it directs economic assistance to local communities, it attracts a unique clientele, and has economic advantages over other uses (among other ecotourism-related variables). Fennell (2000), however, argued that billfishing, or any other type of fishing in the conventional context, could not be viewed as ecotourism because of the failure to recognise some key factors in differentiating ecotourism from other forms of tourism. That is, while there was disagreement with the variables used by Holland *et al.* (1998) to define ecotourism, there was also the belief that some very crucial aspects of human–animal relations had not been considered. These included the intention to entrap the animal (which is not the same as the intentions of ecotourists which should be geared towards minimum disturbance and impact in all cases); the pain and stress which results from catching the animal; consumptiveness (catch-and-release practices still may be viewed as consumptive along a continuum); and values, such that ecotourists have a different set of values related to sport and the intrinsic/extrinsic motivations surrounding participation in these activities. Fennell argues that despite the angler's best intentions to minimise stress on the animal, one can only do so to a point after which he or she must cease to pursue and capture the animal. The argument follows that the treatment of animals cannot be based on an acknowledgement of healthy populations (i.e. it is fine to catch or hunt animals because of the healthy state of the population as a whole), but rather that respect must be shown to the individuals comprising these populations (Taylor 1989). The use of weak definitions or principles behind ecotourism thus opens the door for a great deal of misrepresentation and the prospect of any number of different activities (hunting, as noted above) that places human needs over the value of other species (more on this issue in Chapter 7).

In an effort to shed light on the intricacies of the heightened demand for ecotourism and the propensity for including more consumptive forms of tourism under the ecotourism label, Fennell and Nowaczek (2010) constructed a framework that includes a number of different interactions with fish. The framework is thought to be useful in positioning the type of interaction based on consumptiveness and need, but also according to the differences between anthropocentric ethics and universal ethics. Figure 3.1 shows that there are eight different types of interaction that nature-based tourists can have with fish. These range from the most consumptive and non-essential, and increasingly anthropocentric forms of NBT, to ecotourism defined as non-consumptive, and increasingly ecocentric (perhaps better stated as biocentric.). An example of the former is groups of tourists fishing for sport, in competition, mainly catch-and-release. Here it is suggested that catch-and-release fishing can be morally questionable given that fish are subjected to great suffering and often death as a result of the activity. The stock case of the latter is ecotourism where there is a focus on fish viewing, learning and appreciation, and with no direct physical handling of the fish. Along these lines, Stoll *et al.* (2009) identify fish viewing as a new leisure activity. Their work on sturgeon viewing in northeast Wisconsin provides ample evidence of an emerging industry, although primarily for the local market. Proponents of the fish viewing as ecotourism perspective would argue that this activity is philosophically very different than the more conventional fishing for sport form.

Tremblay (2001) advances our understanding of consumptiveness in wildlife tourism by suggesting that how we view consumptive has generally advocated one form of tourism (non-consumptive wildlife viewing) at the expense of others (consumptive hunting and fishing). He challenges the belief that non-consumptive activities convey morally superior values leading to more desirable experiences, on the basis that these lead to increased understanding, education or respect. This dichotomy between good and bad is potentially damaging based on the fact that tourists might alienate local people because of their utilitarian reliance on wildlife as a form of sustenance. Tremblay argues for complementarity in regard to so-called consumptive and non-consumptive activities in a balanced provision of nature-based services – a stance that is advocated here. This complementarity bridges over to a shared interest in attractions, such that the hunter's experience is enhanced by the quality of wildlife in the vicinity to be viewed. What is subject to debate, however, is the degree to which the wildlife viewing tourist's experience is heightened by the opportunity to hunt or fish (see Fennell and Weaver 1997).

Taking this discussion in the same direction, Meletis and Campbell (2007) argue that ecotourism should not be labelled as non-consumptive because it is in many ways as consumptive as other forms of tourism. The ecological footprint of ecotourists is often quite large because tourists travel many hundreds or thousands of miles to nature destinations. But they also argue that being consumptive often fits with the cultural practices of people in many remote places around the world. Ecotourists could be invited on ecotours where hunting and fishing form part of a community feast or celebration. Canada's high north is used as an example of such a tour.

A potentially fruitful area of research in the future is the notion of the acceptability of using natural resources in conducting ecotours. It may very well be that ecotourists do not mind at all that wild or domestic animals are killed for human consumption, even for meals that they might enjoy as part of their ecotourism experience. It is a very different mindset, however, for ecotourists themselves to be involved in killing these animals; that they see their ecotour operators killing them for immediate consumption; or that the animals killed were done so for personal satisfaction – as part of the ecotour experience. Hunting and fishing by recreationists are activities that are undertaken for purposes of personal satisfaction, and any time you remove an animal from the environment for such reasons, the action may be of justifiable concern for ecotourists. Not unlike the discussion on soft and hard typologies of ecotourism in reference to activities and accommodation, future research may discuss the ecotourist in reference to choices around the consumption

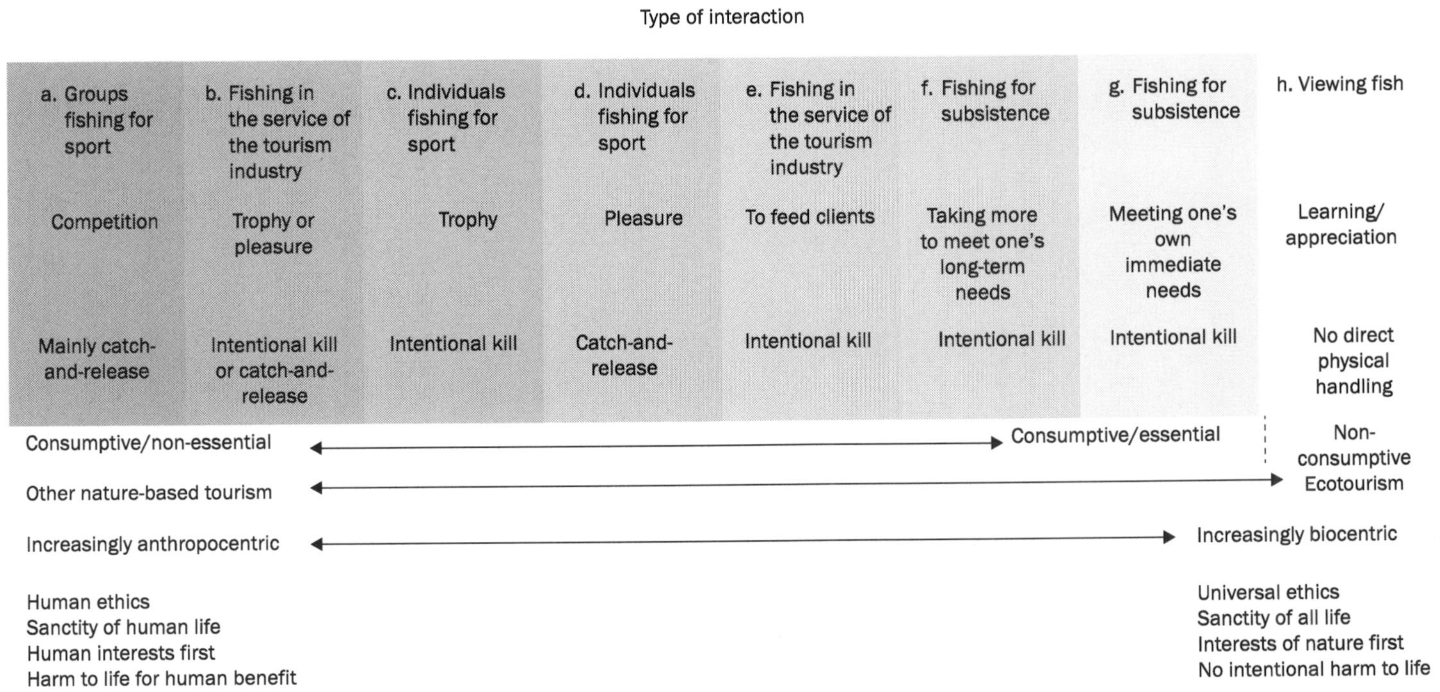

Type of interaction

	a. Groups fishing for sport	b. Fishing in the service of the tourism industry	c. Individuals fishing for sport	d. Individuals fishing for sport	e. Fishing in the service of the tourism industry	f. Fishing for subsistence	g. Fishing for subsistence	h. Viewing fish
	Competition	Trophy or pleasure	Trophy	Pleasure	To feed clients	Taking more to meet one's long-term needs	Meeting one's own immediate needs	Learning/ appreciation
	Mainly catch-and-release	Intentional kill or catch-and-release	Intentional kill	Catch-and-release	Intentional kill	Intentional kill	Intentional kill	No direct physical handling

Consumptive/non-essential → Consumptive/essential

Other nature-based tourism → Non-consumptive Ecotourism

Increasingly anthropocentric → Increasingly biocentric

Human ethics
Sanctity of human life
Human interests first
Harm to life for human benefit

Universal ethics
Sanctity of all life
Interests of nature first
No intentional harm to life

Figure 3.1 Human priorities and actions in recreational interactions with fish

Source: Reynolds and Braithwaite (2001)

of meat. The hard–soft path continuum may run from vegan (the hardest path ecotourist) and to vegetarians, to those who consume abundant bushmeat, and finally to those consuming mass-produced animals in factory farms (Fennell 2013b). How these animals are killed is an obvious concern. An animal that has enjoyed a long, free life in the wild, but killed under conditions of moderate pain and stress, with some suffering, may be a better life than that of an animal reared in factory farms, where it would have to endure continual suffering throughout the course of its life (see Singer 2009).

Wildlife tourism

Wildlife tourism is a relatively new addition to the literature on ecotourism and NBT, and it has pushed the research on human–animal interactions in new and important directions. Higginbottom (2004: 2) defines wildlife tourism as 'tourism based on encounters with non-domesticated (non-human) animals'. This includes wildlife-watching tourism (free-ranging animals); captive-wildlife tourism in man-made confinements, like zoos; and hunting and fishing tourism. The general definition posed above means that encounters with animals for tourism purposes can be both non-consumptive as well as consumptive, suggesting that wildlife tourism can be ecotourism in the first case (non-consumptive), but fall outside the bounds of ecotourism in the latter case (consumptive forms of outdoor recreation). It is there-fore a mistake to treat ecotourism and wildlife tourism as being synonymous on the basis of the consumptiveness variable, because wildlife tourism does not involve other aspects in the natural world like plants, and because of the conflict that goes along with categorising activi-ties that are not based on the same value sets. These value sets or orientations have been investigated by Deruiter and Donnelly (2002), who found that especially powerful determi-nants of wildlife value orientations were socialising agents including family members, as well as early encounters with wildlife. Place of upbringing also had a profound influence on wildlife values. The authors observed that rural children with familial hunting backgrounds tended to be more anthropocentric, while children who spent time contemplating nature – a direct experience with wildlife – maintained stronger ecocentric feelings.

Reynolds and Braithwaite (2001) have developed a framework, which illustrates the relationship between ecotourism and wildlife tourism, as well as other related forms (Figure 3.2; see Novelli *et al.* (2006) for a useful modification of the Reynolds and

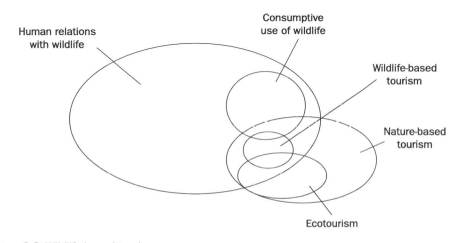

Figure 3.2 Wildlife-based tourism

Source: Reynolds and Braithwaite (2001)

Braithwaite (2001) NBT conceptualisation in reference to wildlife tourism). Based on their examination of a series of wildlife tourism brochures, Reynolds and Braithwaite place the wildlife tourism product into one of seven categories:

1 *NBT and wildlife component*, where wildlife is an incidental part of the overall NBT product.
2 *Locations with good wildlife viewing opportunities*, includes accommodation units that are located in wildlife-rich habitat and may attract wildlife through the provision of food.
3 *Artificial attractions based on wildlife*, which include, for example, man-made attractions where animals are kept in captivity.
4 *Specialist animal watching*, for special-interest groups like birders.
5 *Habitat specific tours*, which focus on areas or regions rich in animal life.
6 *Thrill-offering tours*, where dangerous animals are enticed to engage in spectacular behaviour for the viewing pleasure of tourists.
7 *Hunting/fishing tours*, in natural, semi-captive or farmed environments where animals are killed or released back into the wild.

Some of this research on wildlife tourism has sought to examine the expectations, emotions, and levels of satisfaction that wildlife tourists have in captive or semi-captive environments. For example, Hughes *et al.* (2005) have found that in captive wildlife environments like Barna Mia in Australia, the feeling of being in the wild may be enhanced by limiting the number of visitors, removing certain barriers between tourists and animals, and removing barriers to allow for easier movement of animals within their enclosures. Other studies have found that commercial development of wildlife tourism has at times threatened biodiversity conservation, attempts to develop sustainably, as well as to favour use values over non-use values arising from wildlife tourism (Tisdell and Wilson 2004; see also Lovelock and Robinson 2005).

Curtin (2010) studied the essence of what makes a memorable wildlife encounter amongst a group of serious or dedicated wildlife tourists. Responses to the question 'Describe your most memorable wildlife experiences', respondents gave a number of answers which generally revolved around the following categories: memories in the making, charisma and appeal, large numbers of wildlife, first-time sightings, spontaneity, close proximity and eye-to-eye contact, and embodied experiences. Close proximity, one-to-one or eye-to-eye contact enhanced the experience for three main reasons. It allowed for a closer and more detailed view; the encounter is more intimate; and the experience allows for a totally unmediated view. Some respondents said that the wildlife viewing experience became hard-wired only later after the excitement of the event had passed. Others mentioned that being the first to discover or identify a species provided a sense of achievement.

In other related work, Ballantyne *et al.* (2011a) used a qualitative research design to investigate participants' memories of wildlife tourism experiences. They sent respondents a web survey four months after their marine-based wildlife tourism trip, and found that there were four levels of visitor response to these types of experiences. These include (with examples from the web survey transcripts):

1 Sensory impressions, such as visual, auditory, olfactory or tactile memories (e.g. 'The wonderful colours of the fish on display', and 'Being able to touch the sea creatures in the rock pools' (p. 773)).
2 Emotional affinity, where visitors reported emotional responses or emotional connections to animals (e.g. 'I got quite emotional when I saw the dolphins, they are so

intelligent and graceful', and 'Seeing the big turtle laying was exciting, but I couldn't help but feel that we were getting in the way of nature' (p. 774)).

3 Reflective responses, included new insights by visitors based on cognitively processing their experiences or by reflection of such experiences (e.g. 'I saw the turtles walk to the sea and I felt that humans need to protect them; the world is for all of us', and 'Humans do more damage on this planet than any other creature . . . we all must take responsibility for looking after the planet' (p. 774)).

4 Behavioural responses, such as the actions visitors have taken in response to their wild-life tourism experience, including heightened awareness (e.g. 'I certainly do not use as many plastic bags and I am very careful about what goes down our drains', and 'I refuse to buy anything Japanese until they stop their senseless slaughter of whales' (p. 775)).

CASE STUDY 3.3

Wilderness tourism and outfitting in the Yukon, Canada

The interpretation of ecotourism takes a more general form in the Yukon Territory of Canada. The territory's selection of 'wilderness tourism' as an umbrella term for a variety of different backcountry experiences is very much a function of geography (the remote wilderness setting of the territory) and the existing products and resources operating in the region. The undeveloped wilderness setting is thus a key factor in the delivery of tourism products in the region. Wilderness tourism is defined as 'tourism related to nature, adventure, and culture, which takes place in the "backcountry" and is primarily associated with multi-day trips, although it also includes day trips' (Tompkins 1996). Wilderness tourism is divided into: (1) hunting; (2) fishing; and (3) adventure. This latter category is very much a reflection of the adventure tourism definition and activities of the Canadian Tourism Commission. Such activities include nature and wildlife observation, land adventure products, water adventure products, winter adventure products, air adventure products and native tourism.

Cultural tourism

Cultural tourism is defined as 'visits by persons from outside the host community motivated wholly or in part by interest in the historical, artistic, scientific or lifestyle/heritage offerings of a community, region, group or institution' (Silberberg 1995: 361). Silberberg places primary emphasis on museums and heritage sites in urban settings. He also mentions that while people may visit an urban setting for many reasons like visiting friends and relatives, or business or conventions, they may not be mainly interested in culture, but may visit cultural attractions if made aware of opportunities or convenient.

The profile of the cultural tourist is not dissimilar in some respects to the ecotourist, as observed by Silberberg (1995). The cultural tourist:

● earns more money and spend more money while on vacation;
● spends more time in the area while on vacation;
● is more likely to stay at hotels or motels;
● is more highly educated than the general public;
● includes more women than men;
● tends to be in older age categories.

McKercher and du Cros (2003) provide further scope into different types of cultural tourists through the refinement of a cultural tourism typology based on the experience sought (deep versus shallow) and the importance of cultural tourism in the decision to visit a destination (low versus high). The most highly motivated of the five types identified, the purposeful cultural tourist (high centrality and deep experience), is said to be the exception rather than the norm, and representative of a small minority of those participating in cultural tourism. Other types representing the majority include the serendipitous cultural tourist, the incidental cultural tourist, the casual cultural tourist and the sightseeing cultural tourist.

What is interesting about the discussion on these different cultural tourist types is that they are similar to the commentary on ecotourism types. McKercher and du Cros argue that the majority of all cultural tourists seek cultural attractions on a recreational and pleasure basis, and not for deep learning. These experiences are based on enjoyment and entertainment; with a de-emphasis on experiences that are overly taxing mentally and which do not challenge the tourist's personal ideologies. These latter types of experiences should be presented in an enjoyable, entertaining and easy to consume manner. What is also interesting about the commentary offered by McKercher and du Cros is the belief that for many who participate in cultural tourism, culture may in fact be a secondary motivator to visit a region, and may not play a role whatsoever in the destination of choice. As observed in Chapter 1, this may also be the case with ecotourism in reference to culture – culture is a secondary motivator and may be incidental to the overall trip.

McKercher and du Cros' work is backed-up by Stebbins (1996), who contends that cultural tourism is a liberal arts hobby within the structure of serious leisure theory, and further that it is a genre of special interest tourism. Those who take one or two cultural tours may be viewed as dabblers, while those of a more serious nature are deemed to be hobbyists through 'the acquisition of a broad, profound, nontechnical knowledge and understanding of, for example, an art, cuisine, language, culture, history, or area of the world' (p. 949). Stebbins makes reference to the general cultural tourist, who makes a hobby of visiting different sites and regions, accumulating knowledge and experience. By contrast, the specialised cultural tourist tends to focus on one or a small number of different regions in order to attain a broad understanding of the culture of this (these) region(s).

In Chapter 1, ecotourism was defined using many different variables. One variable that was deemphasised was culture. The rationale for this was that most if not all forms of tourism engage with the culture of the host destination to varying degrees. And further, that if there were an overriding emphasis on culture, it would not be ecotourism but rather cultural tourism. It is important to provide a little more scope into the relationship between culture and ecotourism as we move forward.

Culture is said to be important in ecotourism because of the relationship between humans and nature. That is, interest in other human groups appears to emerge in relation to how they have lived as part of the fabric of the natural world, especially indigenous people. From this perspective it is not difficult to see why scholars emphasise this relationship, and even go so far as to suggest that 'Cultural tourism can be regarded as a subset of ecotourism' (Ryan 2002: 953). These authors follow the lead of some of the earliest work on ecotourism, including Ceballos-Lascuráin (1987), see Chapter 1, as well as Ziffer (1989), who contends that ecotourism is 'a form of tourism inspired primarily by the natural history of an area, including its indigenous cultures'; and also The Ecotourism Society (1990), cited in Wood (1991), which defined ecotourism as 'purposeful travel to natural areas to understand the culture and natural history of the environment . . .' In this last case, the emphasis on culture is placed even over natural history.

But we should recognise, as the Commonwealth Department of Tourism (1994) has in Australia, that a broad interpretation of the natural environment would logically include cultural components. Furthermore, it is essential that the ecotourism industry enables local people to gain economically from ecotourism (and that ecotourism needs to contribute economically to parks and protected areas, and build awareness in people over

Plate 3.3 To some tourists culture is the primary attraction; to others it is merely a secondary feature of the overall experience

the importance of conservation). However, there is still the overriding need to 'put the ecology back into ecotourism' (Valentine 1993: 108). Ecotourism is a concept that needs to be guided by the ecologists and conservationists, Valentine adds, and not necessarily by those in other fields.

Adventure tourism

Another closely related form of tourism to ecotourism is adventure travel, which in some circles is felt to subsume ecotourism. For example, in suggesting that ecotourism is a branch of adventure tourism, Dyess (1997: 2) admits that he did not have a full appreciation of the heated disagreement that 'exists over the semantics of the two terms, as proponents of ecotourism and adventure travel strive to define and sanctify their own approach to travel'. In a general sense, differentiation between the two forms of travel can simply be based on the type of activity pursued (again, with respect to the primary motivation in participating in the activity). However, in cases where activities are broadly categorised as either ecotourism or adventure tourism, there may be problems, as evident in the following example.

Tourism Canada has defined adventure tourism as 'an outdoor leisure activity that takes place in an unusual, exotic, remote or wilderness destination, involves some form of unconventional means of transportation, and tends to be associated with low or high levels of activity' (Canadian Tourism Commission 1995: 5). The following is a list of adventure travel activities developed by the Canadian Tourism Commission (CTC) under this definition: (1) Nature Observation; (2) Wildlife Viewing (e.g. birding, whale-watching); (3) Water Adventure Products (e.g. canoeing, kayaking); (4) Land Adventure Products (e.g. hiking, climbing); (5) Winter Adventure Products (e.g. dog sledding, cross-country skiing); and (6) Air Adventure Products, including hot-air ballooning, hang-gliding, air safaris, bungee jumping and parachuting. The point to be made from this example is that both ecotourism and adventure tourism products may adhere to the CTC's definition, despite the fact that to many there is a clear distinction between nature observation (ecotourism) and kayaking (adventure tourism). Clearly, further analysis of these activities is required in order to detect similarities and differences. However, while the activity classification itself is important in identifying adventure pursuits, Priest (1990) writes that there must be an element of uncertainty associated with the event. The answer to the question of how adventure and non-adventure tourism experiences, like ecotourism, differ may lie in the realm of social psychology, which examines why participation occurs from cognitive and behavioural standpoints (instead of from some arbitrary basis established on the setting and other such variables). Both Ewert (1985) and Hall (1992) write that it is risk that plays a primary role in the decision to engage in adventurous activities. In addition, Hall suggests that it is the activity more than the setting that provides the dominant attraction for pursuit of adventure recreation and tourism. Two other factors in analysing one's motivation to engage in risk-related activities include challenge and skill, the fundamental variables behind Csikszentmihalyi's (1990) model of flow (see also Csikszentmihalyi and Csikszentmihalyi 1990). Individuals who are said to have reached a state of flow have matched their personal skills with the challenges of a particular activity. Flow includes a number of main elements as defined by Csikszentmihalyi. These include:

1 *Total immersion into the activity.* This relates to the elimination of distractions that enable the person to lose touch with his or her surroundings.
2 *Enhanced concentration.* A result of the previous factor that allows the participant to forget about the unpleasant tasks that may be associated with the activity.
3 *Actions directed at fulfilling the goal.* The goals and objectives of the event are clearly understood by the participant, who knows how best to approach the situation.
4 *The activity requires skill and challenge.* The relationship between these two variables is important in that if skill far exceeds challenge, boredom will result, whereas if challenge far exceeds skill, anxiety will result.
5 *Flow involves control.* The participant exercises control over his or her movements and the situation, with a degree of anticipation of the events which will unfold.
6 *A sense of transcendentalism.* Here the participant has the experience of transcending his or her physical being, as rooted on the face of the earth, to reach some higher level of understanding or being. A sense of oneness with the surroundings or objects involved in the experience is felt.
7 *The loss of time.* Frequently participants feel as though they have been involved for a short period (e.g. one hour), when in fact they have been involved for a long period (e.g. four hours).

Csikszentmihalyi (1990) suggests that flow is not necessarily restricted to those who engage in adventurous activities, but also includes those, for example, playing chess or physicians engaged in surgery. According to Hall, it may be the desire or the enhanced desire to experience the state of flow that moves individuals to participate in risk or adventure-related activities.

To Quinn (1990), adventure lies deep within the spiritual, emotional, intellectual and objective spheres of humanity, and is the eternal seduction of the hidden (Dufrene 1973: 398). More specifically, Quinn argues that adventure is a desire for a condition that is absent within the individual. Under rather tenuous conditions, the individual must harbour doubt as to the adequacy of his or her ability. The further one attempts to go beyond one's perceived personal talents, the more intense the adventurous situation becomes. The adventure experience therefore is one that varies in intensity. The result is that in today's marketplace tourists are able to select from a broad range of hard and soft adventure experiences, offering associated degrees of risk and uncertainty. The hard end of the continuum features various forms of extreme tourism for goal-oriented tourists who gain pleasure and satisfaction from putting themselves at risk (e.g. mountain climbing, white water rafting, ice diving, ice climbing, caving).

According to Christiansen (1990), it is the task of the adventure tour operator to provide the client with an adequate perception of risk, while ensuring a high level of safety and security. This is accomplished through an accident-free history, good planning, and the maintenance of the highest level of leadership, skills and experience. Christiansen provides a good description of the difference between soft- and hard-core adventure experiences, as shown in Figure 3.3. The soft adventure activities in the figure are pursued by those interested in a perceived risk and adventure with little actual risk, whereas the hard adventure examples included are known by both the participant and the service provider to have a high level of risk. For those activities that are at the upper end of the risk continuum, there are many ethical questions that need to be asked related to operators and clients. For example, if a client has sufficient money and motivation to climb Mount Everest, should operators take them on if they are unskilled and/or unfit? Where does the responsibility lie – with the client or the operator, or both?

Fennell and Eagles (1990) focused on the element of risk in examining potential differences between adventure travel, ecotourism and mass tourism in the framework shown in Figure 3.4. Here it is implied that preparation and training, known/unknown results and risks, and certainty and safety are all variables that may be used to differentiate between these forms of tourism. The figure demonstrates that the three kinds of activity are not mutually exclusive; ecotourism may share some elements of the other two experiences, while still remaining distinct from mass tourism and adventure tourism.

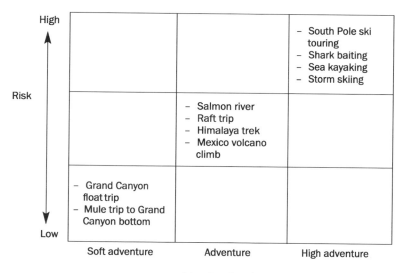

Figure 3.3 Levels of risk in tour packages

Source: Christiansen (1990)

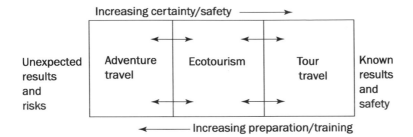

Figure 3.4 Tourism activity spectrum

Source: Fennell and Eagles (1990)

The link between motivation and level of experience in mountain climbers has been analysed by Ewert (1985), who based his research on the premise that the perception of danger is important in the experience of the risk recreation participant. From a survey of climbers at Mount Rainier National Park, Washington, he found that inexperienced climbers participated for recognition, escape and social reasons. Conversely, the experienced climbers were motivated by more intrinsic reasons, including exhilaration, challenge, personal testing, decision-making and locus of control. The implications of the research are such that over time, climbers who continue with a risk recreation activity, such as scuba-diving, mountain climbing and spelunking (cave exploration), may require conditions that are less crowded, more rugged and less controlled than their novice counterparts (see the discussion on specialisation in Chapter 2).

ACE Tourism

Given the foregoing discussion on ecotourism and its variations, the concept of ACE tourism is developed here (Figure 3.5) to illustrate the evolving relationship between three distinct, but related, NBT products; namely ecotourism, adventure tourism and cultural tourism. As suggested earlier, the overlap between these three appears to have become stronger over the past few years, to the point where policy and practice have considered them as almost completely synonymous (this phenomenon is represented by the acronym ACE in the figure). Depending on the setting and situation, ACE either expands or contracts to represent different concentrations of adventure, culture and ecotourism in product content.

Figure 3.5 advances our understanding of multiple and often overlapping types of experiences in two ways. First, it suggests that ACE tourism is different than these other forms of tourism, the latter of which may be seeking some form of homogeneity based on adventure, culture or natural history offerings. For example, those programmes that are classed solely as ecotourism avoid the inclusion of conditions that relate to culture tourism or adventure tourism (or both). In this regard, ecotourism should be considered as unique according to its function and role within the tourism marketplace, a stance which is based on the fact that: (1) there is not enough empirical evidence to demonstrate homogeneity between adventure tourism, culture tourism and ecotourism; and (2) there may be an associated dilution factor or effect on ecotourism if these three types of tourism merge into a combined form. Second, the figure illustrates that there is a form of tourism, which we may refer to as ACE tourism, that contains aspects of adventure, ecotourism and cultural tourism in its programming. The degree or percentage of culture, adventure and ecotourism in the ACE programme offering is subject to the service provider, and the

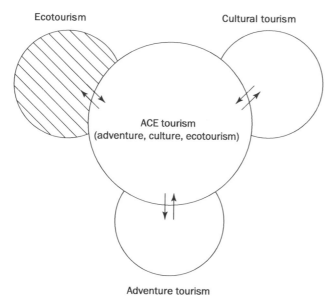

Ecotourism

Cultural tourism

ACE tourism
(adventure, culture, ecotourism)

Adventure tourism

Figure 3.5 ACE tourism

Plate 3.4 Ecotourism programmes may contain aspects of adventure in unique settings. But is this ecotourism?

resources and setting at hand, which means that a number of adventure, ecotourism, culture and ACE service providers can operate in the same setting but with different concentrations of these various components. This would allow each to occupy a niche in the same general setting by virtue of their differential loadings on these components (i.e.

Table 3.1 *Hybrid adventure, culture and ecotourism experiences offered by commercial ecotourism businesses in New Zealand (1999)*

Tourism experiences	Frequency	Per cent
Ecotourism and adventure	111	47.8
Ecotourism and culture	23	9.9
Adventure, culture and ecotourism (ACE)	10	4.3
Total	144	62.0

Source: Higham and Dickey (2007)

all operators would not be able to compete with each other in this setting if they all attempted to offer ecotourism alone, when they could have specialised their offerings on the basis of different concentrations of culture, adventure and nature). Weaver (2002a) has used the ACE framework to discuss mountain trekking in Asia. This activity is characterised by an amalgamation of adventure, cultural and ecotourism in a hybridised form, where the diversity of, 'motivations and impulses increases the challenge of maintaining any focus on the core ecotourism criteria' (Weaver 2002a: 167).

In their comprehensive overview of the New Zealand ecotourism industry, Higham and Dickey (2007) found that relatively few operations ($n = 232$) provided activities related to adventure and culture in their offerings. The lack of culture was especially surprising to the authors given the importance of Maori culture in New Zealand, and the central role that this culture has played in the development of tourism in New Zealand. The types of tourism experiences are broken down by frequency and per cent in Table 3.1.

These results may suggest the inclination towards thinking that the market is generally or mainly interested in ecotourism and adventure, which may in fact be the case. On the other hand, it may also mean that there may be an opportunity to exploit the ACE niche in New Zealand and possibly develop a more diverse set of programmes that emphasise the ecotourism and culture or ACE market in the future.

Ecotourism as mass tourism

An intriguing argument in the literature is that if ecotourism in philosophy and practice is indeed the best form of travel, how can it spill over into other forms of travel, like mass tourism, in making these other types better? As the chief proponent of this line of thought, Weaver (2002b) argues that because scale does not matter in our efforts to be sustainable (i.e. small and large scale can either be good or bad) there is no reason to believe that ecotourism could not occur at a grander scale. Weaver conceptualises this perspective as two ideal types along an ecotourism spectrum, with hard (active, deep) ecotourism at one end, and soft (passive, shallow) at the other. Quite naturally one thinks of larger numbers of soft path ecotourists creating a greater range and intensity of impacts from their activities. Weaver, however, argues that the softer-path ecotourist is more likely to restrict their activities to a small percentage of area in parks (typically hardened sites that can absorb the impacts of numbers), while the hard-path ecotourists are more likely to penetrate deeper into the back regions of protected areas. Furthermore, Weaver notes that it is the softer-path ecotourist who is more likely to contribute to conservation financially – as an essential component of ecotourism – by virtue of revenue generated by larger numbers of tourists for this factor to have an effect. Citing Boza (1993), who felt that ecotourism was the strongest argument for the development of Costa Rica's national park system, Weaver

(1999) observes that it is passive ecotourism assumed by mass tourists and their direct and indirect revenue that provides the strongest argument for the development and defence of protected areas.

Weaver's is not the only voice that echoes this sentiment. Based on his work in Phuket, Thailand, Kontogeorgopoulos (2004a) argues that conventional tourism and ecotourism are conceptually, operationally and spatially linked in a symbiotic relationship. Two of the region's oldest ecotour operators, Sea Canoe and Siam Safari, are said to uphold the principles of ecotourism even though they have structural connections to the package tourism industry that provides their main market. As such, ecotourism in Phuket has emerged not in isolation or opposition, but as a function of the mass tourism industry. But what is especially intriguing is the degree to which these tourists adhere to stringent practices of ecotourism as set out in strict definitions of the term (i.e. environmental conservation, sustainability, low impact and non-consumptive). Kontogeorgopoulos illustrates that Phuket tourists typically stay in 4- to 5-star hotels, visit Phuket en route to other destinations or for short holidays, and book through travel agents and tour operators. Tourists hear about local ecotours (the supply for these trips is a mere 20 km away) in their hotels, with less than three-quarters knowing anything about their ecotour operator prior to visiting the region. Ecotourism is also said to be one of many different niche daytrip options in the region, including entertainment, shopping and health-related activities. In this way, ecotourism benefits mass tourism, 'by offering a chance for visitors to take a break from the beach or shopping to observe and learn about natural attractions' (Weaver 2001a: 110). (See also Diamantis 2000, on ecotourism and the Mediterranean islands, where the vast majority of ecotourists were characterised as occasional by virtue of their interest in other types of activity.) In a subsequent publication, Kontogeorgopoulos (2004b) argues that Phuket is a prime example of the importance of overlapping mass and ecotourism markets, which has helped ecotourism survive financially. This example shows that: (1) package tourists are willing to accept and employ a sense of responsibility in a type of experience that might radically differ from the sea-sun-sand-sex experience; and (2) 'a one-size-fits-all approach is insufficient when defining ecotourism or assessing its potential for promoting sustainability, environmental education, and ethical management' (Kontogeorgopoulos 2004a: 105).

But what is seen as an advantage, above, by some is a disadvantage to others. The problem with ecotourism according to Myles (2003: 185) is that it fails to differentiate between the following two experiences: (1) a large group of people visiting a private game reserve, arriving at their destination by aircraft, staying in luxury accommodations, expecting to be pampered with all the modern conveniences, and spending very little money at the destination because everything is prepaid before departure; and (2) a low impact ecotraveller, backpacking or camping to save money on accommodation, but spending money spontaneously wherever they travel, pursuing a minimal impact experience in a wilderness environment. Myles would likely sit at the other end of the spectrum from Weaver and Kontogeorgopoulos, when he says that because of mass tourism, ecotourism is losing its integrity. This would seem to be the position of Diamantis (1999), too, who argued that there is a tendency for ecotourism to slip with apparent ease into a small-scale form of mass tourism. The challenge according to Diamantis is to identify measures to prevent ecotourism from becoming what he viewed as a 'mass ecotourism' phenomenon.

Assistance in further examining the argument for and against mass-tourism-as-ecotourism may come in the form of resource use and infrastructure support used by tourists to accomplish their ecotouristic ends. The ecological footprint has emerged as a creditable tool by which to calculate the amount of land needed to produce the resources that tourists or other user groups use, and absorb the wastes that are produced from touristic activities (see Rees and Wackernagel 1996; Fennell 2002b; Gössling et al. 2002). As observed by Hunter and Shaw (2006), the fact that some softer path ecotourists stay in much more luxurious forms of accommodation, as is the case with Phuket, above, means that their higher resource demands (use of energy, water, foodstuffs, etc.) will contribute to a larger ecological footprint. The nature of this larger ecotourism footprint has been discussed by Cater (2006), who argues that the economic elite gain considerable social capital by visiting exclusive softer path ecotourism resorts, despite the fact that they might not be ecologically benign. By contrast, the deep ecotourist does, or should, not require a level of western comfort that would place strain on the ecological resource base of the destination (Acott et al. 1998).

The connection to the mass tourism market that companies like SeaCanoe and Siam Safari (noted above) have enjoyed in southern Thailand may ultimately have contributed to some rather significant problems that have evolved more recently. Shepherd (2002) writes that these companies have experienced troubles because it has been relatively easy for competitiors to move in and claim a piece of the pie too. With few legal measures in place to control the development of new companies using the same premise – and the same piece of land – there is a danger of trying to accommodate too many visitors in restricted spaces and periods of time. Such has also been the case in Costa Rica, which has experienced tremendous growth in tourism over the last two decades. Lumsdon and Swift (1998) argue that as the country moves towards a mass tourism market culture – where the volume of tourists can no longer be referred to as eco-specialists, but rather eco-participants – stringent controls will need to be implemented to safeguard the country's cultural and ecological assets from larger densities of visitors. The complexity behind fixing the ecotourism label on mass tourists has been discussed by Johnson (2006) in the context of cruise line ecotourism in the Caribbean. Cruisers could not become anything more than soft ecotourists because of a lack of time afforded to passengers in which to experience the attraction, as well as an emphasis on selling the attraction rather than environmental education. In addition, there appeared to be no meaningful attempt to support or encourage indigenous culture through the purchase of local products – thus minimising 'ecotourism' benefits to local people. Further investigation is required to see if stronger linkages to the mass market contribute to some of the problems identified above. Still, this work has not stopped other researchers from using the ecotourism label even for the conventional cruise market. Beadles Thurau et al. (2007) identified four different markets of cruisers along the Panama Canal. These include cultural discovery tourists, action adventure tourists, conventional tourists and natural discovery tourists. These authors conclude by suggesting that ecotourism attractions were preferred to a degree not expected, and that more marketing should be directed to this area in satisfying this demand.

CASE STUDY 3.4

Urban ecotourism

In one of the classic early papers on tourism, Christaller (1963) wrote that tourism by nature avoids central places in favour of the natural resource base found outside the limits of the city. This, he reasons, is attributable to the notion that cities have become urbanised and overpopulated to the point that quality of living could be better realised through such excursions. Peripheral regions were thus seen by Christaller as the antithesis of cities. By its nature, tourism has facilitated the process for capitalism to achieve further growth in new sectors and in new places. As suggested by Husbands, 'Tourism is . . . an example of such a new activity [in the periphery] and the peripheral space provided by it is an example of the production of space, and the further penetration of the periphery by capitalism' (Husbands 1981: 51). This new space has been most notably used for sun, sea and sand tourism (e.g. Caribbean), but also more recently in regions which support adventure travel and ecotourism.

Given the foregoing discussion, it is interesting to note an emerging focus on eco-tourism in urban regions. This phenomenon may be defined as travel and exploration within and around an urban area that offers visitors enjoyment and appreciation of the city's natural areas and cultural resources, while inspiring physically active, intellectually stimulating and socially interactive experiences; promotes the city's long-term ecological health by promoting walking, cycling, public transportation; promotes sustainable local economic and community development and vitality; celebrates local heritage and the arts; and is accessible and equitable to all (Blackstone Corporation 1996: 2–5).

Such an elaborate definition underscores the importance of nature, culture, physical activity, sustainability and community development in urban areas. Logical places for urban ecotourism include parks, cemeteries, golf courses, sewage lagoons and stormwater control ponds, landfill and waste disposal sites, high rise and other structures, and zoos and botanical gardens (Lawton and Weaver 2001). Good examples of these include El Avila National Park in Caracas, Venezuela, which lies just to the north of the city, containing hundreds of well-kept trails and providing excellent views of the coast and Caracas; Century City in South Africa, and New York's Central Park. Lau and Johnston (2002) argue that Auckland, New Zealand is emerging into an ecotourism destination in its own right through a range of businesses and services that use local resources to sustain the industry at the micro scale. Although the market for ecotourism is not nearly the size of general interest tourism, there has been a deliberate move to cater to the specialised interests of those seeking eco-experiences. Perhaps one of the most intriguing issues concerning this phenomenon is how to reconcile the rather lofty philosophical base of ecotourism and the pace and congestion of the urban environment.

Lawton and Weaver (2001) observe that golf courses (and cemeteries for that matter) are notorious for their use of pesticides in keeping lawns green. Although these regions may harbour keynote species like eagles, alligators and bears, the volume of chemicals introduced to the environment raises serious questions about the ecological health of the entire ecosystem. This can also be the case with the concept of the urban marineland habitat, which contains a variety of marine mammals like killer whales. Unfortunately, however, these animals are unable to live out their life cycles fully because of a variety of issues related to captivity. Conversely, many authors (Lawton and Weaver 2001; Higham and Lück 2002) have identified a number of innovative developments in urban environments which have stimulated ecotourism through the creation of space, much in the way described by Husbands earlier. Municipalities have reclaimed mines as well as waste disposal sites and turned these into green belts, including parks and golf courses. These have become important sites for native flora and fauna, and have contributed to developments which have allowed managers to link green belts (corridors) throughout the urban environment. Higham and Lück (2002) have taken up this issue in suggesting that many groups have made significant contributions to conservation through the restoration of natural areas which have been degraded by human intervention, making ecotourism in the urban environment an option.

As a case in point, Canada's Green Tourism Association (Toronto) argues that the principles of ecotourism applied to urban tourism may actually be better for the environment in certain places because cities are able to absorb the impacts of tourism when compared to wilderness areas (Dodds and Joppe 2001). The city of Toronto has approximately 20,000 acres of green space within the city, including about 370 species of birds. Examples of ecotourism include the Leslie Street Spit, which acts as a migratory flyway; R.C.Harris Filtration plant; the Humber River; self-guided discovery walks throughout the city; the Don River restoration project; and a number of bike, blade and heritage walks provided

continued

by the conservation authority. In addition, the Green Tourism Association, in concert with the city, published the '"Other" Map of Toronto', which is a green map linking tourism with the natural environment. In February 2002, the Green Tourism Association participated in the International Outdoor Adventure Show in Chicago where they conducted a study which revealed that 71 per cent of visitors said they would use businesses, restaurants and accommodation that made provisions to minimise negative environmental and social impacts. In a previous study carried out in July 2000, 91 per cent of respondents said that when they visit a city they 'sometimes or often' visit green spaces and park lands. The study also observes that 83 per cent of those polled felt that ecotourism could take place in a city (Green Tourism Association 1999).

It remains to be seen how effective cities will be in stimulating ecotourism. There is little question that the greening of cities is an urban response to sustainable development, and more specifically sustainable tourism. Perhaps one of the main constraints for ecotourism in the urban environment is overcoming the idea that tourism, from a spatial context, is a place for business and culture, and less for adventure and ecotourism. There is also the argument, although controversial, that tourist dollars might better be spent in peripheral areas, possibly helping to stimulate the economies of marginalised peoples – acknowledging that there are many intervening factors which might catalyse or constrain people from venturing into more peripheral areas. However, if planners are able to integrate key aspects of conservation, environmental education, community welfare (broadly defined), ethics and sustainability, then there may be few reasons to dispute the legitimacy of ecotourism in the urban environment.

Conclusion

The possibility of confusing ecotourism with many other forms of NBT is complicated by the fact that these other forms share the same setting as ecotourism. Adventure tourism, wildlife tourism, NBT and cultural tourism, were examined for the purpose of navigating these often uncertain waters. The discussion on consumptive activities like fishing and hunting and non-consumptive activities like ecotourism provided assistance in moving the debate forward. There continue to be questions about the legitimacy of activities that, while helpful to local people in terms of jobs and other local benefits, have implications on the welfare of individual animals as well as populations. These are discussions that will be examined in more detail as the discussion on the key criteria used to define ecotourism continues in the following chapters.

Summary questions

1 What is a natural resource, and why are natural resources so important to the ecotourism industry?
2 How is wildlife tourism different than ecotourism?
3 How is a consumptive activity different than a non-consumptive activity?
4 Some theorists argue that ecotourism can fit within a mass tourism model. Agree or disagree with this perspective.
5 How can ACE tourism translate into different niche opportunities for tourism operators sharing the same physical space?

4 Sustainability 1: local participation and benefits

This chapter – the first of two chapters on the sustainability criterion of ecotourism – is partitioned into two main sections. The first section deals with the concept of sustainability itself. This includes a discussion of the basic features of sustainable development, as well as a link between sustainable development and tourism. The second section focuses more on the need for local participation and local benefits that are such essential aspects of ecotourism. Chapter 5 deals more with the conservation agenda that is tied to sustainability and ecotourism. Owing to the fluid nature of many topics and issues in ecotourism, aspects of local participation and benefits will be discussed in some of the chapters to follow, including economic impacts of ecotourism (Chapter 9) as well as policy and governance (Chapter 10).

Sustainable development and tourism

The measurement of development (i.e. a nation's stage of socio-economic advancement) has often been discussed via key economic indicators, including protein intake, access to potable water, air quality, fuel, health care, education, employment, gross domestic product (GDP) and gross national product (GNP). The so-called 'developed' world (countries like Australia, the USA, Canada and those of western Europe) is characterised by the existence of these socio-economic conditions, whereby those with more are considered more highly developed (more on development in Chapter 10). Furthermore, one's level of development, either objectively or subjectively, is often synonymous with one's perceived stage of 'civilisation', whereby progress (usually economic) is a key to the relationship between who is civilised and who is not. The *Oxford English Dictionary* defines civilisation as an 'advanced stage of social development', and to civilise as to 'bring out of barbarism, enlighten'. The point to be made is that our perception of what is developed and what isn't, what is civilised and what isn't, is a matter of debate and one that our more recent approaches to development need to better address. For example, the most developed 20 per cent of the world's population (those in the 'West') are thought to use some 80 per cent of the world's resources in achieving development status. If it is our goal to have the entire world 'developed' according to this Western paradigm, the planet will be in serious jeopardy (acknowledging that Western countries are uneven in their use of natural and social resources in their development).

Deming (1996) shares the view that humanity needs to take a good long look at civilisation, by observing that people have an insatiable hunger to see more and more of the planet, and to get closer and closer to its natural attractions. This behaviour surfaces continually in tourism as the tentacles of the tourism industry seek to push the fine line that exists between acceptable and unacceptable human–wildlife interactions. For example, animal harassment regularly occurs in Point Pelee National Park in Ontario,

Canada, as thousands of birders converge on the spring migration of birds in the park. Despite posted warnings, tourists continue to venture off the designated paths in identifying and photographing species. Deming asks: in the face of global warming, diminishing habitat and massive extinctions, what can it mean to be civilised? Her response is a plea for limits, both social and ecological, in facing the enemy within:

> As Pogo said during the Vietnam War, 'We've seen the enemy and it is us.' Suddenly we are both the invading barbarians and the only ones around to protect the city. Each one of us is at the center of the civilized world and on its edge.
>
> (Deming 1996: 32)

Milgrath (1989) talked of values as fundamental to everything we do (see also Forman 1990). He argues that humans have as a central value the desire to preserve their own lives, which has naturally evolved into a concern and value for other people – a social value. The face of this social orientation, Milgrath says, is most noticeably reflected though economic development priorities with serious implications for the long-term sustainability of societies and the resources upon which they rely. This form of instrumentalism (something valued as a means to an end), takes us away from the realisation that non-human entities have value in and of themselves, and should exist in their own right. This 'ethic of nature' perspective is one which is more broadly intrinsic and ecocentric (Wearing and Neil 1999; more on this in Chapter 7).

Sustainable development has been proposed as a model for structural change within society; one that ventures away from a strictly socio-economic focus to one where development, 'meets the goals of the present without compromising the ability of future generations to meet their own needs' (World Commission on Environment and Development 1987: 43). As such, the principles of ecology are essential to the process of economic development (Redclift 1987), with the aim of increasing the material standards of impoverished people living in the world (Barbier 1987). Sustainable development's advocacy of balance between economic, social and ecological systems makes it especially relevant to tourism where there is a wealth of literature that has emerged since the 1980s.

One of the first action strategies on tourism and sustainability emerged from the Globe '90 conference in British Columbia, Canada. At this meeting representatives from the tourism industry, government, non-governmental organisations (NGOs), and academe discussed the importance of the environment in sustaining the tourism industry, and how poorly planned tourism developments often erode the very qualities of the natural and human environment that attract visitors. The conference delegates suggested that the goals of sustainable tourism are: (1) to develop greater awareness and understanding of the significant contributions that tourism can make to environment and the economy; (2) to promote equity and development; (3) to improve the quality of life of the host community; (4) to provide a high quality experience for the visitor; and (5) to maintain the quality of the environment on which the foregoing objectives depend. Although their definition of sustainable tourism development was somewhat non-committal (i.e. 'meeting the needs of present tourist and host region while protecting and enhancing opportunity for the future'), a number of good recommendations were developed for policy, government, NGOs, the tourism industry, tourists and international organisations. For example, the policy section contains 15 recommendations related to how tourism should be promoted, developed and defined, in addition to a series of regional, interregional and spatial and temporal implications. One of the policy recommendations states that, 'sustainable tourism requires the placing of guidelines for levels and types of acceptable growth but does not preclude new facilities and experiences' (Globe '90 1990: 6).

From the perspective of financial prosperity and growth, there is an economic rationale for sustainability; as McCool (1995: 3) asserts, 'once communities lose the character that makes them distinctive and attractive to non-residents, they have lost their ability to vie for tourist-based income in an increasingly global and competitive marketplace'. In addition, McCool quotes Fallon in suggesting that sustainability is all about the pursuit of goals and measuring progress towards them. No longer is it appropriate to gauge development by physical output or economic bottom lines; there must also be consideration of social order and justice (see also Hall 1992 and Urry 1992). McCool feels, therefore, that in order for sustainable tourism to be successful, humans must consider the following: (1) how tourists value and use natural environments; (2) how communities are enhanced through tourism; (3) identification of tourism's social and ecological impacts; and (4) management of these impacts (see also Liu 2003).

Accordingly, theorists have initiated the process of determining and measuring impacts. As outlined above, Globe '90 was one of the initial and integral forces in linking tourism with sustainable development. This was followed by Globe '92 (Hawkes and Williams 1993) and the move from principles to practice in implementing measures of sustainability in tourism. Even so, it was recognised in this conference that there was much work to be done, as emphasised by Roy (in Sadler 1992: ix):

> Sustainable tourism is an extension of the new emphasis on sustainable development. Both remain concepts. I have not found a single example of either in India. The closest for tourism is in Bhutan. Very severe control of visitors – 2000 per year – conserves the environment and the country's unique socio-cultural identity. Even there, trekking in the high altitudes, I find the routes littered with the garbage of civilization.

Although many examples exist in the literature on tourism and sustainable development (see Nelson *et al.* 1993), few sustainable tourism projects have withstood the test of time. An initiative that has received some exposure in the literature is the Bali Sustainable Development Project, coordinated through the University of Waterloo, Canada and Gadjahmada University in Indonesia (see Wall 1993; Mitchell 1994). This is a project that has been applied at a multi-sectoral level. Tourism, then, is one of many sectors, albeit a prime one, that drives the Balinese economy. Wall (1993) suggests that some of the main conclusions from his work on the project are as follows: (1) be as culturally sensitive as possible in developing a sustainable development strategy; (2) work within existing institutional frameworks as opposed to creating new ones; and (3) multi-sectoral planning is critical to a sustainable development strategy and means must be created to allow all affected stakeholders to participate in decision making. (See also the work of Cooper (1995) on the offshore islands of the UK and the work of Aylward *et al.* (1996) on the sustainability of the Monteverde Cloud Forest Preserve in Costa Rica as good examples of tourism and sustainability.) The integration of tourism with other land uses in a region has been addressed by Butler (1993: 221), who sees integration as, 'the incorporation of an activity into an area on a basis acceptable to other activities and the environment within the general goal of sustainable or long-term development'. Butler identified complementarity, compatibility and competitiveness as variables that could be used as a first step in prioritising land uses, where complementarity leads to a higher degree of integration, and competitiveness leads to segregation of the activity relative to other land uses.

Other models have been more unisectoral in their approach to the place of tourism and sustainability within a destination region. These have tended to underscore a range of indicators that identify sustainable/unsustainable approaches to the delivery of tourism. Examples include Canova's (1994) illustration of how tourists can be responsible towards the environment and local populations; Forsyth's (1995) overview of sustainable tourism and self-regulation; Moscardo *et al.*'s (1996) look at ecologically sustainable forms of tourism accommodation; and Consulting and Audit Canada's (1995) guide to the development of core and site-specific sustainable tourism indicators (see also Manning 1996). Table 4.1 identifies the core indicators identified in this document (e.g. site protection, stress, use intensity, waste management etc.) which must, according to the report, be used in concert with specific site or destination indicators. This report identifies two categories of this latter group of indicators: (1) supplementary ecosystem-specific indicators (applied to specific biophysical land and water regions), and (2) site-specific indicators, which are developed for a particular site. Table 4.2 provides an overview of some of these 'secondary' ecosystem indicators.

Research has also discussed tourism and sustainability from the perspective of satisfying basic principles or codes of ethics (codes of ethics are discussed at length in Chapter 7). While indicators are variables that are identified and used to measure and monitor tourism impacts, codes of ethics are often lists of guidelines designed to elicit change in the behaviour of stakeholder groups; a form of compliance for acceptable behaviour at a tourism setting. The *Beyond the Green Horizon* paper on sustainable tourism (Tourism Concern 1992) is a good example of this form of education, which can be seen in the following definition of sustainable tourism and its ten guiding principles (see Figure 4.1). To Tourism Concern, sustainable tourism is:

> tourism and associated infrastructures that, both now and in the future: operate within natural capacities for the regeneration and future productivity of natural resources; recognise the contribution that people and communities, customs and lifestyles, make

Table 4.1 *Core indicators of sustainable tourism*

Indicator	Specific measures
Site protection	Category of site protection according to IUCN index
Stress	Tourist numbers visiting site (per annum/peak month)
Use intensity	Intensity of use in peak period (persons/hectare)
Social impact	Ratio of tourists to locals (peak period and over time)
Development control	Existence of environmental review procedure or formal controls over development of site and use densities
Waste management	Percentage of sewage from site receiving treatment (additional indicators may include structural limits of other infrastructural capacity on site, such as water supply)
Planning process	Existence of organised regional plan for tourist destination region (including tourism component)
Critical ecosystems	Number of rare or endangered species
Consumer satisfaction	Level of satisfaction by visitors (survey-based)
Local satisfaction	Level of satisfaction by locals (survey-based)
Tourism contribution to local economy	Proportion of total economic activity generated by tourism
Composite indices	
Carrying capacity	Composite early warning measure of key factors affecting the ability of the site to support different levels of tourism
Site stress	Composite measure of levels of impact on the site (its natural/cultural attributes due to tourism and other sector cumulative stress)
Attractivity	Qualitative measure of those site attributes that make it attractive to tourism and can change over time

Source: Consulting and Audit Canada (1995)

Table 4.2 *Ecosystem-specific indicators*

Ecosystem	Sample indicators[a]
Coastal zones	Degradation (percentage of beach degraded, eroded) Use intensity (persons per metre of accessible beach) Water quality (faecal coliform and heavy metals counts)
Mountain regions	Erosion (percentage of surface area eroded) Biodiversity (key species counts) Access to key sites (hours' wait)
Managed wildlife parks	Species health (reproductive success, species diversity) Use intensity (ratio of visitors to game) Encroachment (percentage of park affected by unauthorised activity)
Ecologically unique sites	Ecosystem degradation (number and mix of species, percentage area with change in cover) Stress on site (number of operators using site) Number of tourist sitings of key species (percentage success)
Urban environments	Safety (crime numbers) Waste counts (amounts of rubbish, costs) Pollution (air pollution counts)
Cultural sites (built)	Site degradation (restoration/repair costs) Structure degradation (precipitation acidity, air pollution counts) Safety (crime levels)
Cultural sites (traditional)	Potential social stress (ratio average income of tourists/locals) In season sites (percentage of vendors open year round) Antagonism (reported incidents between locals and tourists)
Small islands	Currency leakage (percentage of loss from total tourism revenues) Ownership (percentage foreign ownership of tourism establishments) Water availability (costs, remaining supply)

Source: Manning (1996)

Note: [a] These ecosystem-specific indicators are merely suggested, and act as supplements to core indicators

to the tourism experience; accept that these people must have an equitable share in the economic benefits of tourism; are guided by the wishes of local people and communities in the host areas.

Nothing is measured but 'rules' are stated for the purpose of prompting or reinforcing this appropriate behaviour.

The Tourism Industry Association of Canada (1995) joined forces with the National Round Table on the Environment and the Economy in creating a document that demonstrates commitment and responsibility in protecting the environment through cooperation with other sectors and governments at all levels. Their sustainable tourism guidelines were developed for tourists, the tourism industry, industry associations, accommodation, food services, tour operators and Ministries of Tourism. Each of these sections contains appropriate guidelines that deal with policy and planning; the tourism experience; the host community; development; natural, cultural, and historic resources; conservation of natural resources; environmental protection; marketing; research and education; public awareness; industry cooperation; and the global village.

A final publication that merits attention in this section is the work of the Federation of Nature and National Parks of Europe (1993). Its comprehensive look at sustainable

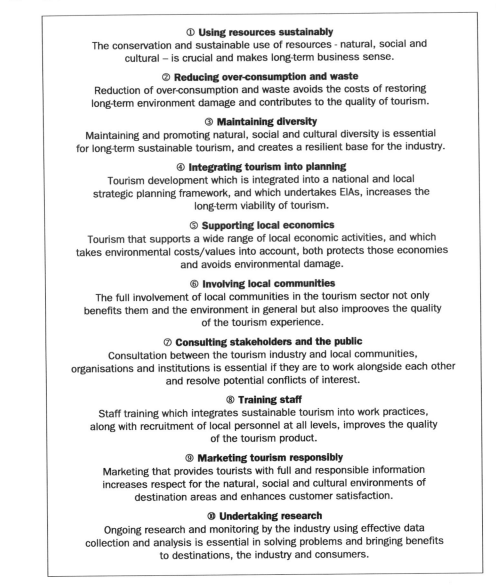

① Using resources sustainably
The conservation and sustainable use of resources - natural, social and cultural – is crucial and makes long-term business sense.

② Reducing over-consumption and waste
Reduction of over-consumption and waste avoids the costs of restoring long-term environment damage and contributes to the quality of tourism.

③ Maintaining diversity
Maintaining and promoting natural, social and cultural diversity is essential for long-term sustainable tourism, and creates a resilient base for the industry.

④ Integrating tourism into planning
Tourism development which is integrated into a national and local strategic planning framework, and which undertakes EIAs, increases the long-term viability of tourism.

⑤ Supporting local economics
Tourism that supports a wide range of local economic activities, and which takes environmental costs/values into account, both protects those economies and avoids environmental damage.

⑥ Involving local communities
The full involvement of local communities in the tourism sector not only benefits them and the environment in general but also improves the quality of the tourism experience.

⑦ Consulting stakeholders and the public
Consultation between the tourism industry and local communities, organisations and institutions is essential if they are to work alongside each other and resolve potential conflicts of interest.

⑧ Training staff
Staff training which integrates sustainable tourism into work practices, along with recruitment of local personnel at all levels, improves the quality of the tourism product.

⑨ Marketing tourism responsibly
Marketing that provides tourists with full and responsible information increases respect for the natural, social and cultural environments of destination areas and enhances customer satisfaction.

⑩ Undertaking research
Ongoing research and monitoring by the industry using effective data collection and analysis is essential in solving problems and bringing benefits to destinations, the industry and consumers.

Figure 4.1 Principles for sustainable tourism

Source: Tourism Concern (1992)

tourism in Europe's nature and national parks provides good insight into the challenge of implementing sustainability in that part of the world. Many of the protected areas in Europe are situated in rural working landscapes (e.g. England, Wales, Luxembourg) and must contend with different pressures as compared with some of the larger and less densely populated areas surrounding the protected areas of Australia, New Zealand, Canada and the USA. However, Europe also contains many large national parks and biosphere reserves that are maintained accordingly. In both cases (rural and wilderness environments) policy-makers and practitioners are charged with the task of implementing sustainable tourism in these varied settings. The European national parks document recognises that people must be able to improve the quality of their lives, maintain jobs, improve their economy, enjoy their cultures and promote harmony between cultures. These must be accomplished with an eye to environmental education, political support for the environment, and the protection of heritage values through restorative projects and direct practical help.

While the principles behind sustainable tourism are sound in theory, researchers are often sceptical about how theory translates into good practice. Hunter (1995) suggests that sustainable tourism development approaches are often flawed because they condone the planning and management of tourism in a manner inconsistent with the design of sustainable development. In particular, tourism does not adequately address issues of geographical scale and intersectoral cooperation which are so important for achieving sustainable development. Furthermore, Macbeth (1994) calls attention to the fact that sustainable tourism is more reactionary than proactive in nature. Macbeth suggests that, 'the history of capitalism is full of examples of how reactionary tendencies are easily coopted by capitalism to sustain its own existence, thus extending the status quo of exploitive relations rather than overthrowing them' (1994: 44). This will continue to occur, according to Macbeth, unless the present form of capitalism is overcome. Liu (2003) argues that sustainable tourism research has been patchy and disjointed because of a critical lack of focus on tourism demand, inter-generational equity, the nature of tourism resources and so on. The author suggests that a transformation of current research should take place according to systems and interdisciplinary perspectives (see Fennell 2003).

McKercher (1993a) feels that tourism is vulnerable to losing sustainability for four main reasons. First, tourism is not recognised as a natural resource-dependent industry; second, the tourism industry is invisible, especially in urban areas; third, tourism is electorally weak, with little support in government; and fourth, there is a distinct lack of leadership driving the industry, which ultimately makes tourism vulnerable to attacks from other land users. McKercher cites the example of resource use in northern Ontario as a case in point. In this region the economy has been dominated politically by the large extractive industries (forestry and mining). The disaggregated structure of the tourism industry in Ontario's north (predominantly outfitters and lodges) prevents it from having any political decision-making influence at all.

Other critical reviews of tourism and sustainability include Goodall and Cater's (1996) belief that sustainable tourism will probably not be achieved, despite the most committed environmental performance; Burr's (1995) work illustrating that sustainable tourism development is unlikely to occur unless the people of rural tourism communities work together to make it happen; and Clarke's (2002) view that no type of tourism can ever properly be sustainable, as sustainability is more typically a process to a desired state rather than an end unto itself. Important in Clarke's message is that sustainability is not just applicable to ecotourism, but rather any form of tourism, including mass tourism depending on how it is planned, developed and managed (see also Laarman and Gregersen 1994). This also means that unique market segments, like ecotourism, can potentially be sustainable based on the use of the natural environment, long-term economic benefits, environmental protection, and stimulates local community development. But they may be equally unsustainable if improperly managed. It is therefore potentially dangerous to look at sustainable tourism as a specific market, instead of from site-specific or regional perspectives.

CASE STUDY 4.1

Sustainable tourism and the Green Villages of Austria

Understanding the vital link between landscape and tourism, Austria has embarked upon a policy of sustainable tourism with the aim of preservation and an overall improvement in the quality of the natural environment. Specifically, the following measures have been proposed:

1 a straightening out of the demand curve to avoid peak demands and burdens;
2 reducing the consumption of space for tourism;
3 preservation of natural landscapes;
4 cooperation with other industries, in particular agriculture and forestry;
5 professionalism within the industry; and
6 a changing of the behaviour of tourists.

One of the most significant programmes in Austria is the Green Village endeavour, which is designed to allow communities to accommodate the growing demands of tourism in a sustainable way. Towns are encouraged to incorporate solar panels in their heating, restrict building height to no more than three storeys, keep parking places a minimum of 80 m away from buildings to eliminate noise and fumes, keep motorways at least 3 km away from Green Villages, restrict vehicular traffic through villages, designate cycle paths, recycle, restrict building to the town site only, eliminate single-crop farming in adjacent farmlands, discriminate in favour of sustainable craftsmen, build hotels using natural products, insist that farmers be able to sell their products locally and use local, natural pharmaceuticals. Such a philosophy, it is thought, will benefit both communities and the tourism industry.

Figure 4.2 illustrates that sustainability has to be more than simply one aspect of the industry (e.g. accommodation) working in a sustainable way. The illustration recognises that in essence the tourism industry experiences a tremendous degree of fragmentation by virtue of the fact that consistency in sustainability is not likely to be found across all sectors. The aim, then, for sustainability is to ensure that all aspects of the industry are working in concert. In addition, the figure incorporates the notion of both human and physical elements working within each of the four sectors; that is, the fact that the people working at a physical attraction very much dictate the extent to which sustainability is achieved at the site. This is a notion examined more recently by Font *et al.* (2006), who observe that tour operators can only be sustainable if suppliers are sustainable in the first instance. But because tour operators have sufficient influence over suppliers throughout the supply chain, and because society is demanding more accountability in tourism operations, tour operators can in fact promote improvements in the performance of their partners. They can do this, according to the authors, through a sustainable supply chain policy and management system; they can support suppliers in reaching sustainability goals; and they can choose suppliers that meet sustainability criteria.

The upshot of the relevance of sustainability to all aspects of tourism is the recent move by mass tourism operators to introduce authentic, less commercialised experiences, 'the discovery of cultures and amazing unspoiled places', in the words of Font *et al.* (2006), in

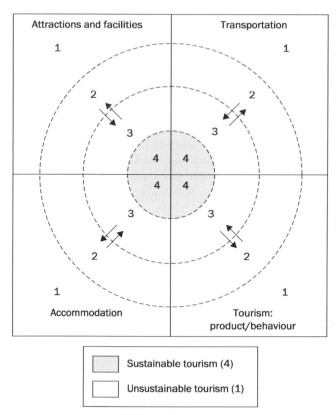

Figure 4.2 Degrees of sustainable tourism

appealing to market demands – keeping pace with society means behaving in a sustainable manner in order to be competitive. It is thought-provoking that this is the case (in reference to unspoiled places) from the mass tourism perspective, given that increasingly there is less of it to be found (see also Welford and Ytterhus 1998). There has also been discussion on striving to attract the 'most desirable type of tourist' in the destination's efforts to be more sustainable (and profitable). This has been addressed by Kaosa-ard (2002), who reports that in Thailand the national tourism authority has been pressured to find better *quality* tourists, as opposed to more *quantity*. The hotel association in Thailand argues that those who stay in the big hotels (ride in chauffeurs and dine in expensive restaurants) are in fact better tourists because they spend more money at the destination. Conversely, civil groups of one form or another argue that the best type of tourist is one who spends his or her money at locally owned hotels and eats at local food stalls, for example, where money penetrates more deeply and widely. In fact what has been found is that the latter group is seen to be more attractive. Based on a study of 1,200 local and foreign tourists, Kaosa-ard *et al.* (1993) found that a very desirable cohort of tourists is females, aged 40 and over, who spend more money per day and have a keen interest in cultural attractions – thereby distributing more money into the hands of local people.

Conceptualising tourism and sustainability

In an effort to better understand and better situate sustainability in tourism, theorists have looked at the relationship of the two (tourism and sustainability) as evolving over time according to different approaches. For example, Clarke (1997) writes that there are four different rather discrete stages in the relationship between tourism and sustainability. The first stage includes a perspective where mass tourism and sustainable tourism were completely separated by a conceptual barrier – polar opposites. Clarke's second approach, *continuum*, places both mass tourism and sustainable tourism on a continuum, based on flexibility of earlier ideas shared between both. The third approach, *movement*, is characterised by mass tourism improving or taking on aspects of sustainability so that it would not be positioned as a villain. The fourth stage, *convergence*, indicates that all forms of tourism, regardless of the scale, could be sustainable in nature.

Some of what Clarke (1997) was referring to in reference to different approaches is found graphically in Figure 4.3, where there is a focus on contrasting mass tourism and AT. In a general sense, the illustration provides a good sense of the relative size of mass tourism and AT according to the corresponding circles in the diagram. Although mass tourism may be said to be predominantly unsustainable, more recently new and existing developments in the industry have attempted to encourage more sustainable practice through various measures, some of which include the controlled use of electricity, a rotating laundry schedule and the disposal of wastes (the arrow indicates that there is a move towards an increasing degree of sustainability in this sector). On the other hand, the illustration indicates that most forms of AT are sustainable in nature (in theory). The AT sphere is shown to comprise two types of tourism, socio-cultural tourism and ecotourism. Socio-cultural AT includes, for example, rural or farm tourism, where a large portion of the touristic experience is founded upon the cultural milieu that corresponds to the

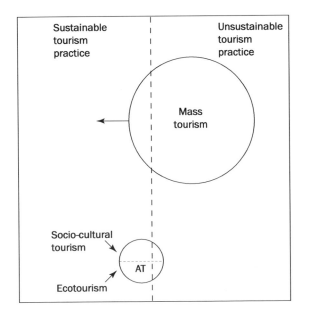

Figure 4.3 Tourism relationships

Source: Adapted from Butler (1996) in Weaver (1998)

environment in which farms operate. Ecotourism, however, involves a type of tourism that is less socio-cultural in its orientation, and more dependent upon nature and natural resources as the primary component or motivator of the trip, hence the division within the AT sphere. What has confounded theorists and practitioners over the years, is the belief that the difficulties in defining ST have led to the inability to properly apply it. This ambiguity has meant that any form of tourism can be termed sustainable, despite the fact that these forms are often not (Butler 1999), as discussed earlier in this chapter.

Supporting local economies, involving local communities

The benchmark for community-based planning and management has for some time been Murphy's (1985) community approach to tourism. Although just as relevant in many respects today as in the 1980s, there has been a great deal of work in the area to expand upon many of Murphy's original ideas. Community development, as described by Smith (1990a), originated in the self-help programmes that were developed during the depression years in Canada, the USA and the UK. A defining characteristic of community development is that it is based on local initiatives by advocating a site-specific approach to finding solutions to community problems using community members and community resources. Bujold (1995: 5) defines community development as, 'the process by which the efforts of the people themselves are united with those of governmental authorities to improve economic, social, and cultural conditions of communities'.

Tourism is increasingly seen as a key community development tool, with the recognition of its economic contribution in bolstering stagnating economies and diversifying existing sectors, and its ability to unify community members. Such is the case in the Shetland Islands, Scotland, where tourism is being relied upon to sustain an economy that was once dominated by North Sea oil development (Butler and Fennell 1994); or the Finnish island of Åland, where all tourism initiatives are owned or controlled by local people (Joppe 1996). Joppe (1996) stresses that it is vitally important to understand that there exists a fundamental division between conventional community development and community economic development (CED) models. While conventional economic development focuses on the attraction of new businesses to the community (seen as an outward-directed approach to development), the CED focus is on being small, green and social, and is more inward in its orientation by striving to 'help consumers become producers, users become providers, and employees become owners of enterprise' (Joppe 1996: 476), through the principles of economic self-reliance, ecological sustainability, community control, meeting individual needs and building a community culture (see also Davidson 1995). What is at stake here is the role that tourism can play in securing quality of life within the community. At the foundation of any tourism development strategy is the realisation that

> If tourism development is to be viable as a long-term economic strategy, these concerns [social and ecological] must be addressed, and the resource base must be protected in the process. The host community is the economic, social, cultural, and infrastructural resource base for most tourism activity, and resident quality of life is a measure of the condition of the resource.
>
> (Christensen 1995: 63)

Christensen proposes a community quality of life framework that addresses both objective and subjective indicators at the individual and community scale. A tourism development project is said to affect changes in the quality of life of members of the community, which in turn cause impacts at different social scales. These impacts need to be evaluated both by individuals and by the community, depending on the scale of reference to the individual. There is a deliberate hierarchy established with respect to agents of change, suggesting that appropriate solutions need to be implemented at the individual social scale first, followed by neighbourhood groups, coalitions, responsible industry, policy and governmental regulation (see also Mabey 1994; Belasco and Stayer 1993; Campbell 2002; Lipske 1992; Williams 1992; Sproule 1996; and White 1993).

The tactical approaches used by regions (local or national governing bodies) will make or break how ecotourism is perceived by local people, according to MacKinnon (1995). She cites a number of examples of communities which have adopted or been identified to develop ecotourism in Mexico. One such community is Tres Garantías in the Yucatan peninsula. This region was initially planned as a hunting reserve by an international development team, but the area's residents found that non-consumptive recreation was more attractive to tourists than consumptive forms. This project was thus conceived using a top-down approach, and even though there have been isolated benefits derived from ecotourism, MacKinnon suggests that social integration within the community is not complete, and will not be complete until benefits are more widespread and the perception of ownership is given to the community. MacKinnon discusses too the example of Los Tuxtlas, Veracruz, to illustrate that ecotourism has been successfully integrated into a community through more of a grassroots approach (initiated through the efforts of a local woman who set out to build empowerment slowly and deliberately). In this latter case, the community has been more quickly able to realise benefits which have been spreading on a more regional scale, along with a better understanding of the advantages of conserving their own natural-resource base – much more so than large developers, according to MacKinnon, who don't live in the area and who perpetuate the widening gap between rich and poor in countries like Mexico.

Another example of how communities cooperate in controlling ecotourism development initiatives is offered by Drake (1991), who defines local participation as, 'the ability of local communities to influence the outcome of development projects such as ecotourism that have an impact on them' (1991: 132). Important in this process is the demonstrated benefit to the community both through community members' participation and through their realisation that some aspect of their community has been treated or protected (e.g. natural resources). The following is proposed as a model of local participation in the development of ecotourism projects (1991: 149–155):

- *Phase 1: Determine the role of local participation in project.* This includes an assessment of how local people can help achieve set goals through efficiency, increasing project effectiveness, building beneficiary capacity and sharing project costs.
- *Phase 2: Choose research team.* The team should include a broad multidisciplinary approach and include people in the social sciences and those within the media.
- *Phase 3: Conduct preliminary studies.* Political, economic and social conditions of the community should be studied in the context of the environment, from existing documents and by other, survey-related, work. Identification and assessment of the following is important: needs, key local leaders, media, the community's commitment to the project, intersectoral involvement, traditional uses of the land, the type of people interested in the project and why, the role of women, who will manage and finance the project, land ownership and cultural values.
- *Phase 4: Determine the level of local involvement.* Local involvement occurs along a continuum from low-intensity to high-intensity. This must be determined in addition to when the involvement is to occur. In cases where government is not supportive of local government, intermediaries (e.g. NGOs) can be used to facilitate local participation.

- *Phase 5: Determine an appropriate participation mechanism.* This is impacted by the level of intensity of participation, the nature of existing institutions (e.g. government, NGOs, citizens' groups), and the characteristics of local people (how vociferous they may or may not be). This phase may include information sharing and consultation, which usually takes the form of a citizen advisory committee with representatives from many groups within the community. The committee is charged with the task of commenting on goals and objectives or other project-related aspects.

- *Phase 6: Initiating dialogue and educational efforts.* The use of the press is important in this phase as a means by which to build consensus through public awareness. Key community representatives can be used in this process. The ecotourism team should explain the goals and objectives of the project, how the project will affect the community, the values of the area, any history of threats and the benefits of the project. Various audiovisual techniques should be used to emphasise these points. Workshops or public meetings could be organised to identify strengths and weaknesses of the project.

- *Phase 7: Collective decision-making.* This is a critical stage that synthesises all research and information from the local population. The ecotourism project team present the findings of their research to the community, together with an action plan. Community members are asked to react to the plan, with the possible end result being a forum through which the team and local people negotiate to reach a final consensus based on the impacts of the project.

- *Phase 8: Development of an action plan and implementation scheme.* In this phase, the team and community develop an action plan for implementing solutions to identified problems. For example, if members of the community express the need to increase the community's standard of living, the team may respond by purchasing agricultural produce from local people at market rates or on a contractual basis. They may also develop a variety of positions to be occupied by local people, including gift shops, research positions, park management positions and private outfitting companies. This local action plan must then be integrated into the broader master plan of the project.

- *Phase 9: Monitoring and evaluation.* Monitoring and evaluation, although often neglected, should occur frequently and over the long term. The key to evaluation is to discover whether goals and objectives set out early in the project's life cycle have been accomplished or not.

More recently, Garrod (2003) suggests changes to Drake's model in a number of subtle ways. For example, with regard to Phase 1 on determining the role of local participation in project, Garrod suggests that focus should be on how this participation might be encouraged, along with the need for such to take place at every stage of the project.

Community development initiatives have a better chance of being accepted by local people if developers acknowledge that different groups within the community want different things, depending on their role in, affinity within, and utilisation of the community. This perspective is discussed by Jurowski (1996), who feels that because the impacts of tourism are not the same for all residents, residents' individual values need to be recognised by tourism developers in order for their projects to be successful.

The first unique group identified by Jurowski is the 'attached resident'. This person is likely to be a long-term resident or an older individual who loves living in the community because of its social and physical benefits. To these people, control over the form and function of their community is important. In general, tourism developers can gain support for their projects from this group by involving citizens in the planning process, establishing a focal point and common theme, developing projects that emphasise heritage themes, and showing that the project has social and ecological benefits for the community.

The second type established by Jurowski, the 'resource user', typically includes people like anglers and other recreationists who, although ambivalent about the economic impacts of tourism, can be won over by developers. The developers gain their support by providing skill opportunities for youth, involving this group in events related to their interests (e.g. bike races), protecting 'their' sites for participation, and allocating tourism funds for the development of facilities and services they desire.

The final group outlined by Jurowski is the 'environmentalist'. Although this group is the most likely to focus on the negative aspects of development, it is in the best interests of development teams to do the following: (1) provide information on how the project will protect the environment; (2) incorporate ecological education programmes; (3) encourage the participation of environmentalists in development; and (4) prompt citizens to develop their own educational programmes for tourists. Projects that reflect the interests and concerns of the community, therefore, are said generally to stimulate volunteer activity and minimise conflict. This research is attractive because it acknowledges that people residing in the local community are indeed different from one another. This is in contrast to much tourism research, which considers the local community as one homogeneous group (see also the work of Weaver and Wishard-Lambert 1996).

CASE STUDY 4.2

Ecotourism and Monarch butterflies

One of North America's great yearly migrations is undertaken by the Monarch butterfly which, in the millions, flies upwards of 8,000 km from Canada/USA to one of 15 small overwintering sites spread over only 50 acres in the Transverse Neovolcanic belt of Mexico. This region contains sufficient quantities of fir trees, moisture, wind protection and nectar, needed for survival. Here the butterflies will stay for about five months before venturing off north. The most astonishing aspect of this migration is that no single butterfly ever makes the entire round-trip. As suggested by Halpern (1998), three to four generations separate those that came from Mexico from those that return. Halpern says that a Monarch born in August in New York flies all the way to Mexico to overwinter, and begins the northward journey in March, laying eggs along the coast from Texas to Florida before dying. Butterflies born in Florida, for example, continue north breeding and laying eggs along the way, as do their offspring. By August, another butterfly, four generations removed from the one born in New York the previous summer, will metamorphosise in New York and repeat the steps. In completing the cycle the butterfly goes to a place, the small parcel of land in Mexico that it has never been to.

While the natural history associated with this story is a great mystery of science, the human component is rather less spectacular. Barkin (2000) writes that the creation of a special biosphere reserve to protect the Monarch from human encroachment (a result of a 1976 article in *National Geographic Magazine* by Urquhart which detailed the journey of the Monarch) had many catastrophic implications for local communities. The NGO responsible for establishing the reserve, through the World Wildlife Fund, did not undertake any consultation with the local community. In establishing protection for the butterflies, no compensation was afforded to the locals who, without the ability to harvest resources from the protected areas, were unable to earn a living. And while some facilities were constructed to cater to tourists, in no way were they able to accommodate the needs of 250,000 visitors (1998–99 figures) – representing a tenfold increase from a decade earlier.

Not only did local people lose the ability to control their lands, the benefits of ecotourism were only realised by a relatively small percentage of the inhabitants of the region. In many cases, the shortage of jobs forced some people to leave their families behind in order to search for other low-paying jobs as far away as Mexico City (three hours distant) to support their families. Still others who traditionally logged the forests of the region were forced through necessity to go back to this way of life illegally. Aridjis (2000) and Halpern (1998) observe that even those in charge of reserves have been involved in illegal logging due to the corruption that has become common in the region. Some individuals have resorted to putting large gashes in the trees so they eventually die in order to take them 'legally'. Added to this is the feeling that some of the reserves are in danger, as lucrative logging concessions are awarded in exchange for political patronage.

While some have suggested that ecotourism is one of the few ways to rectify the problem (e.g. through developing small guest houses and having children and women sell souvenirs), others are not so optimistic. Halpern notes that such a scheme is unrealistic because in the absence of sound planning there is a lack of infrastructure to support a growing number of tourists; it does not take into account the impact from these tourist numbers (trail erosion, garbage, water pollution etc.); and there are no regulations in place to limit the number of tourists in the same way that logging was regulated in the past. There is also the issue of environmental education that appears to be absent in local guides and tourists. Halpern notes that an American 'naturalist' who wanted to go past a barbed-wire fence to get closer to the butterflies paid a guide US$8 to lift the fence. In getting closer to the butterflies, the tourist trampled upon a newly seeded part of the reserve, killing some of the seedlings in the process.

What is needed, according to Barkin, is a new model to break the cycle of impoverishment and environmental degradation. He says that ecotourism can provide opportunities for local people, but only as a complementary activity to other land uses. He proposes the following:

- Government must recognise locals' right to be in control and benefit from the Monarch butterfly, through training, technical assistance and resources.
- Mechanisms are needed to ensure the traditional system of food production is maintained.
- Consider the relevance of sustainable tourism practices for the region, although it is unlikely that the Mexican government will recognise this form of tourism for rural development.
- Exploration of the potential to harness geothermal energy for agro-industrial, recreational and medicinal purposes (spas).
- A way to address the seasonality problem through the promotion of cultural tourism and other aspects of the region's biodiversity.
- To spread the benefits of tourism to many, instead of resting it in the hands of a few.
- A move away from current tourism policies which are environmentally destructive, contributing to social polarisation with a great deal of leakage.
- An attempt to integrate tourism into regional development for a more balanced economy.

This situation is quite similar to the problem in Tres Guarantias, Mexico, where the imposition of ecotourism from a top-down perspective has proven disastrous to local people.

Again, the aspect of control emerges as a critical element in the development of an ethical ecotourism industry. Communities are maintained in an equilibrium state before the imposition of tourism (tourism which will benefit a few). Time, education, control, and the intrinsic nature of involvement need to be factored into the formula for any degree of success within such an enterprise.

Aboriginal ecotourism

Ecotourism as a community development vehicle for Aboriginal people has emerged as an important topic. Too often there are cases where Aboriginal people, often living in remote regions, are margnalised socially and economically with very few options to make a living. Since Aboriginal people have a close connection to the land, there is the willingness – almost the need – of Aboriginal people to work with the land. There are certainly numerous success stories in the literature. For example, Hitchcock (1993) illustrates that the semi-nomadic pastoralist people of Purros in Namibia have been able to develop a successful ecotourism industry largely on the back of transparency and equal benefits (profit) distribution. Other aspects of the ecotourism initiative include: (1) having tourists abide by structured guidelines (developed by the local people) (see Table 4.3); (2) tour guiding and handicraft prices are set to avoid undercutting; and (3) the development of programmes to educate tourists on resource use in the area. This type of community action, according to Hitchcock, is thought to be on the rise in the less developed countries (LDCs).

However, what has emerged from the preceding discussion is that although community development and management is theorised to be more 'people-centred', it has fallen short in certain cases for a variety of different reasons. Tipa and Welch (2006) list these reasons as follows: (1) the state is almost always the dominant partner in such associations; (2) indigenous people frequently lack tools and expertise to complete tasks; (3) the process ties indigenous people to formal institutional and legislative rules that strips their ability to act on behalf of special interest groups; (4) the relationship often incurs costs borne by the community; and (5) the concept of 'community' is often defined more broadly such that Indigenous people are merely one group of several that may be outweighed by the values and expectations of other non-indigenous parties.

The case of coral reef management in Gili Indah Indonesia is representative of the fact that community-based management is not always the most beneficial manner by which to govern resources (Satria *et al.* 2006). Despite the homogeneity of the community (all residents coming from South Sulawesi) a newly implemented zoning rule is seen to be more favourable to the tourism industry because it has restricted access and property rights of fishers, with the end result of marginalising fishers in their own backyard, and allowing the tourism industry to be a majority in holding management and exclusion rights. The authors cite Ostrom (1999), who observed that often local tyrannies will prevail because:

> Not all self-organized resource governance systems will be organized democratically or rely on the input of most appropriators. Some will be dominated by a local leader or power elite who only make changes that will be an advantage to them.
>
> (Ostrom 1999, cited in Satria *et al.* 2006: 113)

Other theorists share this sentiment as well. Investigating the perspectives of three stakeholder groups involved in community-based ecotourism (decision-makers, operators and tourists) in Malaysia, Indonesia and Thailand, Rocharungsat (2004) concluded that:

it is unrealistic to imagine that communities will always be able to successfully and independently implement all stages of CBT: juggling marketing, hospitality, conservation and evaluation. Nevertheless, if all stakeholders are serious in their hopes to realize a successful sustainable model of CBT, we must first share this goal of a strong, empowered community.

(REST 2003: 10)

The view that ecotourism does not always benefit local communities is explored by Stonza (2007). Increased incomes from ecotourism are not always the solution. Income may meet their short-term needs, but there is no guarantee that this income translates into a heightened sense of responsibility for conservation. Furthermore, this extra income may be used to purchase tools (fishing tackle, hunting supplies) that may allow local people to exploit resources further. In other work, Stronza and Gordillo (2008) investigated communities involved in three different ecolodge projects in the Amazon, and found that apart from increases in income, there were various social discords. Interviews and focus group data indicated that there were often decreases in reciprocity among community members. Leaders from Kapawi, for example, argued that *mingas* (traditional gatherings of community members to work on community tasks) have decreased in frequency, especially in places close to ecolodges, because of obligations at the ecolodges and the income generated from employment. Some workers at the Posada Amazonas ecolodge in Peru have even tried to 'buy out' their communal work responsibilities. Kiss (2004) corroborates these findings by suggesting that many community-based ecotourism projects, despite being heralded as success stories, actually involve little change in local land- and resource-use practices. Furthermore, only modest improvements in local livelihoods seem to be the norm, and there is too much dependency on external support over the long run.

Nepal (2004) lists a number of important constraints facing the Tl'azt'en First Nations group in central British Columbia, Canada, in the development of ecotourism. Issues included the belief that most of the commercial ecotourism niches were already filled; poor understanding of the target market; competition with other northern First Nations communities for ecotourism; and the belief that ecotourism markets would hold a stereotypical image of First Nations peoples. For Keller (1987), Inuit, Indians and Metis of the Northwest Territories of Canada were hampered in the development of tourism enterprises because of a lack of control of the industry at local and regional levels. Keller observes that people of this peripheral region must have decision-making control and act to limit development to a scale of growth in tune with the social, ecological and economic climate of the area (see also Seale 1992; McNeely 1993).

For ecotourism development to be successful, Aboriginal people must be allowed to take control despite the social and political forces that often work against them. This has been a problem in the restructuring of South Africa, where there remains a legacy of poor education and lack of training for black people (Swart and Saayman 1997). The result is that the whole touristic experience is one that is inherently foreign to the poor in South African society. Given that part of the government's tourism vision relates to addressing the need to improve the quality of life of every South African, Swart and Saayman recognise the need for significant changes in legislation to benefit all people. Lack of control is a problem for Aboriginal communities of the Kimberley region of northwest Australia, when the cruise line industry is unwilling to seek permission to gain access to tribal lands and waters and not following cultural protocols amounts to a lack of respect for the custodians of these lands (Smith *et al.* 2009). Control over decision-making is a problem for the Sami of northern Sweden because decision-makers see ecotourism as one of only a few options for employment and economic development. There is top-down pressure on the Sami to conform (Müller and Huuva 2009).

The issue of control in the development of ecotourism was at the centre of an incident between the Kuna Indians of Panama and Panama's national government. Chapin (1990) writes that in the 1970s, the government had the specific aim of working with multinational corporations in the development of tourism in the region of the Kuna (around Carti on the Caribbean side of the country), 'In return for their services as tourist attractions, the Kuna were to be given employment as service staff' (Chapin 1990: 43).

According to Chapin, the Kuna were unwilling to comply with the feasibility team and collectively forced the tourism industry influence out of the region. While plans for the development of casinos had to be discarded, one operator had his boat confiscated and another individual had to endure being shot, set on fire, a hanging attempt and a pummelling at the hands of Kuna youth. In their efforts to remain autonomous and in control, the Kuna own and manage the many small hotels of the region and cater to a small and manageable number of tourists on the basis of Kuna ownership and law. The Kuna are mindful of the importance of sustainable development and are striving to achieve it, but under the following conditions: (1) they must have control of the industry; (2) projects must be small in size and therefore manageable; and (3) there must be equity and dialogue among the various stakeholders (see also Wearing 1994).

To the aforementioned list we could also add skill development. A pressing need is for Aboriginal people to continue to develop skills that will enable them to be successful in an industry of global proportions. In this regard, De Lacy (1992) discusses the development of a number of Aboriginal employment programmes in Australia during the 1980s, including the initiation of many courses and skills that are currently available at selected universities within the Australian system. Similar training opportunities are reported

Table 4.3 *Code for the indigenous-sensitive ecotourist*

- Who operates the programme. Is it indigenous-run? If so, is it operated communally or do only a few individuals or families profit?
- If not indigenous-run, do local communities receive an equitable share of the profits or any other direct benefits, such as training? Or do only a few individuals/families benefit?
- Learn as much as you can about the local culture and customs. Visit local indigenous federation offices for information and materials with an indigenous perspective.
- Do not take photographs without asking permission.
- If you want to give a gift, make it a useful gift to the community rather than to an individual (e.g. gift to a school). Most indigenous communities function communally.
- Refrain from tipping individuals. If you are with a group, everyone can contribute to a gift to the community.
- Be aware of the boundaries of individual homes and gardens. Never enter or photograph without permission.
- Bring your own water purification tablets. Don't rely on boiling water exclusively as it depletes fuelwood and contributes to forest destruction.
- Pack out what you take in, and use biodegradable soaps.
- Be sensitive to those around you.
- Don't make promises you cannot or will not keep – for example, sending back photos to local people.
- Do not collect plants or plant products without permission.
- Wear appropriate and discreet clothing (e.g. many cultures are offended by women in shorts even though they may go topless).
- Respect local residents' privacy and customs. Treat people with the respect you would expect from visitors to your own home.

Source: Colvin (1994)

in the American Affiliated Pacific Islands, where support of indigenous ecotourism entrepreneurs is thought to be critical in achieving sustainable economic growth. Liu (1994) suggests that government needs to take the lead in providing financing, development and operational support, and training programmes to develop appropriate marketing and business skills for this population. National level controls are needed, in concert with local involvement, as a means by which to coordinate ecotourism development with the aim of reducing the effects of commodification on local indigenous cultures (King and Stewart 1996).

The stewardship issue

The most prevalent discourse in the literature on Aboriginal people and ecotourism is one that stresses the strong relationship that indigenous people have with the land. This relationship with the land is rather like a kinship, ethic or special bond, which is holistic and spiritual (see, for example, Zeppel 2006). This relationship is referenced by Higgins-Desbiolles (2009), who argues that the close bond that indigenous people have with nature may be used to educate non-indigenous people about how to better approach the natural world. Higgins-Desbiolles argues that indigenous ties to ecotourism would be better stated as indigenous *cultural–ecological tourism*, emphasising the strong bond between culture and the land. A parallel motive is to help eliminate the destructive nature of Western thinking – which is suggested to be antithetical to the indigenous worldview.

An important addition to the literature on the holistic and spiritual link between Aboriginal people and the land comes from Hinch (1998). Hinch feels it is a mistake to believe that just because Aboriginal people occupy the same wilderness or natural environment spaces as the ecotourism industry, both share the same views on human-nature relationships. Indigenous tourism is defined as 'tourism activity in which indigenous people are directly involved either through control and/or by having their culture serve as the essence of the attraction' (Hinch 1998: 121). And while it may be that indigenous people consider themselves to be at one with nature, this 'oneness' takes on a character that may be inconsistent with the tenets of ecotourism. Hinch uses the example of the climax of a whale hunt in which the whale is drawn up on the shore, and butchered in front of tourists. This would be a difficult spectacle for tourists who have as their mission the pursuit and enjoyment of animals in their wild and unmolested state – tourists who might be contributors to organisations that are involved in the protection of endangered or threatened animals like whales. The traditional practice argument used by indigenous people may not be strong enough to outweigh the brutality of the actual kill, especially if these tourists are more interested in whale watching than whale eating.

Another contribution along the same lines comes from Buckley (2005), who provided a first-hand account of the ethical challenges facing ecotourists and Aboriginal people in the Arctic region of Canada. Canadian Inuit people who transport tourists to the marine habitat of the narwhal often involve themselves in the hunting of this endangered species, for the purpose of selling narwhal products to wealthy Asian cruise boat passengers (non-subsistence). It is illegal to do so, and the issue becomes one of whether the operator ought to stop running tours even though they know this practice continues (and the operator indirectly helps Aboriginal people by employing them to shuttle tourists to these regions). The direct impact on the ecotourism industry is that narwhal disappear as soon as they sense tour boats approaching. They have learned to avoid human humans because of the hunting threat. Buckley writes that because the government of Canada provides the Inuit with lifestyle and material goods, including boats and firearms, they have an obligation to adhere to Canadian law. This is evidently not the case.

King and Stewart (1996) observe that one of the chief motivating factors for indigenous involvement in ecotourism is the declining state of natural resources, suggesting that ecotourism is a rational income possibility in the face of, for example, severe deforestation or over-grazing. This has been discussed by Young (1999), who found that ecotourism involvement in Baja Mexico evolved on the basis of more extrinsic factors. In investigating the state of the small-scale fishery, Young discovered that the absence of fisheries conservation techniques led to resource conflicts. Based on yearly declines in fish stocks, whale watching was proposed to balance economic benefit and conservation. However, local interest in grey whales increased only in response to the recognised economic potential of the resource as an ecotourism attraction. Furthermore, whale watching has not been so lucrative as to alleviate the pressures on the in-shore fishery. As such, conflict over the right to use marine resources has only intensified as a result of ecotourism. The same scenario has unfolded in regards to the Sami people of northern Europe. Pettersson (2006) writes that modern reindeer herding practices, along with lower reindeer meat prices, have prevented the Sami from taking part in this traditional work, and forced the Sami to look for alternative forms of income, with tourism emerging as an important option. Coupled with this is the observation that financial incentives are often required to get indigenous groups to take part in conservation. Local people will conserve for a reward, especially if the resource has economic importance (see Wells *et al.* 1992; Gullison *et al.* 2000). It also means that ecotourism may be a secondary choice among local groups, with less of an intrinsic motivational base to conserve, spread economic benefits, and to see tourists as something more than a vehicle for profit.

Questioning the conservation motives or practices of Aboriginal people, as above, is a touchy subject. We have been conditioned to believe that Western civilisation is the seed of our past and current ecological discontent, and that Aboriginal people are simply casualties of this Western way of life. Fennell (2008) investigated this discourse through a survey of the broad literature (biology, archaeology and anthropology) on Aboriginal stewardship. He found that there are a number of studies pointing to the fact that traditional societies have had difficulty managing resources in a sustainable way, just like non-Aboriginal people, with over-utilisation as the norm rather than sustainable practice. This is demonstrated through the work of Krech (1999) on North American Indians, who observed that:

> For every story about Indians being at the receiving end of environmental racism or taking actions usually associated with conservation or environmentalism is a conflicting story about them exploiting resources or endangering lands.
>
> (Krech 1999: 227)

Low (1996) has also investigated the lack of a stewardship ethic in Aboriginal people. Based on the results of a study of 186 cultures, Low argued that if ecological impact is found to be low within a traditional society it is not because of conservation mindedness, but rather from conditions of low population density, poor technology and/or the absence of a market from which to profit. This prompted Diamond (2005: 9) to recognise that the main error is in, 'viewing past indigenous peoples as fundamentally different from (whether inferior or superior to) modern First World peoples'. His point is that we (i.e. *Homo sapiens*) have always found difficulty in managing the natural resources of the world in a sustainable fashion. This means that because we share a common evolutionary heritage, we are really no different (acknowledging the myriad cultural differences that make us distinctive in many unique ways) when it comes to how we use, and have used, resources across cultures and time.

An interesting parallel to the myth of stewardship discussion, above, is offered by Waite (1999), in the context of the Australian Tourism Commission's representation of

Australia's Indigenous peoples in five international television advertisements. The 'otherness' of Aboriginal people, and the context in which they live (colonial representations), is discussed as an emerging drawing card for international tourists. One of the major themes explored by these advertisements is the concept of Aboriginal people as eco-angels. Waite feels that it is wrong to suggest that Aboriginal people are separate from civilisation and just as wrong to suggest that they have had little impact on the natural world. Waite lists massive burning of the landscape and the extinction of mega-fauna as cases in point.

Conclusion

This chapter initiated the discussion on the importance of sustainability in ecotourism through an analysis of the discourse on local participation and benefits. If local people are allowed to play a part in the planning, development and management of ecotourism, there may be some incentive to embrace ecotourism if this participation leads to benefits. Often this is the case – ecotourism projects do in fact lead to participation and benefits. But it is also the case that the benefits promised by ecotourism do not necessary come to fruition. Blersch and Kangas (2013) have pushed this debate forward through the construction of a simulation model for ecotourism developed over a 100-year period for Belize (time step of one year). The authors found that there would be a decline in the income generated from ecotourism due to rising oil prices (cost of travel) and because of the environmental impacts of ecotourism. Conservation could slow these declines, but in the end the system was found to be unsustainable. Benefits have been positive in Belize over the last 20 years, but these benefits are temporary in view of the long term. This study provides an excellent introduction to the next chapter, by emphasising the fact that sustainability is a concept that emphasises both development and conservation.

Summary questions

1   Aboriginal interests have often been overlooked in development initiatives. Identify a few examples used in the chapter, which help to explain how Aboriginal people might benefit from ecotourism.
2   How do core indicators of sustainable tourism differ from ecosystem-specific stem-specific indicators?
3   What is the myth of stewardship?
4   Why is it important to have all sectors of the tourism industry working in a sustainable way?
5   Is mass tourism always unsustainable? Why or why not?

5 Sustainability 2: conservation

> We abuse land because we regard it as a commodity belonging to us. When we see land as a community to which we belong, we may begin to use it with love and respect. – Aldo Leopold

The above quotation aptly describes the dilemma that has existed for some time in humanity's struggle to balance natural resource utilisation with conservation. Globally, we have been slow to recognise how our development activities and patterns of consumption have impacted the resource base. In this chapter, a general discussion on conservationism is used to describe the polarisation of thought on the value and role of resources within society. This chapter is a continuation of the previous chapter in reference to sustainability. While Chapter 4 dealt with the local participation and local benefits aspects of sustainability, this chapter examines why conservation is important, and many of the complex issues tied to the use and management of natural areas and biodiversity. The human component is thus never far removed from the discussion on nature and natural resources. An attempt is made to highlight some of the main considerations of the development of a conservation philosophy in light of the pressures that humans continue to place on the natural world. Parks and protected areas are discussed as settings that have an important role in balancing ecological integrity and tourism industry demand.

The exploitation of the natural world

The earth's bounty

In an influential treatise on the historical roots of our present-day ecological dilemma, White (1971) argues that where formerly humankind was part of nature, over the past several hundred years our species has become the exploiter of nature and natural resources. The strength behind this conviction lies in how Western Christianity and capitalism have polarised humans and nature by insisting that 'it is God's will that man exploit nature for his proper ends' (White 1971: 12). Historical accounts charge that many scientists of the Middle Ages and Renaissance periods were tempered by this Christian philosophy. During the great scientific renewals of the 1500s and 1600s, a mechanistic and static perception of nature developed, replacing the passive mystic and organic views of earlier times. The push was towards a type of science that had designs on enlarging the bounds of the human empire in the endless pursuit of knowledge and control in order to 'regain the level of understanding and power once enjoyed in the Garden of Eden' (Bowler 1993: 85).

Another significant element that contributed to human dominance over nature was fear of the unknown. Although scientists in medieval times had begun to unravel some of nature's mysteries, their lack of understanding of the intricacies and interconnectedness of the environment contributed to the feeling that marginal areas – like wilderness – had to

be subdued (Short 1991). Land and animals were dichotomised as settled or savage, culti-
vated or uncultivated, domesticated or wild. Through Christian eyes, wilderness areas
were settings that contained pagans exercising pagan rites. In order to bring about religious order, such lands had to be cut down and cleansed.

The abundant literature that exists on human–wilderness relations illustrates that there
is no specific material object or space that one can identify as being 'wilderness' (Nash
1982). Wilderness is a concept that produces specific feelings or moods in people, and
occurs within the mind as a perceived place. Such a place may be wilderness to one but
not to another, as identified by Saarinen (2005), who illustrates how wilderness areas
differ in the minds of local people and tourists. This attitude prevailed in the Puritan mind
as the first settlers set foot on North American soil, where settlers were religiously and
morally obligated to spend the bulk of their time in work-related activities. Leisure was
severely curtailed and strictly regulated in accordance with religious protocol. The implications for nature were such that the wilderness represented both a challenge and an
obstacle to a sustained European colony in the New World.

CASE STUDY 5.1

Natural history travel in Shetland, Scotland

Butler (1985) writes that up to 1750, the Highlands and Islands of Scotland were virtually
terra incognita to the people of the rest of Britain. The Shetland Islands were no exception.
However, as a consequence of its unique location, Shetland provided both a geographical
and a cultural link between Scandinavia to the east and Great Britain to the south. Flinn
(1989) suggests that before 1850 the people who travelled to Shetland were artists, geolo-
gists, naturalists, physicists and surveyors. In 1814 Sir Walter Scott visited Shetland and
used the region as the setting for his novel *The Pirate*. As Simpson (1983) illustrates,
Scott's novels suffused a romance and drama of the book's region in the minds of the
readers, motivating them to want to visit these places. In 1832, an ironmaster and naturalist, G.C. Atkinson, travelled to Shetland, perhaps as a result of Scott's influence. He
wrote, 'I have long felt the greatest interest in descriptions of novel and extraordinary
scenery and of the inhabitants and natural productions of regions that have been little
known, either from their difficulty in attainment . . . or from their being so near.' Shetland's
natural history, particularly the bird life and the physiography of the region, were main
attractions for these intrepid travellers. By 1850, access to Shetland improved, according
to Flinn, who wrote, 'by 1859 Shetland had a frequent steamer service from the south
during the summer months, and a trunk road system which made the islands more accessible to tourists of a less hardy and enterprising breed' (Flinn 1989: 235). Tourism, it
appears, had displaced the traveller bent on exploration and challenge with a breed less
willing to endure the hardships of travel.

The roots of conservation

As the modification of the natural world began to intensify, scientists in France and Britain
were discovering that excessive land uses, like deforestation, were the principal cause of
soil erosion and poor productivity. These discoveries prompted early attempts at conservation in Britain, and in British India, where forested lands were set aside for the purpose
of shipbuilding. Such areas were referred to as *conservancies*, and the foresters in charge
of these areas as *conservators* (Pinchot 1947). (There is evidence, however, to suggest that

reforestation policies were being implemented in Britain in the seventeenth century – see Bowler 1993.)

In North America, conservation evolved on three fronts (Ortolano 1984). The first on the view that conservation should entail the maintenance of *harmony* between humans and nature; the second that conservation related to the *efficient* use of resources; while the third was based on the feeling that conservation – preservation – could be attained from the standpoint of religion and *spirituality*. This latter view of conservation engendered a philosophy with the aim of saving resources *from* use rather than saving them *for* use (Passmore 1974). Each of these perspectives is discussed in further detail below.

Harmony

In the USA, a former Minister to Turkey and founder of the Smithsonian Institute, George Perkins Marsh, became an instrumental figure in illustrating to Americans that their actions (commerce and lifestyles) were uniquely potent. Marsh wrote that:

> The earth is fast becoming an unfit home for its noblest inhabitant, and another era of equal human crime and human improvidence . . . would reduce it to such a condition of impoverished productiveness, of shattered surface, of climatic excess, as to threaten the depravation, barbarism, and perhaps even extinction of the species.
>
> (Marsh, cited in Bowler 1993: 319)

His *Man and Nature* or *Physical Geography as Modified by Human Action*, originally published in 1864, recognised that in the changing conditions of nineteenth-century America, harmony between human influences (modifications) and the natural world could be achieved only through society's commitment to a moral and social responsibility to future generations. (This view was rekindled over 80 years later by the land ethic philosophy of Aldo Leopold.) Marsh identified the collective power of large corporations as the factor most responsible for the increasing negative impacts to the natural world. Politically, the accountability of the large firms did not become an issue until the turn of the century. The Roosevelt administration recognised that the federal government had been too generous in granting favours (leases, land rights, etc.) to such corporations (Hays 1959), and as a result attempted to: (1) find an adequate means of controlling and regulating corporate activities; and (2) resist the efforts of corporations to exploit the natural resources of the nation on their own behalf.

At the time, changing technology, industrialisation, urbanisation, population growth and transportation all contributed to the feeling that the American frontier was diminishing. This fact was most ardently illustrated by the direct relationship between the advancement of the transcontinental railroad and the decline of American buffalo. The combination of a loss of habitat and the shooting of buffalo for sport (from the trains themselves) contributed significantly to the downfall of this species. Space was also a critical factor in the frontier mentality. Americans had reached the supposed limits of their manifest destiny by encountering the Pacific to the west, Canada to the north and Mexico to the south. In Canada, the frontier attitude lingered on because of the challenges encountered in settling and harnessing the conditions of the north and because of a much smaller population base.

Efficient use

By the beginning of the twentieth century, Americans started to exercise concern over the fate of their resource base (Nash 1982). Conservation became a vehicle to represent the new frontier; the vehicle to allow American society to maintain vitality and prosperity. However, despite conservationism's apparent simplicity – the wise use of natural resources – fierce debate surfaced over how such resources ought to be utilised, if at all.

The efficient use of resources perspective represented the opposite end of the conservation spectrum from the perspective of preservationism, and was championed by the American, Gifford Pinchot. Pinchot developed a concept of natural resources for the greatest good of the greatest number for the longest time (Herfindahl 1961). More directly, conservation was to engender direct control over natural resources on the basis of three principles: (1) to develop the continent's existing natural resources for the benefit of the people who live there now; (2) to prevent the waste of natural resources; and (3) to develop and preserve resources for the benefit of the many, and not merely for the profit of a few (Pinchot 1910).

This philosophy recognised the need for people to acknowledge that natural resources were finite. Individuals and industries had to be more accountable for their actions under the premise that what was utilised today might have repercussions for those in the future. From this standpoint, conservationism was indeed progressive, as suggested by Hays (1959: 264):

> The broader significance of the conservation movement stemmed from the role it played in a transformation of a decentralised, non-technical, loosely organised society, where waste and inefficiency ran rampant, into a highly organised, technical, and centrally planned and directed social organisation which could meet a complex world with efficiency and purpose.

Spirituality

A more romantic view of wilderness was developing in response to the technological and industrialised transformation of Britain and Europe. Based on the works of Erasmus, Darwin, Wordsworth, Coleridge and Carlyle, Romanticism developed from the more regressive view that society had declined from past, more harmonious times. Romanticism embodied a deeper spirituality and awareness that a simpler life was attainable without the complications of a society blemished by materialism, and could be accomplished under the following conditions: (1) untouched spaces had the greatest significance; (2) these spaces had a purity which human contact degrades; (3) wilderness was a place of deep spiritual significance; and (4) the conquest of nature was a fall from grace (Short 1991).

The first proponent of the Romantic philosophy in North American society was Ralph Waldo Emerson, who had met and been inspired by the romantic poets of Britain in the early 1830s. The main doctrine of Emerson's interpretation of Romanticism surrounded the belief that although mankind was firmly rooted to the physical world, people had the ability to 'transcend' this condition (spiritually) in searching for deeper philosophical truths. Emerson's transcendentalism was a spiritual doctrine linking humans to nature. To Emerson, humanity was divided into materialists and idealists, with the first class founded upon experience and the second on consciousness. The materialists were to insist on facts, history, circumstances and animal instincts; the idealists on the power of thought, will and miracle. His major work, *Nature* (1835), had a significant impact on many other writers to follow, including Henry David Thoreau, Herman Melville, John Burroughs, Walt Whitman and John Muir.

Nature in North American society thus began to hold and inspire a small but articulate core of advocates paving a clear path out of the downward-spiralling pattern of a materialistic, consumptive society. Although the works of Emerson were largely idealistic, the transcendentalist movement also provided the rationale for practical change within American society. Thoreau, for example, campaigned to have the American government establish national preserves for the purpose of ensuring the future well-being of animals (Finch and Elder 1990). Such a call was well ahead of its time, as the United States did not endeavour to establish protected areas until a number of years later.

The emergence of a second tier of conservation (environmentalism or the green movement) occurred in the 1960s in response to the rapid and overwhelming increase in the impact of technology in society. Publications such as Rachel Carson's *Silent Spring* were effective in calling attention to the effects of chemical use in society. In addition, ecology and the growing importance of science became institutionalised as a mechanism to evaluate the many social and ecological ills of the time. Bowler (1993) writes that the most militant supporters of the environmental movement opposed the entire economic structure of society in favour of a reversal to a simpler, more natural state. Examples of organisations/movements advocating a more radical approach to environmentalism include Greenpeace, Deep Ecology (founded by Arne Naess) and the Eco-Feminist Connection. Those environmentalists considered less militant called for the protection of selected parcels of land that held natural and cultural significance, with the request that industrialisation and development occur outside the realm of such areas. According to Bowler, a fundamental difference existed between the two camps. Those less enthusiastic (e.g. those in favour of the development of nature reserves) recognised that minor changes could occur within the current system, while the most enthusiastic environmentalists wanted to destroy the existing social order. It is the former group that, in a practical sense, has made the most significant gains in the context of the global arena. Ecotourism, to some, is merely an extension of this philosophy of 'working within the system' and one that, at least conceptually, attempts to knit the elements of economy and ecology together (via parks) through the philosophy of sustainable development.

Parks and protected areas

The concept of 'park' is one that is firmly established in Western civilisation (Smith 1990a; Wright 1983). The Greeks and Romans met at designated open spaces *(agorae)*, while in medieval times the European nobility used their private lands as hunting reserves. With very few unmodified open spaces left in Britain by the nineteenth century, parks were recognised for their role in securing outdoor recreation opportunities in the countryside. To accommodate demand, the nobility began to lease open space for summer and winter recreation of all classes within British society. The importance of parks was more formally established by the Municipal Corporations Act in 1835, allowing for the creation of municipal parks and for the secure right of public recreation.

In Britain, the parks movement stemmed largely from the response of the British public to the effects of urbanisation, pollution and loss of leisure equated with the Industrial Revolution. The same could not be said for the evolution of parks in Canadian society. Wright (1983: 45) illustrates that urban parks in Canada were created to satisfy a concept rather than a reality:

The establishment of local parks in the 1850–1880 period in Ontario was merely the continuation in Canada of a program and philosophy adopted in Britain and maintained by the elite in their own environment, not because the conditions in Canada demanded its implementation but because the colonial settlers wished to preserve the values and beliefs inherited from their ancestral homes.

Ontario was the first of the provinces in Canada to enact legislation governing municipal park development in 1883 (Eagles 1993). This occurred at a time when there was a heightened awareness in North America of the need for parkland. The world's first national park had been created in Yellowstone in 1872, and Canada's first national park, Banff, was created in 1885.

Parks and protected areas (the terminology used in this section to refer to public lands held in trust with both a recreation/tourism and conservation/preservation mandate, and owned and usually operated by a public agency) have a certain mystique to travellers interested in some of the best representative natural regions. In fact, it was Johst (1982) who suggested that visitation to parks may increase by virtue of their designation as such. Simply stated: parks and protected areas often generate more recreational use simply because they are recognised as parks (although McCool 1985 contests this point).

Harroy (1974) has shown that Yellowstone was created to satisfy a broad mandate of concerns that had emerged from the USA's frontier mentality. Foremost, the park was set up to prevent the exploitation of wildlife and the environment, for the purpose of recreation, and finally as a means of scientific study. Banff, on the other hand, was established for political and economic reasons, including the generation of tourism dollars (largely from the therapeutic and recreational benefits of the hot springs) for the purpose of offsetting the cost of building the transcontinental railway (Lothian 1987). Banff's popularity quickly escalated to the point where the park was absorbing a wide variety of recreational demands including fishing, horseback riding, hunting, mountain climbing and so on, but also other resource demands such as mining, logging, grazing and a town site. Rollins (1993) suggests that as a case study, Banff is representative of the many significant problems that Parks Canada has had to address for over a century regarding the management of natural features, with many conflicting stakeholders harbouring consumptive and non-consumptive designs on the use of park resources.

National parks are broadly mandated with the dual purpose of protecting representative natural areas of significance, and encouraging public understanding, appreciation and enjoyment. Historically, the preservation ideal within parks was not fully developed or emphasised. However, as the system of protected areas continues to grow (as illustrated by the increasing circles over time in Figure 5.1), park management philosophies have become better integrated, recognising that parks do not exist as ecological islands, but must be managed according to environmental conditions both inside and outside their boundaries (Dearden 1991).

In general, the threats to parks have evolved, having been primarily internal but now more external in their orientation (Dearden and Rollins 1993). This evolution has coincided with the fact that the role of parks has changed significantly over time, from a primarily recreational purpose, to one that maintains ecological functioning first and foremost. Section 5(1.2) of the Canadian Parks Act (1988 amendments) suggests that maintenance of ecological integrity through the protection of natural resources is to be the first priority when considering park zoning and visitor use in the management plan (Canada, Parliament 1993). Park zones are established on the basis of natural resources as well as the need to absorb recreational use. An overview of the basis for zoning in national parks in Canada (Environment Canada 1990) is illustrated below (for more of a history on the development of outdoor recreation land classification, the reader

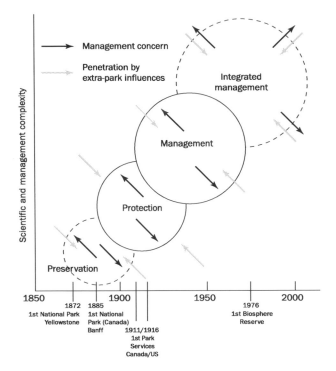

Figure 5.1 The evolving role of parks

Source: Dearden and Rollins (1993)

should consult the American Outdoor Recreation Resource Review Commission reports of the early 1960s):

- *Zone 1: Special Preservation.* Specifies areas or features, which deserve special preservation because they contain or support unique, rare or endangered features or the best examples of natural features. Access and use will be strictly controlled or may be prohibited altogether. No motorised access or human-made facilities will be permitted.
- *Zone 2: Wilderness.* Extensive areas which are good representations of each of the natural history themes of the park and which will be maintained in a wilderness state. Only certain activities requiring limited primitive visitor facilities appropriate to a wilderness experience will be allowed. Limits will be placed on numbers of users. No motorised access will be permitted. Management actions will ensure that visitors are dispersed.
- *Zone 3: Natural Environment.* Areas that are maintained as natural environments, and which can sustain, with a minimum of impairment, a selected range of low-density outdoor activities with a minimum of related facilities. Non-motorised access will be preferred. Access by public transit will be permitted. Controlled access by private vehicles will be permitted only where it has traditionally been allowed in the past.
- *Zone 4: Outdoor Recreation.* Limited areas that can accommodate a broad range of education, outdoor recreation opportunities and related facilities in ways that respect the natural landscape and that are safe and convenient. Motorised access will be permitted and may be separated from non-motorised access.

- *Zone 5: Park Services.* Towns and visitor centres in certain existing national parks which contain a concentration of visitor services and support facilities as well as park administration functions. Motorised access will be permitted.

Rollins (1993) observes that zoning is primarily natural resource based, and does not define the types or levels of recreational opportunities that can occur within such regions of the park. Visitor management within parks in North America is addressed through a number of preformed planning and management frameworks, including the Recreation Opportunity Spectrum (ROS), the Limits of Acceptable Change (LAC) and the Visitor Activity Management Process (VAMP) (see Diamantis 2004; see also Hadwen *et al.* 2008).

Ecosystem management

As identified previously in Figure 5.1, parks management has evolved significantly over time. One of the finest examples of this evolution is the development of the ecosystem management philosophy, which has blossomed as a consequence of the realisation that in order to effectively safeguard an environment one must scientifically understand the relationships and processes that exist within these settings. The biodiversity crisis, new ecological theories and dissatisfaction with governmental regulatory measures also contributed to the birth of this mode of thinking (Grumbine 1996). Foremost, biological and social systems theory became the foundation of ecosystem management once it became clear that ecological sustainability can be attained only through substantial societal change. Indeed, as Francis (n.d.) suggests, we must begin to see ourselves as integral components of complex ecosystems; components, which are in turn related to one another over a variety of spatial and temporal scales. The feeling is that the species with which we share ecosystems have their own inherent value, and should not be judged on the basis of their ability to provide us with resources, but rather as important elements within these complex systems.

In definition, ecosystem management, according to the Canadian Environmental Advisory Council (CEAC) refers to the, 'integrated management of natural landscapes, ecological processes, wildlife species and human activities, both within and adjacent to protected areas' (CEAC 1991: 38). The following definition, put forward by Johnson and Agee (1988), emphasises the inclusion of both social and ecological processes that help shape and transform ecosystems. Hence, we can never separate human and biophysical elements within an ecosystem; they are inseparable.

> Ecosystem management involves regulating internal ecosystem structure and function, plus inputs and outputs, to achieve socially desirable conditions. It includes, within a chosen and not always static geographic setting, the usual array of planning and management activities but conceptualised in a systems framework; identification of issues through research, public involvement, and political analysis; goal setting; plan development; use allocation; activity development (resources management, interpretation); monitoring; and evaluation. Interagency coordination is often a key element of successful ecosystem management, but is not an end in itself. Success in ecosystem management is ultimately measured by the goals achieved, not by the amount of coordination.
>
> (Johnson and Agee 1988: 7)

An interesting perspective that fits with the ecosystem management philosophy is one advanced by Chipeniuk (1988), who says that the management of parks from an ecological perspective is not complete, and ecosystems are not natural, because they lack a principal component of these ecosystems that existed at the time of European contact: human hunters and gatherers. Chipeniuk argues that this is a niche, which, for biological and socio-cultural reasons, must be filled by human surrogates. In a biological sense, the environment would be returned to more of a natural state (i.e. pre-contact); while from the socio-cultural context the return of people as part of the functioning of park ecosystems would cease to delude us about the proper place of human beings in the natural world. This point is touched on by Pretty and Pimbert (1995: D8), who argue that when local indigenous people are excluded from conservation the goals of conservation are at risk. They write that, 'some "pristine rain forests," assumed to be untouched by human hands, are now known to have once supported thriving agricultural communities. The "pristine" concept of the wilderness is an urban myth that exists only in our imagination'.

CASE STUDY 5.2

Last Mountain Lake National Wildlife Area

This Saskatchewan wildlife area is significant in that 1,013 hectares of it was designated as the first federal bird sanctuary in North America on 8 June 1887. According to the Canadian Wildlife Service (1995), over 280 bird species have been recorded at Last Mountain Lake during peak migration, including 50,000 cranes, 450,000 geese, and several hundred thousand ducks. In addition, the area provides habitat for 9 of Canada's 36 vulnerable, threatened and endangered birds, including the whooping crane and peregrine falcon. Last Mountain Lake has been designated as a Wetland of International Importance along with approximately 30 other sites in Canada and 700 locations world-wide (as of 1995). Although conservation is the primary purpose at Last Mountain Lake, more consumptive activities such as hunting and fishing are allowed in adjacent areas under strict regulation.

Although many of the more developed countries like Canada have been world leaders in the policy and management of protected areas, many of these countries are still far from completing their national parks systems because of a complex array of land and human use issues (as noted above). Despite Canada's low productivity in the establishment of parks, public opinion is in favour of the creation of more wilderness areas. Reid *et al.* (1995) illustrate that on the basis of a survey of over 1,500 residents of British Columbia, respondents, on average, would be willing to pay between $108 and $130 annually in taxes for a doubling of designated wilderness areas, and between $149 and $156 for a tripling of wilderness areas. Comparatively, many developing countries have been far more conscientious in the development of their parks systems. Panama, Nicaragua and Costa Rica, for example, have set aside much higher percentages of their national territories as parks and protected areas. In fact, Costa Rica and Panama have cooperated in the development of an international park called La Amistad, as has Malaysia and Indonesia in a transboundary conservation initiative called 'Heart of Borneo', which is an effort to preserve the rich and abundant natural resources of the region. Ecotourism is an important industry in the region through recognition that conservation must be linked with sustainable livelihoods. Hitchner *et al.* (2009) identify a number of challenges for making the initiative successful. These include: protection of forests and cultural sites for ecotourism; improved communication between villages, guides and lodges; increased promotion of transboundary trekking options; preparation for more tourists and more equitable distribution of ecotourism income; improvements in tourism infrastructure; improving complications arising from international border; and maintaining local control over ecotourism in the Heart of Borneo.

In general, parks are a function of the socio-political and economic conditions in which they are situated. The 15 national parks in Britain (England, Scotland and Wales), for example, are managed from a very different perspective from those in other Western societies, like Canada. According to Henderson (1992: 397), conservation in Britain is based on a, 'steady state of human intervention designed to maintain a given habitat at a particular successional stage in perpetuity', which he feels is the most unnatural conservation policy possible. British national parks are essentially living, working landscapes, and administratively it is the responsibility of a National Park Authority (which operates within the local government system and whose members are appointed by the government in consultation with the Countryside Commission) to: (1) preserve and enhance park natural beauty; (2) promote access, use and enjoyment; and (3) ensure proper practices of agriculture, forestry and economics (Exmoor National Park 1990). The result is that there is a significant amount of socio-economic activity occurring within the national parks of Britain, because virtually none of the land in Britain is excluded from human activity (Phillips 1985). Exmoor National Park, the second smallest park in the system, for example, has a resident population of approximately 11,000 as of 2001, hosts about 3 million visitors per year, and is almost 80 per cent privately owned (http://www.exmoor-nationalpark.gov.uk/index/learning_about/general_information/moor_facts.htm; accessed 24 November 2006). The park plan explains the role that economic development, agriculture, forestry, housing, services and facilities, and tourism play in the context of the park, which is consistent with the types of activities that occur in British parks in general (Phillips 1985). Based on this philosophy of use, tourism has a significant role to play within the national parks and is not

Plate 5.1 On Canada's west coast, loggers and environmentalists have fought over some of the world's largest trees. Since being saved these trees have generated much interest among ecotourists in the region

Plate 5.2 Many wilderness areas have both natural and cultural heritage value. This petroglyph (painting of a moose) in the Clearwater River Provincial Park, Saskatchewan, was painted by Aboriginal people many hundreds of years ago

Plate 5.3 The wilderness character of Clearwater Park, Saskatchewan, Canada. This park is well known for canoeing and hiking

questioned to the point that it is in other national parks. However, the tourism industry is still expected to operate within guidelines established by the Countryside Commission and the English Tourist Board (Countryside Commission 1990).

The national parks of Britain, as suggested above, are somewhat of an anomaly in that they do not meet the category II criteria of the International Union for the Conservation of Nature and National Resources (IUCN) on national parks. Internationally, guidelines have been established in an effort to both control and direct countries in how best to set aside specific lands and water. Phillips (1985) feels that the national parks of Britain would be better classified as protected landscapes, or category V of the IUCN guidelines (see Table 5.1).

The IUCN's categories for conservation management illustrate the diversity of protected areas developed internationally, each of which focuses on different aspects of development (the level of tourism infrastructure and use) and preservation. The extent of the work of the IUCN goes well beyond the categorisation of protected areas, though, to include establishing a system of biogeographical provinces of the world; publishing lists and directories of protected areas; publishing conceptual papers on protected areas; publishing the quarterly journal *Parks*; cooperating with United Nations agencies (e.g. UNESCO); holding international meetings, such as the World Conferences on Parks and Protected Areas; and supporting field projects for the establishment and management of protected areas (Eidsvik 1993: 280).

In addition, a number of other units have been developed for conservation purposes. The World Heritage Convention of 1972 (UNESCO) provides for the establishment of natural and cultural heritage sites of outstanding universal value. These sites are woven into the fabric of existing protected areas and, although the system does not impose any new management criteria on existing parks, it does impose an element of symbolism and prestige for countries that maintain such sites. As of 2006 there were 830 sites (644 cultural, 162 natural and 24 mixed) in 138 member states (http://whc.unesco.org/en/list/; accessed 24 November 2006).

The biosphere reserve system (category IX), was developed for the purpose of alleviating many of the park–people problems that have traditionally constrained other protected area types. The biosphere reserve concept grew out of the 1970 UNESCO general meeting which provided the impetus for the development of the Man and the Biosphere programme meetings of 1971, and later the development of the first reserves in 1976. This protected area system is founded on three themes: development, conservation and research. Spatially, these reserves incorporate three distinct zones: (1) a core area, which is minimally disturbed and strictly protected; (2) a buffer, situated around the core and allowing certain types of resource use that do not disturb the core; and (3) a transition zone, which extends outwards into the adjacent territories with no fixed boundary and allowing a full range of human uses. Seven guidelines characterise the establishment criteria of biosphere reserves. They: (1) work as a linked network of natural areas; (2) are representations of the 227 biogeographical provinces; (3) are examples of special environments (i.e. natural biomes); (4) are large in size to ensure effective conservation; (5) act as benchmarks for research, education and training; (6) have some form of legal protection; and (7) have incorporated within them existing protected areas.

Eidsvik (1983) observes that the biosphere is a unique concept relative to other types of protected areas in that it operates under the premise of an inverted pyramid, where the decision-making does not necessarily occur from a centralised federal authority, but rather from the grassroots level with support at various other levels. Eidsvik (1983: 230) described this concept as follows:

The system is designed to support and cherish the participant, operating at his or her own level ... Those services that cannot be provided by individuals or their

Table 5.1 Categories for conservation management

Category I – Scientific Reserve/Strict Nature Reserve
Areas with some outstanding ecosystem features and/or species of flora and fauna of national scientific importance, representative of particular natural areas, fragile life forms or ecosystems, important biological or geological diversity, or areas of particular importance to the conservation of genetic resources. Concern is for continuance of natural processes and strict control of human interference.

Category II – National Park
A relatively large area where one or several ecosystems are not materially altered by human use, the highest competent government authority has taken steps to prevent or control such alteration, and visitors are allowed to enter, under special conditions for inspirational, educative, cultural, and recreative uses.

Category III – Natural Monument/Natural Landmark
Area normally contains one or more specific natural features of outstanding national significance which because of uniqueness or rarity should be protected. Ideally little or no sign of human activity.

Category IV – Nature Conservation Reserve/Managed Nature Reserve/Wildlife Sanctuary
A variety of areas fall into this category. Although each has as its primary purpose the protection of nature, the production of harvestable renewable resources may play a secondary role in management. Habitat manipulation may be required to provide optimum conditions for species, communities, or features of special interest.

Category V – Protected Landscape or Seascape
A broad category embracing a wide variety of semi-natural and cultural landscapes within various nations. In general, two types of areas, those where landscapes possess special aesthetic qualities resulting from human–land interaction and those that are primarily natural areas managed intensively for recreational and tourist uses.

Category VI – Resource Reserve (Interim Conservation Unit)
Normally extensive, relatively isolated, and lightly inhabited areas under considerable pressure for colonisation and greater exploration. Often not well understood in natural, land use, or cultural terms. Maintenance of existing conditions to allow for studies of potential uses and their effects as a basis for decisions.

Category VII – Natural Biotic Area/Anthropological Reserve
Natural areas where the influence or technology of modern humans has not significantly interfered with or been absorbed by the traditional ways of life of inhabitants. Management is oriented to maintenance of habitat for traditional societies.

Category VIII – Multiple Use Management Area/Managed Resource Area
Large areas suitable for production of wood products, water, pasture, wildlife, marine products and outdoor recreation. May contain nationally unique or exceptional natural features. Planning and management on a sustained-yield basis with protection through zoning or other means for special features or processes.

Category IX – Biosphere Reserve
Intended to conserve representative natural areas throughout the world through creation of global and national networks of reserves. Can include representative natural biomes, or communities, species of unique interest, examples of harmonious landscapes resulting from traditional uses, and modified or degraded landscapes capable of restoration to more natural conditions. Biosphere reserves provide benchmarks for monitoring environmental change and areas for science, education and training.

Source: Nelson (1991)

communities then become the responsibility of local agencies. More specialized services come from the provincial governments – and finally, highly specialized residual services are provided by federal agencies.

In practice, there are good examples of the success of the biosphere reserve concept in the developed and developing worlds. In Canada, for example, the Long Point Biosphere Reserve and region in Ontario operates in partnership with a variety of municipal, provincial and federal agencies (Francis 1985) including the Canadian Wildlife Service, Transport Canada, the Ontario Ministry of Natural Resources, the private Long Point Company, a conservation authority, a regional municipality and other private land holdings, all in an area of approximately 33,000 ha. In Mexico, the Sian Ka'an Biosphere Reserve in the Yucátan Peninsula was developed as a result of local efforts to safeguard this 1.3 million acre territory. A non-profit organisation, the Amigos de Sian Ka'an, was established to mediate between the private sector and various levels of government. Tourism, owing to the proximity of the reserve to Cancún, has grown significantly since the 1980s, posing both a threat and an opportunity. Crucial to the viability of the reserve is cooperation between levels of government, the ENGO and local people, in coming together to solve common dilemmas and improve the quality of lives. This has prompted the executive director of Amigos de Sian Ka'an to remark that, 'If the people who live in the reserve support it, they will take care of it. If not, no amount of guards will stop them' (Norris 1992: 33). Despite the positives of the biosphere programme, there are those who have suggested that it still exists as a top-down approach to conservation, without the commitment to localism that should exist in such areas (Janzen, cited in Chase 1989).

A more recent addition to the bank of protected areas, and in recognition of the many different types and levels of pressure that humans place on natural systems, as noted above, is Europe's Natura 2000 programme, spanning the European Union's (EU) member states. Based on the Bird Directive of 1979 and the Habitat Directive of 1992, the programme is designed to stimulate sustainable economic development, while at the same time protecting the environment and maintaining biodiversity (recognising that humans are a fundamental component of nature). The programme is important because there are 250 habitat types recognised across the EU, including 450 animals and 500 plants that are rare or threatened (Aperghis and Gaethlich 2006). However, it is also important in recognising that with such diversity in social, political and economic conditions among the various member states, there is little consistency with respect to the importance of biodiversity conservation. This seems to be the case in Greece, where Aperghis and Gaethlich (2006) argue that there has been a significant degree of political indifference to environmental protection. Without the demands of the EU and the Natura 2000 programme, the authors contend, the environment, and biodiversity conservation more specifically, would still rank much lower politically. Two successful Greek Natura 2000 sites, the Forest of Dadia and the Profitis Elias Monastery, are renowned for their birding opportunities, and are good examples of ecotourism sites where jobs have been created in rural areas, residents keep close attachment to the land and the natural environment is being protected.

Biodiversity conservation: issues and challenges

In Chapter 3, conservation was discussed as one of several variables that have been used to define ecotourism. Examples included some of the more consumptive forms of outdoor recreation such as fishing for marlin or sailfish, where catch-and-release techniques, coupled with the attachment or insertion of tracking devices, could help with the scientific investigation of these species and, in the process, further efforts to conserve these species.

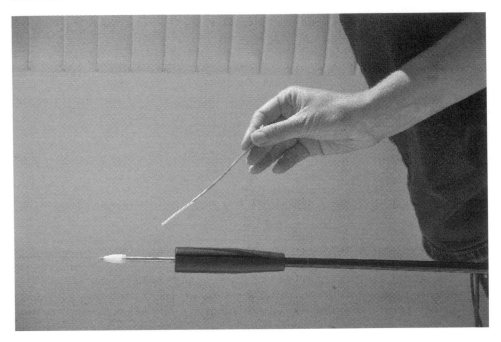

Plate 5.4 Tracking device for sharks

An example of these sorts of practice can be found in Plates 5.4 and 5.5. In this example, shark fishing off the coast of St. Andrews, New Brunswick, Canada, accompanied by a biologist, aids in research on porbeagle sharks. There are innumerable examples in the literature on how ecotourism has been used to support conservation initiatives (see, for example, Funnell and Weaver 2005).

Here the discussion takes a different approach in looking at many of the broader biodiversity conservation issues and challenges. Examples include the major reasons for biodiversity loss, and how best to conserve these lands and waters, along with some of the challenges local people have in balancing conservation priorities from external sources with the local subsistence needs.

The Natura 2000 programme, above, has placed a premium on the viability and maintenance of biodiversity across the European landscape. The pace at which the programme has evolved is testament to the importance of biodiversity to the overall health of the region. Biologists and conservationists organise biodiversity, short for biological diversity, into three levels: (1) ecosystems, like grasslands and rainforests; (2) individual species of plants and animals, with estimates ranging from between 2 to 100 million on the planet at present (May 1992); and (3) genes, that compose the heredity of the individuals that make up these species (Wilson 2002). We value biodiversity on a number of different levels according to Ehrlich and Ehrlich (1992): directly, through food, medicine and industrial products; indirectly through ecosystem services such as water purification and climate regulation; aesthetically, in that every individual of each species is an irreplaceable treasure that we often value through ecotourism; and ethically, from the perspective that we have a stewardship responsibility to protect these species and individuals from our deleterious actions. Ecosystems and their component parts have only recently been factored into the costs of doing business; that is, the services that the environment does for the planet – free of charge – is estimated to range between US$16 and 54 trillion – or 1.8 times the global GNP in the late 1990s (Costanza *et al.* 1997). (See Iturregui and Dutschke 2005 who report that 60 per cent of the 24 ecosystem services

Plate 5.5 Insertion of a tracking device in a porbeagle shark

analysed are being degraded, including fresh water, fisheries, air and water purification and climate regulation.)

Major contributors to human-induced biodiversity losses include habitat destruction, the introduction of invasive species to susceptible environments, pollution, population increase and over-harvesting of species. In the Barra de Santiago region of El Salvador, for example, Ramírez (2005) writes that population increases have put pressure on local natural resources in a number of different ways. Iguanas, doves, squirrels, ducks and rabbits are illegally hunted for subsistence; parakeets, parrots, crocodile skins and sea turtle eggs are exchanged illegally; tourism has developed in the absence of regulations; habitat has diminished through massive deforestation; and agricultural runoff has contaminated water and soil.

Not unlike the importance that habitats have as ecosystem services, some species play key roles in how ecosystems function. An example of one of these key species is bats. There is the recognition that the conservation of bats (loathed in the past) is important not only because their numbers are dwindling, but because they are vital in sustaining ecosystems. Pennisi *et al.* (2004) write that bats pollinate plants, they eat many different types of insect pests and they distribute seeds through their droppings. Bat tourism is seen to be important in that it supports the conservation of these animals (the construction of bat houses and research into disease and habitat loss), and tourism programmes have the potential to change attitudes towards bats.

The recognition that we have to do a better job at protecting biodiversity has generated an intense debate over methods of conservation. Up until recently, conventional wisdom regarding conservation was based on what has been termed the 'fortress mentality', 'fences and fines' or 'coercive conservation' (see Adams and Hulme 2001). This philosophy evolved from the American model of protected areas management, defined in the Wilderness Act of 1964, which declared that, 'man himself is a visitor who does not remain' (Siurua 2006). As such, because human populations have been so disruptive to the proper functioning of ecosystems through encroachment, habitat destruction, poaching and so on, they had to be removed. The problem with this approach, however, is that those who lived in or adjacent to protected areas, and subsequently removed in the name of conservation, have been stripped of their rights as well as any stake in the traditional utilisation of resources that had been so important before the establishment of protected areas.

Not surprisingly, this philosophy, which has worked relatively well in the USA and other industrialised countries, has been roundly criticised for its advocacy of 'animals over people'. The exclusion of local people from their traditional lands has provoked an intense level of hostility towards protected areas. In Nepal, for example:

> With the declaration of parks and reserves in such areas many people have been legally restrained from using their traditional rights to these resources. Those people living outside the boundaries have no legal recourse to procure compensation for their lost benefits. They ask themselves why they have been deprived of this inexhaustible natural resource which is a common property of the community. They think that it is unjust that 'outsiders' impose these restrictions and they express their feelings of discontent in various ways . . . To demonstrate their anger they sometimes vandalise park property by damaging bridges, signposts and boundary pillars.
>
> (Upreti 1985: 20, cited in Siurua 2006: 75)

(See also Hough 1988, who feels that restrictions of access to traditionally used resources, and the disruption of local cultures and economies by tourists etc. have led to hostility, resentment and damage to park property.)

This problem has been identified in the creation of new national parks in Mongolia (Maroney 2006). Although local people were found to have a strong conservation ethic, they were unwilling to discontinue grazing certain areas without compensation. In the absence of any efforts by government to work with herders or local government in any way, Maroney argues that management plans for the parks need to be developed collaboratively through integrated participatory approaches which clearly outline how local people can benefit directly from conservation.

The problems inherent in the fortress mentality philosophy have prompted theorists to develop a different approach deemed the 'community conservation model', which Siurua describes as being inclusive of, 'local people in the design and operation of conservation schemes, and prioritizes the generation of tangible benefits from such schemes in order to link conservation with local development needs and thus give communities an economic stake in conservation' (Siurua 2006: 76). Community-based conservation (CBC), community-based natural resource management (CBNRM), integrated conservation and development projects (ICDPs) and collaborative management (CM) are just a few of the participatory or bottom-up approaches that have been implemented to sustain this new philosophy (see Chapter 4). These approaches are founded on the belief that the best way to conserve biodiversity is to ensure that the people who share these spaces are fed. The trick, therefore, is to somehow make conservation a profitable enterprise in the face of competing interests like forestry, fishing, mining, slash-and-burn agriculture and so on. This has been the rallying cry of ecotourism pundits for years, which support the notion that ecotourism, if developed and managed appropriately, can generate more revenue per unit area than other more extractive and damaging industries (see Fennell and Weaver 2005). In this regard, Kaae (2006) writes that residents of Doi Inthanon National Park in Thailand are generally in favour of tourism in the park because of the potential for jobs (see also Ormsby and Mannle 2006). What they are opposed to are the existing park restrictions on their traditional activities, and there appears to be a significant level of fear that externally based tour operators will report any actions that fall outside these restrictive policies. (See also Carrier and MacLeod 2005 who observe that managers of the Montego Bay Marine Park restricted access to fishing grounds, but failed to compensate the anglers for their loss.)

But far from being a silver bullet, and despite mounting support for this approach, Siurua says that the community conservation model has many theoretical and operational problems based largely on the differences of power between various groups involved as stakeholders, including local people, government, industry and so on (discussed more at length in Chapter 10). Spinage (1998) argues that there is evidence to suggest that the newer approach to conservation is more about political agenda and the desire for monetary reward than proper conservation. Spinage illustrates that studies designed to discredit the Western or fortress approach to conservation were found to be misrepresentative of the fact that there was much more at stake in regard to the denial of rights to lands, over-harvesting and the loss of traditional knowledge. To claims that parks have been the bane of indigenous people, he argues that:

> One asks why are these resources [wildlife] assumed to be found only in the protected area? The answer must surely be because they have disappeared where people have freely exploited them . . . so . . . why are there greater densities of animals in national parks than outside them? [This is a problem that is not solved] by abandoning the small parks to those interests that created the unsustainable conditions around them.
>
> (Spinage 1998: 271)

In South Africa, for example, the creation of Richtersveld National Park came about in response to the belief that the indigenous people of the region (Nama) were unable to conserve their tribal lands properly. Over time, extensive over-grazing in the Richtersveld has necessitated the implementation of limits to the number of stock in the park in line with the capabilities of the region to absorb such use (Boonzaier 1996). Similar observations have been made by Janzen, (cited in Chase 1989), who illustrates that the establishment of African game parks has been self-defeating because they are slowly becoming consumed by poaching and adjacent farmers (see also Lovejoy 1992).

An interesting addition to the debate on how best to conserve biodiversity (i.e. the diametrically opposed positions of the authoritarian approach and integrated conservation and development project approach) is one offered by Brechin *et al.* (2002). These authors argue that past attempts have failed to incorporate social and political processes that would greatly assist conservation efforts. Six key issues for biodiversity conservation include human dignity, legitimacy, governance, accountability, adaptation and learning, and the effects of non-local sources. These are further articulated into a number of recommendations for enhanced international conservation efforts on the basis of: (1) conservation with social justice; (2) conservation in context; (3) knowledge about conservation; (4) increased capacity for organisational coordination; (5) conservation performance based on indicators; and (6) an ongoing dialogue on conservation. Such an approach takes a decidedly moral stance on biodiversity conservation acknowledging that there are many key questions regarding benefits, rights and environmental justice that past methods have not broached. The following few examples of biodiversity conservation and protected areas touch on many of the aforementioned themes.

The importance of property rights for local populations is demonstrated through the work of Urich *et al.* (2001) who document the processes behind the establishment of two protected areas in Bohol, Philippines. Protected areas were established based on the effects of intensive deforestation, agricultural exploitation and quarrying. Given limits to how intensively they can participate in traditional activities, local people have become hostile in accusing policy-makers of ignoring their rights because of the top-down approach employed in the establishment of the protected areas (See also Badalamenti *et al.* 2000 for a discussion on the hostility towards marine protected areas in the Mediterranean because of poor local consensus.) The foregoing has led many theorists to contend that environmental degradation is driven primarily by poverty. However, more contemporary studies indicate that wealth and the people behind large-scale development are more likely responsible for poverty (Gray and Moseley 2005). Loss of rights and access to land were at the heart of the discussion on protected areas and indigenous people by Hill (2006), who argues that marginalisation of groups should be a thing of the past. Coexistence between indigenous people and those who manage protected areas should be developed through adaptive models that are based on shared involvement and more inclusive objectives. In this regard, Balint (2006) writes that commons-related issues in general are not the same as commons issues in or adjacent to protected areas. In the general sense, resource users, as one of many different stakeholder groups with separate power and standing, have the opportunity to devise and change rules as well as bargain as a governance tactic (citing Stern *et al.* 2002). But because protected areas are owned by government, and the state has its own set of rules and regulations, local people have limited ability to devise and change rules and to bargain. Balint argues that CBC would be far more effective by giving closer consideration to rights, capacity, governance and revenue (Table 5.2; see also Chapter 10).

Conservation efforts are often challenged in less developed countries (LDCs) by political and economic turmoil which often contribute to the cessation of conservation almost completely (see Hart and Hart 2003). Heinen and Shrestha (2006) discuss the challenges to conservation in Nepal given the disorder that has taken place in this region for over a decade. Problems include policies that continue to stem from Kathmandu despite efforts to decentralise decision-making, and integrated conservation development programmes that emphasise more development than conservation. But despite these issues, the authors argue that the history of conservation organisations, the practice of deriving employment through tourism, and the willingness to explore new conservation approaches has

Table 5.2 *Commons issues in protected areas*

Variables	Definitions	Indicators	Interactions
Rights	Include formal property rights, traditional usufruct rights, statutory and common law authority for natural resource management, and rights to representation and participation.	Assessment of statutes and common laws regarding decentralisation, land tenure, and representation; measures of the practical enforceability of these rights.	Improved governance and capacity can strengthen local rights and participation and promote devolution of authority for natural resource management.
Capacity	Individual and institutional competence, ability, knowledge, experience and skills relevant for specific or general tasks.	Levels of relevant education, training and work experience; measures of the strength of the enabling social and institutional environment.	Good governance depends on adequate capacity; good governance in turn helps ensure that improved capacity provides general benefits.
Governance	Decision-making processes and institutions.	Measures of transparency, accountability, corruption control, community participation and power-sharing.	Good governance has two-way relationships with rights and capacity and can contribute to increasing revenue.
Revenue potential	Income per household that a given project can reasonably be expected to generate for involved communities.	Projected (and actual) revenue per household.	Expanded and strengthened rights, capacity, and governance allow communities to increase revenue potential.

Source: Balint (2006)

insulated conservation in Nepal. This resiliency means that conservation, according to the authors, will remain intact despite the threat of political vagaries over time.

The disconnection between government and industry in regard to tourism development is demonstrated in the case of Bahia, Brazil, where local and state governments have had little institutional capacity to control development (Puppim de Oliveira 2005). This has resulted in rapid and chaotic tourism development in the region up to the 1990s leading to excessive environmental impacts. In an effort to offset this negative trend, two protected area development processes were put in place that acknowledge the importance of tourism development on one hand, and the need for conservation on the other. In the first case, protected areas were created in regions showing heightened ecotourism demand based on pressure from tourists and local stakeholders. This bottom-up approach to protected area establishment is an example of how local people can effectively mobilise around environmental issues, with the help of government decision-makers. In the second case, protected areas were created for the purpose of acting as a safeguard against the overwhelming impacts of major tourism developments. Puppim de Oliveira argues that external actors such as financial donors can play a part by withholding money for these projects in the event that environmental considerations are not instituted by development teams. In this regard, different participatory approaches to tourism planning and development are providing local people with an opportunity to have their voices heard. PAGIS (an acronym for Participatory Approach and Geographical Information Systems) is one such outlet that integrates expert knowledge (GIS) with local knowledge (PA). The core impetus behind this system is to facilitate a better appreciation of local knowledge and to improve the level of participation of local people tourism planning (Hasse and Milne 2005).

CASE STUDY 5.3

Gorilla tourism in Africa

The brutal 1994 campaign of genocide between the Tutsis and Hutus of Rwanda is a graphic reminder that conservation is very difficult to achieve without an understanding of the socio-economic context in which it takes place. Cultural differences prevent us from truly understanding another culture: their motives, lifestyles and moral bases. Stanford (1999) writes that the people of Central Africa have long suffered from too many governments, geographical divisions, colonial powers and dictators, and physical displacement. This situation, he suggests, has set the stage for an inherently flammable region, with Western aid and tourists adding fuel to the blaze. The 1998 capture of eight tourists in the Congo (four never heard from again), and the 1999 kidnapping of 14 tourists at the ecotourism centre in Buhomoa, Uganda (eight tourists murdered), brings to light the political and economic realities of this region. But despite the war, the importance of gorillas to the regional economy is clear – although millions of people were killed or displaced over the period of genocide in Rwanda, only five gorillas were killed (Stanford 1999).

The Economist reported in May 2002 that for the first time since 1985 poachers are trafficking in Rwanda's mountain gorillas, of which only about 600 remain. They note that two females were killed in order to capture their infants. An infant gorilla can fetch up to US$125,000 on the black market. The report acknowledges that the long-term benefits of gorilla tourism have always been substantial, with this form of tourism ranking as Rwanda's third foreign currency earner. The political instability put an end to gorilla tourism in Rwanda and the Congo; in Uganda, it has begun to rebound only after some significant initiatives by the Ugandan government to control poaching, preserve habitat, give 20 per cent of tracking permits to adjacent farmers, and use other funds to support the conservation budget of the Ugandan Wildlife Authority. In all cases, tour operators have invested substantial sums of money in the region, but this will not be a profitable area unless the government actively enforces policies which instil confidence in tourists and operators alike.

The International Gorilla Conservation Programme (IGCP) was developed with the goal to ensure the survival of the mountain gorillas and their habitat. The mountain gorilla is found only in protected afro-montane forests in northwest Rwanda, southwest Uganda and eastern Democratic Republic of Congo. In these regions there are two populations. One is in the Bwindi Impenetrable National Park in Uganda, and the other in the ecologically homogeneous region of three parks (Mgahinga Gorilla National Park in Uganda, Volcano National Park in Rwanda and Virunga National Park in the Congo). The programme is an initiative between three organisations, including the African Wildlife Foundation, the World Wildlife Federation, and the various protected areas authorities of the three respective countries. The IGCP helps to monitor and protect the gorillas, train staff and advise governments on policy and enforcement. They also help with habitat protection which, after poaching, is one of the biggest threats to the mountain gorilla in an area with tremendous population growth.

Gorilla tours cost about US$280 for the opportunity to visit a gorilla family for a period of one hour. Tourists who have a cold or other illness are not allowed to make the journey. But despite precautions, primatologists argue that the health of the mountain gorillas is in danger because of the volume of tourists they are exposed to. They note that tourists can inadvertently expose animals to human diseases, which may very quickly spread throughout a family (studies of gorilla faeces show that they have indeed picked up new parasites since the introduction of tourism), but it is not known to what extent wild populations are susceptible to diseases which have infected captive groups. The following from

Tourism and Travel in Uganda (www.visituganda.com/wildlife/gorilla-rules.htm) are the rules and procedures that tourists must follow in their interactions with the gorillas:

> A maximum number of six visitors may visit a group of habituated gorillas per day.
> Voices are to be kept low.
> Do not leave rubbish in the park.
> Guides will inform tourists when to take pictures. Flash photography is not permitted.
> Always wash your hands before venturing out to the gorillas.
> Keep a minimum of 5 metres from the gorillas, to avoid exposure to diseases.
> Tourists must stay in a tight group whilst viewing gorillas.
> Do not smoke, eat or drink while near the gorillas.
> If a gorilla charges, crouch down slowly and do not look the gorilla in the eyes.
>   Running will increase the risk of harm.
> Never attempt to touch a gorilla.
> If the gorillas become agitated, the one-hour tour may be cut short.
> After a visit with the gorillas, keep voices down until 200 metres from the family of
>   gorillas.

Unfortunately Butynski (1998) observes that many of these directives are often overlooked in the process of conducting a gorilla ecotour. He reports that:

> The official number of tourists visiting gorillas has escalated from 6, to 8, with 10 being strongly considered, despite the cautions by scientific experts.
>
> All gorilla programmes (tours) suffer from a lack of risk assessments, impact assessments and programme evaluations.
>
> Tourists and guides have been found frequently to bribe park staff to ignore the rules. The benefit to the guide is a larger gratuity at the end of the tour, and likely a kick-back to park staff.
>
> Extended visits with gorillas have gone well beyond the one-hour limit.
>
> Sick tourists are included on tours and unauthorised visits to non-tourist gorilla groups have been allowed.
>
> Conservation has taken a back seat to political power-struggles and short-term financial gains.

Butynski also observes that the absence of sound empirical data on gorilla tourism is the partial cause of so many ethical transgressions, and that information on the extent to which tourists affect gorilla behaviour, ecology, health and survival is essential. The sustainability of gorilla tourism is questioned because there is too much disparity between what needs to be done, and what the most important stakeholder groups (governments and non-governmental organisations (NGOs)) are willing or able to accomplish. This means putting conservation first over economic prosperity; conducting research; stronger regulations, which are enforced; and more money to support conservation. While tourism can be directly affected by war, conservation can be aided by it. A common conservation objective between nations – such as gorilla conservation – often forces these disparate parties to work together. Citing the director of the IGCP, Annette Lanjouw, Snell (2001) writes that peace is not a prerequisite for conservation. Lanjouw suggests that even though she works in a war-zone, conservation can contribute to regional stability.

Websites:
http://www.fauna-flora.org/around_the_world/africa/gorilla.htm
http://abcnews.go.com/sections/science/DailyNews/gorillas990305.html

Private reserves

Sustained growth of ecotourism since the 1970s has generated heightened demand for the use of wilderness spaces. In many regions, however, public land agencies that own and run national parks and other reserves do not have resources to provide the facilities to run ecotourism programmes effectively (e.g. trails, accommodations, interpretive signs etc.). In some cases, the threat from illegal operations such as poaching on public lands, as noted above, curtails the development of a viable ecotourism industry. Eagles (1995) suggests that private reserves have a role to play in the provision of ecotourism services, with parks being flexible enough to entertain new models of financing. He cites the fact that such parks will be ill prepared to accommodate visitation in the future as they increase in number with continued decreases in funding.

A good example of the creation of space for ecotourism purposes is the development of private reserves. In some regions like South Africa and Texas there is a long legacy of private reserves. In the former case, Dieke (2001) explains that much of the tourism industry was based on consumptive activities, principally hunting by sport hunters, commercial hunters and subsistence hunters. As resources became scarcer, it was the latter group who were refused hunting rights, creating tension between those with access and those without, and also the need to control hunting through private parcels of land. In many cases these landscapes remain in the hands of the economic elite, some of whom have incorporated ecotourism programmes in their operations alongside hunting with the purpose of balancing different forms of land use on properties. (As an aside, Dieke notes that the historical removal of wildlife resource rights from rural peasants has given way to an emerging policy of redistribution of resource rights on the basis of ecotourism. Here, it is argued, decision-making control and benefits are greatly sought after by the economically marginalised from such a history of injustice.)

Private reserves are defined by Langholz and Brandon (2001) as lands which are not owned by governmental bodies, they are are larger than 20 ha and maintained in a mostly natural state. While this topic continues to be one that is under-researched, some have provided important inroads. For example, Alderman (1992) suggests that private reserves are developed according to profit, habitat protection and research. She noted that many, as suggested above in the case of South Africa, are in fact hybrids which attempt to balance ecotourism, extractive industry like forestry, education and agriculture. About 25 per cent of the private reserves sampled by Alderman were developed for tourism alone.

One of the world's most noteworthy private reserves for ecotourism is the Monteverde Cloud Forest Reserve in Costa Rica. This parcel of land was originally protected during the 1950s by American Quakers who purchased 1,200 ha of land with the purpose of protecting the rich forests from human encroachment. Initially this protection came about to maintain sufficient lands for the dairy operation undertaken by the tenants, but later expanded with a conservation mandate because of pressure by biologists and other conservationists who recognised the biological value of the land. Honey (1999) notes that tourism was initially quite low key during the 1970s in Monteverde, with only 471 visitors in 1974, climbing to 3,257 in 1980, 11,762 in 1985 and later to 49,580 in 1992, coinciding with the growth of ecotourism throughout the 1980s and early 1990s. Through land purchases financed by various conservation organisations, the reserve has grown to encompass eight ecological zones. Langholz and Lassoie (2002) estimate the number of private reserves in Costa Rica alone to be 250, with a total area of approximately 63,832 ha (see also Langholz et al. 2000). Table 5.3 (as cited in Langholz and Lassoie 2001) outlines the range of private reserve types and their respective ownership structures and

Table 5.3 Private reserve types

Category (type)	Management objective	Example
Formal park (Type I)	Protect nature in a formally recognised national protected area system. Must be legally gazetted through legislation or executive decree.	More than two dozen 'private wildlife refuges' have qualified to be a legally recognised unit in Costa Rica's protected area system.
Programme participant (Type II)	Participant in a formal, voluntary incentive programme designed to promote biodiversity conservation on private lands.	The Natural Heritage Program in South Africa has 150 sites, protecting 216,332 ha. Most owned by private citizens.
Ecotourism reserve (Type III)	Combine conservation with tourism. Tourism the principal revenue generator.	Tambopata Jungle Lodge, Peru.
Biological station (Type IV)	Combine conservation with scientific research. Reserve serves as an outdoor laboratory.	The Jatun Sacha Biological Station protects 2000 ha of forest in Ecuador, while supporting rainforest research, university field courses and tours.
Hybrid reserve (Type V)	Protect nature as one component of a diverse land-use policy. Includes ranches that combine agriculture, forestry or cattle production with reserve providing watershed protection.	The 80,000 ha Hato Pinero operation in Venezuela combines biodiversity conservation, nature tourism and cattle ranching.
Farmer-owned forest patch (Type VI)	Safeguard water and other resources at the individual or family level. Usually informal, small (<20 ha) and not involved in tourism. The least formal type of private conservation area.	Largest category in terms of amount of land protected and number of owners. The category about which the least is known, representing thousands or millions of patches worldwide.
Personal retreat reserve (Type VII)	Maintain natural area as a personal haven at individual or family level. Frequently owned by urbanites who own or inherit land in rural areas; don't rely on reserve for income generation.	Sixty per cent of Adirondack State park in New York State consists of private property, much of it owned by urbanites who use the area for second homes.
NGO reserve (Type VIII)	Protect nature under the auspices of a local, national or international not-for-profit conservation organisation. Includes land trusts, foundations and associations.	School children around the world raised money to protect habitat in Costa Rica's Monteverde Cloud Forest Reserve.
Hunting reserve (Type IX)	Maintain natural areas for sustainable wildlife utilisation. Animals collected for trophies or meat production, or both. Include game ranches owned by hunting clubs.	South Africa has more than 9,000 game ranches protecting 8 million ha. Also supports non-game biodiversity.
Corporate reserve (Type X)	For-profit organisations protect nature as a tool for creating positive public relations, as a result of court orders or a conservation ethic.	Developing country examples usually owned by large multinational corporations (e.g. forestry industry).

Source: Langholz and Lassoie (2001)

management objectives. The types include reference to the IUCN categories of protected areas outlined earlier in the chapter.

In a study of privately owned protected areas in southern Africa, Krug (2002) concluded that a minimum of 14 million ha of private land is associated with the ten types identified above – an area one-half the size of the UK. Krug observes that while the private reserve system is growing, in a self-sustaining economic fashion with zero cost to tax payers, state-managed protected area systems are challenged to increase in size in the face of declining budgets (see, for example, Saayman and Saayman (2006) in reference to declining protected area budgets in South Africa). In South Africa, contractual national parks have been a recent initiative whereby national parks may be established on privately owned land, thereby alleviating government responsibility to purchase large tracts of land (Reid 2001). An essential component of these parks is a joint management agreement where the rights of landholders are formally defined and dictated by a management committee made up of democratically elected landholders and members of the parks authority. Reid suggests that the formalisation of dual responsibility in such parks places landholders in a position of power which paves the way for high levels of participation, thus eliminating any coercive factors that have plagued park–people relationships in the past. Australia has also shown an interest in private protection of lands through their National Reserve System programme. Figgis (2004) notes that private organisations and NGOs can place bids for 2:1 funding for land acquisition and short-term management costs. Ongoing costs must be borne by the group, and the property must be secured either by covenant or legal means, and be managed according to IUCN I-IV categories (see also Buckley, 2004).

Private lands in the form of large farms, ranches or stations can also play a part in biodiversity conservation through ecotourism. Moskwa (2010) writes that, in Australia, those living in the rangelands viewed ecotourism as a means by which to enhance conservation in two ways. First, by generating additional income that could contribute to ecological recovery. Second, ecotourism is deemed important in educating tourists about the value of rangelands. Both are important because some of Australia's most demanding conservation problems are found in the pastoral regions of the country. This research parallels earlier work by Fennell and Weaver (1997) and Weaver and Fennell (1997) who investigated the vacation farm industry in Saskatchewan, Canada. The authors found that ecotourism provided supplementary income to the family, but it also educated not only tourists but also owners of these properties about the value of natural resources. The educational component of ecotourism is discussed at length in the following chapter.

Conclusion

Practical and philosophical issues related to the place and role of humans in the environment will continue to be played out in the conservation arena, with protected areas acting as the most concrete example of the interplay between nature and people. As such, stakeholders must continue to employ new strategies to enable people to strike a fair and equitable balance between use and preservation in a world that emphasises the value of human beings at the expense of other life forms, and which continually encroaches upon the earth's most sensitive and significant regions. The principles that have helped to define biodiversity conservation and ecosystem management may help us to better understand the place of humans in the natural world. However, this technical and scientific information must be analysed in the context of appropriate philosophical and operational questions that allow us to manage protected areas in the context of the people who must also share these spaces.

Summary questions

1   Discuss the differences that exist between conservation perspectives based on efficient use, spirituality and harmony. Name some historical figures who were representative of these positions.

2   Discuss how park management has evolved over time. What are the differences between internal and external park threats, and what are the implications to park management?

3   Why are park zones so important, and what are their implications for the human use of parks and protected areas?

4   What is the difference between public reserves and private reserves? Why are private reserves a better alternative for ecotourism in some countries?

5   What is biodiversity, and how can ecotourism be an effective agent in biodiversity conservation?

6 Learning

I recall being at a conference in Rio de Janeiro in 1992 where a discussion took place on the value of learning as part of ecotourism. One of the delegates argued that 'Learning is what makes it different'. He meant that learning about the natural history attractions we visit is an essential part of the overall experience. Without learning – this includes the emphasis placed on learning by service providers and the willingness on the part of ecotourists to be informed – the experience is diluted. But there are questions. Is it learning about how to be a better climber in Alaska? Is it learning about how to enter a temple in Indonesia? It all depends on how ecotourism is defined.

This chapter starts with a discussion on ecotourism operators and guides, as these are the main facilitators of ecotourism experiences. The chapter goes on to examine interpretation, including signs and viewing platforms, as well as types of learning. Important in this discussion is why learning has situational value: ecotourists may or may not change their attitudes and behaviours *in situ*, while converting back to conventional behaviours at home. The chapter ends with a detailed discussion on ecolodges. This material is located here, because ecolodges ought to be another mechanism by which to educate the eco-tourist through more sustainable designs and practices.

I choose the term learning over education because learning is a more general term. Education is often said to be a process whereby knowledge is passed from one person (a teacher or guide) to another (student or tourist). Knowledge is gained from an outside source. Learning is continual or ongoing in the way we potentially 'learn something new everyday'. For Garavan (1997), learning is an umbrella concept that includes different forms of knowledge acquisition like education and training. This means that, 'Training . . . can be associated with "learning by doing" whereas education is more synonymous with "learning by thinking"' (1997: 42). Ecotourism would involve education especially in how guides and interpreters pass along knowledge to the ecotourist. But ecotourism might also involve learning more broadly in how there may be a range of devices that could be used, either formally or informally, that stimulate the learning process at a destination. These may include other people through word of mouth both at generating and destination regions, guides, brochures, newspaper articles, ecolodges or personal observations.

Operators and guides

A tourism operator is defined as an owner or manager of a place of business (Metelka 1990: 110). In ecotourism, the operator occupies a central position in the facilitation of ecotourism experiences. Higgins (1996) helps us better understand the structure of the nature tourism industry by suggesting that ecotourism is based on the existence of: (1) outbound nature tour operators, located in large key cities in industrialised countries; (2) inbound nature tour operators, who are centred in non-industrialised countries and usually provide services in one country; and (3) local nature tour businesses, which include hotels, restaurants, ecolodges, souvenirs, guides, and so on. It is the outbound nature tour operators who play the greatest role in linking clients with other businesses and destinations around the world, according to Higgins, particularly those in non-industrialised countries. In a survey of 82 American nature tour operators, Higgins found that such operations have increased in number by 820 per cent between 1970 and 1994 (from 7 to 83). Higgins also discovered that client lists for nature tour operators ranged from 25 to 15,000, with an average of 1,674 – 35 of which served over 1,000 clients in 1986 – with the largest five serving 40 per cent of the total market. Research by Ingram and Durst (1989) found that ecotourism firm size ranged from 20 to 3,000 clients, with three serving over 1,000 clients. The majority of their client base was described as 'outdoor enthusiasts, retired couples or students, equally male and female, over 30 years old, and usually travelling as individuals in late summer' (1989: 12). The companies had been in operation for seven years on average.

In other research, Eagles and Wind (1994) used content analysis to analyse advertising of 347 guide-led ecotours in 50 different countries. The average guide-to-participant ratio in these tours was 1:13, which is consistent with the outdoor recreation literature where recommended guide-to-participant ratios are 1:12 for activities like hiking (Ford and Blanchard 1993; see also Hendee *et al.* 1990; Hammitt and Cole 1987). In a survey of 24 North American ecotour operators, Yee (1992) found that while most operators (63 per cent) had been in business for 2–15 years, 17 per cent had been operating between 15 and 20 years, with many suggesting that they were offering ecotours long before the term originated. Ninety-two per cent of respondents followed a code of ethics, and 38 per cent considered ethical conduct in pre-trip orientation; 29 per cent addressed ethics in printed information packs; 13 per cent of those surveyed used videos or slides on ethics, while 33 per cent used lecture formats. The use of trained interpreters and naturalists was also addressed in Yee's research, where it was found that 75 per cent of the respondents had 'ecologists, naturalists or other experts on staff to aid in conducting their tours' (1992: 11).

Researchers have also analysed the degree to which ecotour operators perceive and promote themselves as being environmentally friendly. Weiler (1993), for example, found that about 40 per cent of operators were promoted as being environmentally friendly; 66 per cent of operators felt that their tours were either beneficial or very beneficial to the environment (only one was said to be harmful or very harmful); 70 per cent said that their tour educated tourists about impacts; and only 7 per cent said that their tour enhanced the environment through, for example, removing others' rubbish. Lück (2002a) reports on how Germany's two largest mass tourism operators, TUI and LTU, have taken it upon themselves to control their impacts in destination regions. LTU's operation in the Maldives is a case in point, where tourists are asked to collect inorganic waste during their stay in eco-bags to be shipped back to Germany. Lück notes that after six years of the programme, 300,000 eco-bags had been handed out with 80 per cent of all passengers participating in the programme, amounting to 400 tons of rubbish removed from the local environment.

Moore and Carter (1993) investigated the relationship between ecotourism operators and protected area managers in concluding that it has not been a compatible relationship. Based on interviews of 16 nature-tour operators in Australia, operators felt that resource managers failed to understand certain aspects of their business, including the profit motive, costs associated with marketing, costs of operations, the motivations and needs of tourists, and the need for minimal infrastructure in remote areas. On the other hand, operators were concerned over the fact that managers did not have a conservation ethic, failed to provide visitors with appropriate information, had little understanding of the need to control visitors and did not understand resource fragility. The fundamental differences between the two camps illustrate in part why ecotourism needs to be developed and implemented on the basis of a sound philosophical platform, as noted earlier in this chapter. Allowing profit to overwhelm all other values leads to unhealthy relationships with others who will not or cannot have the same enthusiasm for commercial interests based on their own value set.

The literature on guiding in the outdoor recreation and adventure tourism is comparatively more abundant than what has emerged in the ecotourism field to date. Despite the relative dearth of research, we have come to realise that a good guide is essential to the overall success of an ecotourism venture. Defined, an ecotour guide is, 'someone employed on a paid or voluntary basis who conducts paying or non-paying tourists around an area or site of natural and/or cultural importance utilizing ecotourism and interpretation principles' (Black et al. 2000: 3; see also Black et al. 2001). Priest (1990) suggests that guides need a number of broad skills which can be loosely organised into hard, soft and meta skills. The first of these include the solid and tangible skills that make guides effective, including technical skills (e.g. paddling or climbing competencies), safety skills (e.g. advanced wilderness first aid) and environmental (e.g. minimum impact camping skills). Soft skills include organisational (e.g. planning, preparing and executing skills), instructional (e.g. appropriately teaching how to paddle) and facilitative (e.g. those skills which foster productive group dynamics) proficiencies. By contrast, meta skills are explained by Priest as those talents related to decision-making, problem-solving, experience-based judgement, and the knowledge of when to employ various styles of leadership, including democratic (allowing the group to make certain decisions), autocratic (taking the lead in emergency situations) and abdicratic (abdicating or delegating responsibility to the group).

In view of the fact that ecotourists are a discerning group who may be quite vocal and active in environmental matters, guides and other front-line employees must be educated on natural history as well as any other issues related to environmental policies of the region and methods which have been or are being employed to ensure preservation of species or landscapes (see Mitchell 1992). A study by Weiler and Davis (1993) on the roles of nature-based tour leaders reported that the main roles of the tour leader (on a 5-point scale with 5 being most important) were organiser (mean = 4.6), group leader (4.5), environmental interpreter (4.3), motivator (4.2), teacher (4.0) and entertainer (3.4). As Weiler and Davis suggest, in reference to the leader:

> S/he must be an organizer, a group leader, a teacher, and even an entertainer. In nature-based tourism, the tour leader must also be responsible for maintaining environmental quality, by motivating visitors to behave in an environmentally responsible way during the tour, and by interpreting the environment in such a way as to promote long-term attitude and behavioural change.
>
> (Weiler and Davis 1993: 97)

Weiler and Ham (2001) observe that in Australia there still does not exist industry-wide legal specifications regarding the rights, working conditions, pay and so on of ecotour guides. Although the Ecoguide Certification Program (a voluntary, industry-led certification program developed by the Ecotourism Association of Australia), does not fully address these issues, it has come about with the aim of providing a qualification that is designed to reward guides who achieve various competencies. The program is not designed to train guides, but rather to assess their skills, knowledge, actions and attitudes (Crabtree and Black n.d.), with an overall purpose of instilling a philosophy of best practice among practitioners. Table 6.1 provides an overview of some of the benefits that different tourism stakeholders might realise as a result of better-prepared ecoguides, as outlined in the EcoGuide Program.

Table 6.1 *EcoGuide Program benefits*

Benefits to nature and ecotour guides

1 A recognised industry qualification.
2 An opportunity to promote guiding services as genuine nature/ecotourism.
3 A defined competitive edge rewarded through factors such as better job opportunities.
4 Access to relevant, appropriate and reduced-cost training materials/networking options.

Benefits to nature and ecotourism operators

5 A simple method of recognising and recruiting quality guides.
6 Identification of training gaps/needs within operations.
7 Improved guiding practices leading to fewer negative environmental and cultural impacts and increased client satisfaction.

Benefits to nature and ecotourism consumers

8 An assurance of guides that are committed to providing quality nature or ecotourism experiences in a safe, culturally sensitive and environmentally sustainable manner.

Benefits to protected area managers

9   Improved guiding practices that lead to fewer negative environmental impacts.
10 Guides who role model and ensure good environmental/cultural behaviour.
11 A framework of standards applicable to interpretive rangers.

Benefits to the environment

12 Guides providing relevant and appropriate interpretation that inspires clients and workplace employees and encourages minimal impact actions and a conservation ethic.

Source: Crabtree and Black (n.d.)

Black and Ham (2005) have critically evaluated the EcoGuide Program as a basis for constructing a model for guide certification in improving, overall, tour guiding as a profession. Using a mixed methods approach in their investigation and including six research populations, the authors derived five key principles that could be used in developing a tour guide programme. These include: (1) the importance of research in assessing the need and demand for such a programme; (2) to ensure representation and consultation of key stakeholders; (3) to develop clear programme goals and objectives while simultaneously determining programme owner-ship; (4) to secure funding for the programme; and (5) to establish a timeframe for programme development and implementation (see also Fennell 2002b).

Research on guide training has been advanced by Christie and Mason, who argue that while most training is usually competency-based by stressing the transmission of know-ledge and various skills, it ought to be much more than this. Guiding should be trans-formative such that it not only works on changing skills and knowledge, but it must also positively change values, attitudes and behaviours of those who participate in ecotourism or conventional tourism experiences. As noted by these authors:

> Competency-based guide training courses rarely ask the bigger, philosophical ques-tions that go to the heart of what has been referred to in this paper as transformative tourism. It is suggested that any educative experience (and the tourism experience fits into this category) results in some form of change. Such a change might be slight – the acquisition of some new facts or a new insight into the way other people live. On the other hand the change could be considerable – a rejection of certain stereotypes and attitudes or a new way of viewing the world. It is argued . . . that the . . . guide has a role to play in transformative tourism.
>
> (Christie and Mason 2003: 9; Mason and Christie 2003: 28)

An especially important element of their work is the practice of critical reflection as an option, rather than something that must be imposed. Journal writing, life histories, case studies, critical incidents, idea writing, small group discussion and role playing may be important in allowing the tourist (or ecotourist) to meet head-on their own way of looking at the world in making a decision as to whether or not it is valid.

An innovative guide programme has taken shape at the Brazilian Sea Turtle Conservation Programme in the village of Praia do Forte, Bahia, Brazil. Called the mini-guide programme, it trains local children aged 10–14 to be guides over a one-year period. Along with receiving a monthly stipend, the children are provided with an in-depth education about the marine environment. Results of studies on the programme point to community-wide support and the value of greater environmental awareness on the part of children, which leads to new aspirations and interests in pursuing higher levels of education (Pegas *et al.* 2012).

A simple fact of the matter, however, is that the implementation of tour guide training programmes comes at a cost. This is acknowledged by Black and King (2002) in their work in Vanuatu, where training courses and programmes had to be funded by foreign aid in the absence of public funds, especially in outlying areas. Another sad reality of ecotourism is that it often concentrates economic impacts into the hands of a few, especially in the less developed countries (LDCs). Guiding is one of the aspects of ecotourism that benefits local people directly. This has prompted Ham and Weiler (2000) to outline the following six principles for tour guide training which are designed to help LDCs develop sustainably through the ecotourism industry (see also Weiler and Ham 2002 with respect to their work in Panama, Patagonia and the Galapagos Islands, in regard to these six stages for sustainable tour guiding):

1   The initiative for training should come from the host country and ownership should remain with the host country.
2   Training content and methods should be informed by the literature on what constitutes good or best practice ecotour guiding and the adult training literature, with appropriate customisation to meet local needs and the target trainee group.
3   Training efforts must be systematically evaluated, and lessons learned from these evaluations must be documented and disseminated widely.
4   Training can and should be made accessible, both logistically (location, time and cost) and intellectually to residents in peripheral and rural areas.
5   Delivering cost-effective guide training by mobilising resources for multiple 'ends' is an important sustainable development strategy.
6   Training and supporting in-country trainers is essential for ensuring that ecotourism is of benefit to host economies.

It is often the case that guides wear many hats. This is the case in Madagascar, where guides have responsibilities both inside and outside the parks. Ormsby and Mannle (2006) found that guides are important in promoting conservation awareness to local communities living outside or adjacent to protected areas. The transmission of information has been taken one large step beyond in Indonesia, according to Dahles (2002), where the Indonesian government uses tourism as a medium to spread state propaganda. Regulation of tourism guides is thought to be essential in an effort to maintain the state's manufactured imagery of a tamed cultural heterogeneity through professionalism based on central control and uniformity.

While operators and guides typically facilitate on-site experiences in ecotourism, service provision also includes the 'outfitter', who has strong ties to outdoor recreation, where such providers have been providing adventure (e.g. hunting and fishing) experiences for tourists for some time. The types of experiences of interest to natural-based tourists have necessi-tated the inclusion of businesses that are prepared to offer equipment and other specialised services for tourists. Outfitters, then, are those 'commercial businesses that provide a person with the equipment necessary for an activity or experience' (Dahles 2002: 111). As Tims (1996) observes, outfitting traces its roots back to the early explorers who provided a service to people, usually in a manner that placed humans against nature. More recently, outfitting has progressed to the point where it has taken on professional status through America Outdoors and the Professional Guide Institute (PGI). The PGI provides training for people in areas such as wildlands heritage, back-country leadership, interpretation and outfitter operations, with the mission to, 'identify, enhance and disseminate the natural interpretive and educational resource of the outfitting industry so that outfitters and guides can offer the highest quality of experience to the public' (2002: 177).

Interpretation

While interpretation is often thought of solely in the context of persons working within parks and protected areas (i.e. interpreting cultural or ecological histories), according to Barrow (1994) interpretation is far more elaborate, and based on three types of planning: (1) town and country, (2) marketing and (3) education. In the town and country approach, appropriate land use philosophies and techniques are used to protect valuable land from development, provide access to the public, and to help create an effective tie with the resource base. Marketing interpretation deals with how best to understand various user groups and their particular wants and needs. The marketing approach to interpretation is often product oriented, but must temper this with other concerns of the site (e.g. resource protection). Finally, Barrow suggests that interpretation employs educational theory in its plan development in order to understand how people learn, and what to teach them. It is this third component that is most frequently considered in ecotourism theory and practice.

For guides to be truly effective they must use their training to communicate knowledge (facts, anecdotes and so on) about ecotourism destinations and attractions in ways that will positively impact the attitudes and behaviours of ecotourists. Kuo (2002) writes that interpretive information has the following functions. It provides: (1) friendly or welcoming messages to visitors on, for example, signposts; (2) general information to guide visitors safely and quickly; (3) information on alternative routes and attractions, for the purpose of directing visitor flows; (4) information on congestion and queuing times, in allowing visitors to better plan their trip; (5) educational information on the natural, cultural, histor-ical features of the area; and (6) information on appropriate visitor behaviour in the area (See Burgoon et al. 1994, who say that interpretation needs four components if it is to work properly: a sender, a medium, information content and a receiver.)

In order to manage effectively, interpretive planners must be sensitive to the learning, behavioural and emotional sides of interpretation both as art and science. Veverka uses the following example of an archaeological site to illustrate the relationship between these objectives in interpretation:

'The majority of the visitors will be able to describe three reasons why protecting archaeological sites benefits all visitors'. But the real – or most important – aim the manager may have in mind for interpretation to accomplish is to prevent visitors from picking up 'souvenirs' such as pottery shards at archaeological sites. So the Behavioural Objective might be: 'All visitors will leave alone any artifacts that they may find at the site and not to take them home'. It is the job of the Emotional Objectives for the interpreter to get the visitors to appreciate the value of artifacts left in place, and feel that they are doing a good thing by not removing anything. So an objective of this kind might be: 'The majority of visitors will feel a sense of responsibility for not touching any artifacts they may find on the ground'.

(Veverka 1994: 18)

The behaviour is the action that the interpreter would like to occur as a result of the interpretive programme, whereas the emotional objective prompts the interpreter to strive to realise how best to get the visitor to feel as though he or she is behaving in the appropriate manner at the site and beyond; that is, the behaviour is something that they should want to do. Veverka illustrates that advertisers use the same philosophy today by making you feel as though you want or need a particular product. The behavioural response is to actually go out and purchase the product.

In places like the Great Barrier Reef Marine Park in Australia (GBRMPA), interpretation plays a vital role in education and protection (Weiler and Davis 1993). Hockings (1994) found that while staff–recreationist interactions are not very high as a consequence of the level of use and size of the GBRMPA, tourists have come to rely on tour operators in securing information. Of a sample of 170 operators, 72 per cent of respondents indicated offering interpretation as part of their programme. A higher proportion of this number was found to be day reef trip, diving and sailing operators, while a lower percentage of fishing operators provided interpretation. A significant finding of the study was that 56 per cent of staff involved in interpretation had no relevant formal qualifications, while only 4 per cent of all staff had formal qualifications. It was suggested that this overall lack of training could not be linked to type of activity offered (types of activities included scenic flights, fishing, camping, day reef trip, seaplane rides, dive trips and so on), as those indicated as having the appropriate interpretive qualifications were evenly spread across operator types and sizes. Operators, according to Hockings, relied more on practical experience as a means by which to inform clients.

We get a glimpse of the importance of interpretive programmes in ecotourism through empirical studies of the nature of Lück's (2003), who found that not only were ecotourists to swim-with-dolphin tours exposed to relatively good interpretation, but they wanted more of it. He found that tourists were clamouring for more information on the broader environmental issues and regulations in which the operations (and animals) were immersed. This is important, Higham and Carr (2002) observe, because the experiences that ecotourists have on-site may have an influence on their environmental values and behaviours down the road. This finding prompted Higham and Carr to suggest that the reliance on environmental education as a foundational characteristic of ecotourism is a strong differentiating factor in separating it from other forms of nature-based tourism (NBT).

Whether interpretive messages are actually getting through is open to debate. Armstrong and Weiler (2002), for example, investigated the types and frequency of conservation messages communicated by guides in Victoria's park system (Australia), as well as the percentage of these received by park visitors. These researchers found that the number of environmental messages delivered by guides and the amount of time spent delivering such were quite small given the opportunity presented in park tours (messages pertained to minimising visitor impacts, the significance of heritage values, minimising tour operator impacts and conservation action by individuals). One of the major findings was that despite the number of messages relayed, many visitors could not identify at least two messages of any kind at the conclusion of their day-long experience. This has implications both for the content of the message as well as the style of interpretation. What appears to be more important in interpretive design in protected areas is not necessarily

Plates 6.1 and 6.2 Experienced multilingual interpreters are often central to the ecotourism experience

the intensity of use of media at these sites (i.e. high intensity versus low intensity) in influencing environmental attitudes, but rather the association between activity and interpretation objectives. In a study of two sites in Western Australia (The Treetop Walk and Penguin Island) employing different levels of interpretive intensity, Hughes and Morrison-Saunders (2005: 175) concluded that, 'attempts to communicate a strong conservation message may be hampered by the character of activities offered if those activities communicate messages undesirable to interpretive themes'. Matching more consumptive forms of outdoor recreation (i.e. conservation for human use) with high priority areas dedicated to ecological integrity may produce negative feelings about the site and how it is managed.

Kohl (2005) has been critical of interpretive approaches because the majority of conservation and visitor management programmes lack a systems approach. The Conservation Measures Partnership's (2004) *Open Standards for the Practice of Conservation* is said to be well suited because of its iterative properties. The basic project management cycle is founded upon conceptualisation, planning, implementation, analysis, adaptation, communication and iteration. Kohl uses this approach to model visitor impacts in a park, the control of local impacts outside of a park, interpretation as part of a larger environmental education programme and interpretation in visitor fundraising. While the standards (i.e. industry-wide norms generating measurable results fed back into the cycle) built into the practice of interpretation may be new (although these are not apparent in Kohl's examples), the strategic programme planning approach that surrounds this process is not (see Fennell 2002b; see also Chapter 11).

Signs and platforms

An important form of interpretation in parks and protected areas comes in the form of signs and viewing platforms. Hughes and Morrison-Sanders (2002) investigated interpretive signage of the Tree Top Walk (TTW) in Western Australia, where the operator intentionally minimised signage in an effort to restrict distractions. A subsequent visitor survey indicated that participants of this tour were frustrated with the lack of interpretive information along the course of the tour. This promoted the development of an interpretive sign trail in 2001 for the purpose of assessing how signs impact visitor learning. Hughes and Morrison-Sanders found that trail-side interpretive signs provided no improvement in the visitor knowledge (the only sub-group to demonstrate improvement in knowledge was the repeat visitor group), but interestingly there was a positive increase in the perception of learning because of the addition of more signage.

Referencing back to the work of Hughes and Morrison-Sanders (2005), these authors compared the Tree Top Walk (TTW; referred to as a low intensity use of on-site interpretation) against a high intensity attraction (Penguin Island). The authors found that the intensity of interpretation had little affect on visitor attitudes and perceptions, but the character of the site itself did. For example, the TTW attraction, which is more observational in nature, effectively complements the mainly aesthetic reason for visitation, where the tradition of use and the conservation agenda nicely coincide. This suggests that interpretation efforts should coincide with the type of activity sought by visitors. Too much interpretation in a rich and detailed manner may not be effective for those visitors looking for an aesthetic and shallower experience.

Given the frequency of their use, it is surprising that there is a lack of research on understanding the role that viewing platforms play in educating ecotourists. Duffus and Dearden (1990) highlighted the dearth of knowledge on viewing platforms in their comprehensive work on wildlife tourism. In particular they focused on work needed in the areas of site users, focal animals, as well as the environments in which these animals and sites are located. Using Duffus and Dearden's conceptual framework, Higham *et al.* (2008; see also Curtin 2005) identified a number of key research issues to pursue in the future on these static or mobile sites that are normally located in critical wildlife habitats for breeding, socialising or feeding, as follows:

Key questions relating to site users

- How are discrete groups of tourists affected at locations where multiple viewing platforms exist?
- Do viewing platforms have social carrying capacities?
- Can elements of platform design alter social carrying capacities?
- How do the behaviours of visitors (and perhaps staff) impact visitor experiences at viewing platforms?
- How do the behaviours conducted at one platform impact visitors on other platforms?
- How does visitor origin influence experiences at viewing platforms *vis-à-vis* feelings of stewardship for focal animals?
- Do multiple viewing platforms create symmetric or asymmetric social impacts?

Key questions relating to focal animals

- What aspects of platform design bear upon the impacts of site use?
- How can these impacts be mitigated?
- Should different types of platforms proliferate at a site without careful management?
- What are the implications of mobile viewing platforms for animal biology and energy budgets?
- Should mobile viewing platforms be accompanied by strict management guidelines (or permit conditions) on locations and duration of contact with focal species?

Key questions relating to site ecology

- What are the spatio-ecological elements of the tourist–wildlife interaction?
- What are the critical and important behaviours that focal animals engage in at the wildlife–tourism setting?
- Where do members of an animal population most commonly engage in critical and important behaviours?
- Do specific areas of spatial ecology require complete protection?
- Do mobile viewing platforms have implications for non-focal species?
- What are the consequences of feeding wildlife at viewing platforms?
- How does feeding at viewing platforms impact the abundance of focal and incidental animal populations, and interactions between species within an ecology?

Types of learning

Studies on learning and environmental education have advanced steadily in ecotourism through the advancement of new approaches that explore cognitive, affective and behavioural domains. One of the first models of interpretation comes from Forestell (1993), based on his work on the whale watching industry. Forestell suggests that tourists experience different cognitive states during a whale watching excursion that require the naturalist/interpreter to articulate information in three distinct periods: pre-contact with whales, contact and post-contact. Pre-contact information involves learning on how participants should interact with the whales. This may be supplemented with geographical information and marine environmental education. In the second, or contact, stage, interpreters focus on the need for participants to seek answers to a variety of questions related to the observation of the animals. Questions typically relate to whale identification, whale behaviour, verification of knowledge and safety. The goal of this stage is to, 'generate motivation to learn by creating or uncovering an imbalance between an individual's initial knowledge base, and some current perception of the world . . . [and] provide sound knowledge to allow the participant to regain cognitive balance' (Forestell 1993: 274). The final stage, post-contact, is characterised by the participants' need to make pre- and post-trip comparisons of whales and the marine world, and to consider broader environmental issues (including active and financial support for conservation programmes or issues). Participants also question leaders about the well-being of the whales. Forestell suggests that the ecotourism experience has the ability to empower people to synthesise unscientific observation with scientific fact, which he feels is not an inferior substitute for hard data, but rather one that is experiential and practical. This model demonstrates the need for interpreters to be sensitive to the different stages of an ecotourism experience. In situations where guides are providing more personal knowledge of a setting only, Forestell would perhaps argue for more of a trained approach to the delivery of the interpretive product, which would ultimately enable the tourist to benefit further from the experience. (See Orams 1996, for an alternate model of interpretation.)

Like Forestell, many theorists emphasise the difference between affective (subjective feelings and emotions) and cognitive (objective knowledge acquisition) domains; that is, interpretation can affect both domains differently. Hill *et al.* (2007) found that the circular ropewalk at Crocodylus Rainforest Village was satisfying to ecotourists irrespective of whether they were given interpretive biodiversity sheets or not. As observed by the authors, 'visitors were satisfied with gaining a purely affective experience, where the value of their visit lay in the pleasure they gained from a sensual and immersive encounter with a novel environment' (2007: 82). But further than this, those who received biodiversity information sheets were found to rate their subjective or perceived experience higher than those receiving no information. And further, those given the biodiversity information sheets scored higher on a post-tour quiz than those without the sheets. Still, despite the new biodiversity knowledge, the study found that biodiversity interpretation did not augment the likelihood that these visitors would care more or behave better in rainforest environments in the future. It may well have been that these tourists felt that their behaviour was already ethical enough.

Walter (2013) provides a detailed overview of the many different ways in which we should theorise visitor learning in ecotourism. He organises ecotourism into three different categories: wildlife (soft ecotourism), adventure (hard) ecotourism and community-based (mixed) ecotourism. He suggests that learning has been different in each of the three types. With regard to wildlife ecotourism, Walter contends that learning has been tied principally to behaviouralist and liberal philosophies, changing visitor behaviour, with a focus on educational inputs and behavioural outputs. An example would be the attainment of new knowledge on wildlife biology, habitat and behaviour. Here there is a focus on biodiversity and conservation. Visitor learning in adventure ecotourism is termed 'progressive environmental education', where 'learners are seen as uniquely talented, knowledgeable and skilled, and possessing of an untapped, but unlimited potential for personal growth' (2013: 25). Curriculum involves wilderness experiences, physical and mental challenge, problem-solving and teamwork. Walter positions the final type, community-based ecotourism, within humanist, progressive and radical traditions of education, which centre on the culture and lifeways of mainly indigenous peoples. This form of education is holistic, emotional, spiritual, physical and social in orientation.

Another strategy for enhancing ecological awareness through interpretation is described by Coghlan and Kim (2012) as interpretive layering. This process involves the combination of many sources of interpretation for the purpose of echoing an interpretive message. This method was used at the Great Barrier Reef with good results. Multiple sources were found to reinforce the usefulness of other sources; increased as more sources were used; and the nature of understanding changed for the visitors. This approach has rich potential especially at complex attractions where there are many different messages offered to visitors.

Ballantyne and Packer (2011) investigated the concept of free-choice learning experiences as they apply to wildlife tourism. Free-choice learning includes sources like the Internet and media, as well as attractions like zoos, aquaria, museums, science centres and so on, where learning is under the control of the learner. Learners have the capacity to engage or not engage with these types of sources at will. As observed by Falk and Dierking (2000), the choice to learn is fully under the control of the learner – the will to learn – but also the what, where and when they will learn. Free-choice learning is a perspective that is generally used on-site (see, for example, Ham and Weiler 2002; see also Powell and Ham 2008 in the context of the Galapagos Islands).

Learning drop-off

One of the most persistent issues in ecotourism surrounds the question of whether environmental education programmes have a lasting effect on participants; that is, if ecotourists practise what they learn later on in their lives. Most of this literature examines the attitudes of ecotourists, as well as behaviours and, according to Powell and Ham (2008), results of these studies are mixed. In general, knowledge from interpretive programmes may increase (although Markwell 1998; and Ryan *et al.* 2000 argue that knowledge does not increase), but environmental attitudes and behaviours do not (see, for example, Tubb 2003; Wiles and Hall; 2005). This has prompted Welford *et al.* (1999) to argue that we should not expect nature-based tourists to support sustainable practices, because their prime motivations surround entertainment, consumption and comfort (as discussed in Chapter 2).

In a study on the messages delivered by tour operators in protected areas, Armstrong and Weiler (2002) found that there was not a good match between the content of the environmental messages given by guides and those received by park visitors. Respondents reported fewer key messages ($n = 35$), than the overall number of messages conveyed ($n = 108$). These authors also discovered that some types of messages are better received than others. Information on on-site impacts, the role of protected areas managers and heritage values of the region were better received than, for example, what individuals could do in regard to conservation actions.

Furthermore, there are relatively few studies that evaluate the effects of interpretation on the behaviour of visitors. Munro *et al.* (2008) attribute this to the range of different evaluative techniques and the consequent complexity of choosing and applying evaluation techniques at all. Their findings suggest an emphasis on evaluating interpretative programmes from the perspective of knowledge gain and attitude change (quantitatively), with far fewer efforts that focus on the measurement of behavioural change. They conclude by suggesting that managers ought to isolate a core group of evaluative techniques that could be applied across the interpretation continuum.

These findings appear to be consistent with the empirical work of Ballantyne *et al.* (2011a) and Ballantyne *et al.* (2011b), who investigated the 'drop-off' effect, or the decline in conservation action over time, despite initial intentions. Thirty-three per cent of a sample of 240 respondents indicated in a post-visit survey that although they expressed a strong desire to protect/conserve the environment after their trip, only 7 per cent had actually done so. The authors found that the pattern was the same at each of four different sites. Powell and Ham (2008) report that 70 per cent of tourists on a tour to the Galapagos Islands indicated a strong or moderate intention to donate money to a Galapagos conservation initiative in the future. However intentions, the authors observe, do not always lead to outcomes. In efforts to overcome this drop-off effect, Ballantyne and Packer (2011) advocate for the use of action resources, which include web-based learning materials like Internet forums, weblogs, podcasts and email and instant messaging to reinforce the on-site messages of the original experience. Initial research into the idea has proven useful in instituting long-term behavioural change in tourists, where the experience itself and action resources can work in concert (Hughes *et al.* 2011).

But is it not just tourists missing the boat on learning and education? Jackson (2007) examined attitudes, using two scales, towards the environment of several stakeholder groups: ornithological tour operators, generalist ecotourists, specialist ecotourists and members of a conservation group. Jackson found that although all groups rated highly on both scales, it was the tour operators who fell short of transferring generally favourable environmental attitudes into pro-environmental behaviours. Jackson concludes that there is a lack of a willingness to embrace the principles of ecotourism if these get in the way of business success. There may be many reasons for this. According to Tilley (1999), the expression of positive environmental behaviour may be constrained in small businesses because of economic barriers, poor ecoliteracy, inadequate institutional support, limited business support and low environmental awareness. This corresponds to the work by Malloy and Fennell (1998a) who contend that most businesses operate at the market culture level of moral development instead of the more advanced principled level.

This discussion points to what might be referred to as a 'Dr. Jekyll and Mr. Hyde Syndrome', where individuals change their attitudes and behaviours according to the situation they find themselves in – perhaps as a socialising mechanism. For most of the hard-path ecotourists the environmental ethic already may be present, so pro-environmental attitudes and behaviours are second nature, both during the trip and after. For the vast majority of others, however, there may be a modification of attitudes and behaviours, suggesting that there is a social or normative dimension that induces people to conform to the expectations they have in front of them (during the ecotourism episode). Once they get home, they slip with apparent ease back into regular ways of behaving that may be less ecologically sensitive.

We need more studies on this important area of research. I am curious to see how females and males differ on the emotional aspects tied to these wildlife tourism and eco-tourism experiences. Females are said to reason and make moral decisions on a more emotional level, while males are more objective, rational and prone to morality consistent with normative theory. It would also be interesting to better understand the level or amount of *unlearning* that is taking place. By this I mean the extent to which ecotourism programmes change how individuals have been taught, either through formal or informal mechanisms, to perceive the world (see for example Cochran-Smith 2003). This perspective amounts to a further disruption or disequilibrium of the ecotourist's cognitive (and likely affective) mindset beyond the accumulation of facts from the interpretive experience.

The Aboriginal effect

Relevant to our discussion here is the cultural context of interpretation. When ecotourists visit protected areas they are provided with the opportunity to learn about, for example, species, habitats and ecological relationships. However, Staiff *et al.* (2002) argue, at least in the context of Australian protected areas, there are other narratives that ought to be communicated beyond Eurocentrically based scientific names and descriptions of place names. The authors argue that Aboriginal knowledge has been excluded or marginalised in interpretive schemes such that it is, 'not on equal footing with the universal discourse and truths of the natural sciences' (Staiff *et al.* 2002: 103). Their work uncovers many interesting questions about the representation of what, how, who and why in protected areas. And if we are truly motivated to symbolise or characterise the natural and cultural history of an area, there appears to be much we have yet to learn about interpretive content.

Interpretation can thus be richly enhanced through the involvement of Aboriginal people, who may have a completely different interpretation of the natural world than Western approaches (as above). Zeppel and Muloin (2008) observe that wildlife tours (e.g. viewing animals in captive environments like zoos) can be enhanced through the use of Aboriginal guides, and through the use of signs and displays that accurately describe the Indigenous cultural knowledge of wildlife according to moralistic (spiritual), utilitarian (food) and aesthetic (totems and other aspects of symbolic significance) dimensions of the relationship between animals and the land. In an effort to advance the development of Aboriginal interpretation at wildlife tourism sites, Zeppel and Muloin (2008) offer the following guidelines in Table 6.2.

Table 6.2 *Guidelines for Aboriginal cultural interpretation of wildlife*

Employ local Indigenous guides/wildlife keepers at wildlife attractions.
Use selected local Aboriginal names for Australian wildlife species (where known).
Identify the Aboriginal name (language group) of well-known Australian animals (e.g. kangaroo, koala and kookaburra).
Include quotes and stories from local Aboriginal people about Australian wildlife.
Refer to Aboriginal guides who accompanied historic or recent wildlife expeditions.
Acknowledge Aboriginal totemic/spiritual links with Australian wildlife (beyond 'Dreamtime' stories about wildlife) like Aboriginal custodians of wildlife species.
Contrast Indigenous and non-Indigenous perspectives of Australian wildlife (e.g. wildlife as an Indigenous food resource versus wildlife viewing by tourists).
Describe Aboriginal involvement in contemporary Australian wildlife management (e.g. fauna surveys, wildlife research, endangered species, hunting, wildlife farms).
Interpret contentious issues in Indigenous wildlife use (e.g. traditional hunting).
Develop policies on Indigenous cultural interpretation at captive wildlife attractions.

Source: Zeppel and Muloin (2008)

Sustainable design and ecolodges

The tourism industry has long been recognised as a formidable agent of social and ecological change: some say in a positive context through increased education, renewed pride in culture, the conservation of heritage and, of course, economics (Lickorish 1991); while others say in a negative sense, at the expense of people and the resource base. For example, hotel developments in Spain and Turkey have been identified as being excessive in their impacts on beach environments. In other cases, improper development and the lack of sewerage systems hamper the tourism industry to such an extent in places in Mexico and Brazil that tourists are told not to swim in the ocean owing to the pollution. In such cases tourism development is strongly impacted by what does not exist (i.e. no sewerage systems) or because such systems are unable to keep pace with the ongoing increase in tourism visitation and associated development.

The style and extent of tourism development in the 1990s has been tempered by the trend towards the increase in mega-development projects designed to cater to a growing market of travellers looking for self-contained, hassle-free vacations. Fennell (1989) writes that this mega-resort boom began in the mid–1980s, and includes sites such as the 31,000 acre resort community of Waikoloa in Hawaii and the 1,200 acre Sanctuary Cove mixed-use resort between Brisbane and the Gold Coast. But while most of the development has been centred in what have been politically stable democracies, management consultants have been looking to the untapped markets of the Far East (Fennell 1989; see also Ayala 1996).

In the midst of this mega-development push there has emerged the move towards more responsible travel and development, in response to the green movement (pressure from various groups including academics, non-governmental organisations (NGOs), and the general public), and in keeping one step ahead of the competition. The response to the changing emphasis has meant that even some of the major hotel chains have experimented with ecologically sound rooms through recycling, washing sheets only at the request of visitors, cutting down on the use of electricity and so on (Andersen 1993; Wight 1995). However, where this changing focus has become the most noticeable is through the variety of small-scale initiatives that have developed in urban (bed-and-breakfasts), rural (farm tourism) and back-country settings.

Over the past few years, several documents have emerged that analyse the development of ecolodges and ecotourism facilities on the basis of principles of sustainability. Still, there appears to be some confusion on the scale and magnitude of development of such areas according to Epler Wood, who reports that ecotourism continues to be misunderstood in resort advertising schemes, while the term 'ecoresort' has yet to be appropriately defined (Shundich 1996). For instance, a tourism industry trade publication recently wrote that, 'Another recent and perhaps surprising region for ecoresort development is India . . . There is interest in genuine cultural experiences within the country. One man has proposed a historical tour of India by rail'.

Interest in sustainable ecotourism design took flight in the early 1990s in part as a result of an American national park service publication dedicated to the principles of sustainable design. Two events took place in the USA that prompted the development of this book. The first was the National Park Service Vail Symposium in October 1991 which discussed the severity of environmental stresses within parks; the second was the Virgin Islands National Park initiative in November 1991, which specifically dealt with issues related to sustainable design in parks and protected areas (US Department of the Interior 1993). The latter publication explores interpretation, natural and cultural resources, site design, building design, energy management, water supply, waste prevention, and facility and maintenance operations in its overview of sustainable design. Of particular importance is its extensive checklist for sustainable building design that emphasises and respects the many interrelationships of all parts of the natural and cultural aspects of the site. These were not the first attempts to integrate such developments with the natural world (especially parks), as is evident in the seminal work of Miller (1976, 1989) on ecodevelopment.

Although the U.S. Department of the Interior book is devoted to facility design and construction *in parks*, there is a more generic section that deals specifically with sustainable tourism development. Here the authors advocate using the principles of Aesculapia, the Greek place of healing. From this perspective, 'nature is respected for its restorative qualities. The human experience is set in harmony with the environment, and an opportunity is created to allow a reconnection of human needs to the natural systems on which all life is based' (U.S. Department of the Interior 1993: 58). This type of design principle would then strive to meet the following criteria, according to the authors:

1   provide education for visitors on wildlife, native cultural resources, historic features or natural features;
2   involve indigenous populations in operations and interpretation to foster local pride and visitorexposure to traditional values and techniques;
3   accomplish environmental restoration;

4   provide research and development on, and/or demonstration projects of, ways to minimise human impacts on the environment;
5   provide spiritual or emotional recuperation;
6   provide relaxation and recreation; and
7   educate visitors to realise that knowledge of our local and global environment is valuable and will empower their ability to make informed decisions.

Russell *et al.* (1995: x) define an ecolodge as, 'a nature-dependent tourist lodge that meets the philosophy and principles of ecotourism'. Although they underscore the importance of the ecolodge concept from an educational and experiential perspective, they suggest that it is the philosophical tie with ecological sensitivity that must define these operations. The importance of ecology is also underscored by The Ecotourism Society (now The International Ecotourism Society), which suggests that these lodges promote an educational and participatory experience while at the same time being developed and managed according to the local environment in which it exists (Epler Wood, in Shundich 1996). Russell *et al.* (1995) differentiate between traditional lodges and ecolodges using the 12 points shown in Table 6.3. They further distinguish between the ecolodge and the nature-based lodge, with the latter being associated with fishing, skiing, and luxury retreats.

The issues discussed above have been analysed by Andersen (1993, 1994), who feels that along with environmental codes of ethics, a low-impact approach to the design of ecotourism facilities needs to be employed for such facilities to be truly sustainable – an approach that he suggests would entail a complete reworking of the conventional design of architects. Andersen's work is well publicised in the field of ecotourism, especially his advocacy of a number of principles which, broadly defined, include organisational issues (has an analysis been done of the area's ecological sensitivity?), site planning issues (minimise trail crossing points at rivers and streams), building design issues (maintenance of the ecosystem should take priority over view or dramatic design statements), energy

Table 6.3 *Traditional lodge versus ecolodge*

Traditional	Ecolodge
Luxury	Comfortable basic needs
Generic style	Unique character style
Relaxation focus	Activity/educational focus
Activities are facility based (e.g. golf, tennis)	Activities are nature based (e.g. hiking, diving)
Enclave development	Integrated development with local environment
Group/consortium ownership common	Individual ownership common
Profit maximisation based on high guest capacity, services, prices	Profit maximisation based on strategic design, location, low capacity, services, price
High investment	Moderate/low investment
Key attractions are facility and surroundings	Key attractions are surroundings and facility
Gourmet meals, service and presentation	Good/hearty meals and service-cultural influence
Market within chain	Market normally independent
Guides and nature interpreters non-existent or minor feature of operation	Guides and nature interpreters focus of operation

Source: Russell *et al.* (1995)

resource and utility infrastructure issues (consider the use of passive or active solar or wind energy sources wherever practical), waste management issues (provide facilities for recycling) and evaluation (whether accommodation is made for older guests and physically disabled individuals).

Andersen has been able to operationalise many of these design features in his Lapa Rios Resort development on the Osa Peninsula of Costa Rica's Pacific coast. This development has a main lodge, in addition to 14 private bungalows, which were built by local engineers with the removal of only one tree. (See also Hadley and Crow 1995; Gurung 1995; and Wight 1995 for good discussions on the guidelines and requirements for the design of ecotourism facilities.) Almeyda Zambrano *et al.* (2010) have studied Lapa Rios' contribution to the community and suggest that it fulfils ecotourism's definitional promise along economic, socio-cultural and ecological lines. It remains an important contributor to reforestation in the area, actively buys products from local farms, has made most positions full-time and year round, and continues to support high-quality training and education. Rara Avis in Costa Rica has also been shown to be an effective ecolodge model, especially in reference to the amount and quality of environmental education provided to visitors (Sander 2012).

Given the low-capacity and low-impact philosophies of ecotourism development, a logical question is: At what scale does the ecolodge model *not* apply? That is, when is a development considered too big? This question, at least peripherally, has been addressed by Ayala (1996). As an international ecoresort developer, Ayala believes both that ecotourism is becoming a segment of mass tourism, and it is not incompatible with beach tourism (see Chapter 2). On this account, there is the need to diversify the vast number of sea, sun and sand products through the addition of ecotourism (but not at the exclusion of other hard-path experiences). Ayala feels that the line between ecotourists and conventional tourists is least detectable in long-haul travel, citing similarities in their levels of education, household income and occupation, and that there appears to be another connection by virtue of the monetary donations made by visitors to the mega-resort chains in Africa and Central America. (See also the work of Thomlinson and Getz 1996 on scale in ecotourism development.)

CASE STUDY 6.1

Greylock Glen ecotourism resort

This project is a consortium of state, local, not-for-profit agencies, and private interests which have come together to build a demonstration site for the concept of sustainable development in the Berkshire Mountains region of Massachusetts. The developers hope that this site will be a prototype for similar developments throughout the world. The resort will operationalise the concepts of sustainable development in offering educational, recreational and conference-oriented programmes. In particular, visitors will be able to enjoy a golf course (developed from the Audubon Society's guidelines for environmentally sensitive golf courses), a nature centre (a year-round environmental education facility), hiking, Nordic skiing, biking, camping, tennis, swimming, horseback riding and other recreational activities. In addition to a 200-room resort, 800 acres will be designated as open, recreational land. The developers (Greylock Management Associates) have specified that the following approaches to natural resource management will be followed:

The maintenance of forest integrity and animal preservation will guide all development decisions at Greylock; a natural resource inventory will be compiled. Cultural resources: All cultural artefacts (e.g. farming fields, old roads, orchards) will be preserved and highlighted. Energy management: Energy meters will be present and interpret energy usage at the centre. Water and sewerage: Water-efficient toilets and showerheads will be used. Visitors will be presented with guidelines to help them cut down on water consumption. Waste management: A recycling programme will be implemented, in addition to composting.

(a)

(b)

Plate 6.3 (a) there is no generic style for the construction of ecolodges. Main issues for consideration for ecolodges are the degree to which they are 'green' and the type of ownership (foreign and multinational, or local; (b) an example of a foreign multinational ecolodge, along with an environment which appears to be less ecologically authentic (manicured lawns etc.)

Not unlike the ecotourism operator who misrepresents ecotourism in their programme offerings, accommodation must also be subject to scrutiny in deciding on whether these units are in fact environmentally sensitive or not. Will ecotourism's popularity be directly attributable to the belief that it fits nicely as a form of mass tourism, and that mega-ecoresorts are the solution to bring ecotourism to the masses? Is this the rationale for their development – the intrinsic need to not deny the masses their place in the sun (or rather rainforest)? Or rather is it simply that from an extrinsic motivational perspective it is big business making money the old-fashioned way? Having seen some of these developments first hand, one merely has to dig a bit deeper to see many of the unsustainable features of development. For example, an ecolodge in Mexico recently developed by a large international company used grass and other materials not indigenous to the area in the design of the facility. In addition, no interpretive programmes had been developed to enable tourists to learn about or experience the natural attractions of the area. Such developments are merely cosmetic, often with few educational and sustainable features, which presumably are important to this segment, and undermine what Hawkins (1994) and others have suggested about ecotourism as being minimum density and low impact. Even in cases where ownership is in the hands of one or a handful of individuals, local ownership is rare.

Despite its importance as an integral component of the ecotourism experience, research on ecolodges continues to be sparse. Russell *et al.* (1995) undertook an international ecolodge survey of 28 operators in nine regions around the world (Belize, Costa Rica, Peru, Brazil, Ecuador, the state of Alaska, Australia, New Zealand and Africa). Results indicate that many of the lodges were found in or adjacent to protected areas, with outstanding natural beauty acting as a key to success of the operation (see Chapter 5). Most of the ecolodges sampled were small in size, accommodating about 24 guests, with some successful operations in Amazonia catering up to a hundred guests. The authors felt that although most of the ownership had been typically small scale and independent, corporate ownership was becoming more common. They cited the P&O line in Australia and the Hilton in Kenya as two examples of this recent phenomenon. Finally, they suggested that very few such lodges exist in North America owing to the existing reliance on camping and other, more comfortable recreational lodges – a finding supported in a study of ecotourism in Alberta and British Columbia (ARA Consultants 1994).

Wight (1995) has developed a useful accommodation spectrum that outlines a number of different options according to hard and soft dimensions (see Figure 6.1). The hard aspects of ecotourism can be identified, at least in part, by non-permanent types of dwellings, including tents and hammocks. The ecotourism experience grows continually softer through fixed-roof units (cabins and lodges) on-site, and through fixed-roof dwellings off-site, usually in the form of hostels, hotels and resorts. This hard–soft accommodation continuum relates to the work of Laarman and Durst (1987), at least in part, who suggested that one of the hard and soft dimensions of ecotourism was the physical rigour of the trip (see Chapter 1). Those who were prepared to walk extensively in back-country regions, sleep in camps or crude shelters, or tolerate poorer sanitary conditions were said to have been interested in more of a hard-path ecotourism experience.

If this continuum is to apply to regions like North America, it is logical to ask whether or not such operations are applicable to consumptive or non-consumptive forms of recreation. This is of the utmost importance as outfitters and operators wrestle with how to refit their enterprises to accommodate the changing recreational focus in the north, because, as Anderson/Fast (1996a: 54) suggest, 'depending on the characteristics of the package and the participating tourists, outfitter camps [the only form of accommodation in many remote regions of Canada's north] may or may not be appropriate'.

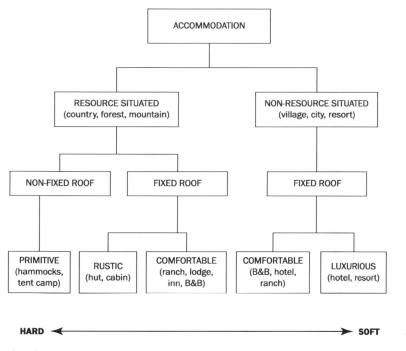

Figure 6.1 Ecotourism accommodation continuum

Other studies on ecolodges have focused on identifying levels of satisfaction among visitors and assessing whether performance goals have been realised. For example, Osland and Mackoy (2004) investigated 21 ecolodges in Mexico and Costa Rica according to type: casual, dedicated, scientific and agri-ecolodges. Based on the use of 84 performance goals, they found that sustainable development goals were mentioned most frequently (although difficult to measure objectively), and scientific ecolodges were more likely to express goals related to the education of ecotourists. Success to managers and owners of ecolodges was typically measured according to financial criteria.

A study by Kwan *et al.* (2010) investigated the demographic and trip characteristics, and motivations of ecolodge visitors, and employed an importance–performance analysis of ecolodges in Belize. Most visitors were from the USA followed by Canada; most were between the ages of 36 and 55; they were highly educated (following other studies on the ecotourist); most worked full-time or were retired; and about one-fifth had household incomes in excess of US$140,000. The majority of visitors stayed in Belize for 8–11 days, and at one ecolodge for three to four nights. Their main travel motivation was to learn and explore nature or explore a new country or culture. It is significant that performance scores exceeded importance scores for all of the 41 attributes used in the study. The five highest ranked attributes for performance were:

- friendliness of staff;
- scenery;
- value for money;
- decent sanitary conditions; and
- quality of the environment or landscape.

The five highest ranked attributes for importance were:

● friendliness of staff;
● scenery;
● quality of the environment or landscape;
● decent sanitary conditions; and
● staff provide efficient services.

Lodges were performing at a level, in areas like hospitality, personal service, and in view of the environmental conditions, above the level of importance as indicated by the patrons.

Nelson (2010) explored the extent to which Eco-Certified accommodations promoted energy use strategies in their units. Specifically, this included the type of information provided, how information was presented and the objectives of promoting this information. Nelson found that only a few of the accommodations were active in educating tourists about energy strategies. The problem with some of these attempts, however, was that tourists were often unfamiliar with the terminology used to explain energy conservation. For example, one accommodation stated that their annual energy savings amount to 480,000 kWh of electricity. Unfortunately, for the uninitiated, there is no basis for comparison. One of Nelson's main conclusions is that eco-accommodations have the potential to expand their conception of ecotourism beyond conventional themes like protection of the visible environment, to include non-visible energy consumption and climate change.

Two studies on the marketing of ecolodges point to some of the problems identified later in the book (see Chapter 9). Lai and Shafer (2005) employed content analysis methods in studying the online marketing information of ecolodges in Latin America and the Caribbean. Their first major finding is that there were a large variety of different products offered online to meet the many interests of the ecotourist market. This points to the heterogeneity of the market, which suggests activities that are offered for the hard-path ecotourist along with many that are more generalised for a wider market. The second main finding of Lai and Shafer is that most of the ecolodge marketing efforts only partially aligned with the main tenets of ecotourism. While aspects of sustainability and local benefits were evident in the marketing of ecolodges, troubling was the lack of emphasis on education in most of these accommodations.

Similar conclusions were drawn by Donohoe and Needham (2008; see Chapter 1) who used content analysis to investigate the marketing of 25 ecolodges in Canada according to six main ecotourism tenets: nature-based, conservation, environmental education, sustainability, distribution of benefits, and ethics and responsibility. They found that 40 per cent of the 25 ecolodges were congruent with less than three tenets; and 8 per cent were congruent with only one of the six tenets. The authors conclude that genuine ecotourism, or the idea of genuine ecotourism, is contentious because the lack of marketing consistency contributes to the misuse of the term. There is the danger that the line between ecotourism and mass tourism is becoming too ambiguous – too blurred, as noted above (see Chapter 2, especially the discussion of Sharpley).

Finally, literature on ecoresorts seems to indicate that accommodation, although important, is not the overriding motivation for visiting regions. ARA (1994), for example, placed accommodation below the natural setting and wildlife viewing (of primary importance), and other secondary activities related to the experience. To this, Wight (1995) adds support by suggesting that accommodation is an enabler of the overall ecotourism experience; basically tourists select the experience, and then choose the accommodation. Often, though, pre-arranged package tours dictate much of the on-site experience, including choice of accommodation, restricting the opportunity for ecotourists to choose lodging, and reinforcing the point that there is an element of ambiguity as to just what ecotourism entails. According to Grenier *et al.* (1993: 9), 'one person's ecotourism dream may be another's touristic nightmare'.

Volunteer ecotourism

In Chapter 2, Sharpley (2006) argued that there was not a separate ecotourist market because ecotourists behave in much the same way as the mass tourism market. Just like this larger group, ecotourists pay large sums of money to relax, have fun and be entertained (see Fennell 2006), and that 'it is highly unlikely that tourists will be motivated to "work" at tourism' (Sharpley 2006: 17). However, there is a group of tourists who do deliberately choose to work at ecotourism. These volunteer ecotourists or research ecotourists or dark green ecotourists (as they have been called) are interested in working with research scientists to, for example, help preserve endangered species or rehabilitate natural areas, which makes this section relevant to the discussion in Chapter 5. The main motivation appears to be altruistic, where the betterment of the environment of the tourism region is a primary goal. Volunteer tourists represent an important sub-sector in ecotourism, and there are many organisations that rely on these volunteer ecotourists to accomplish innumerable socio-cultural and ecological ends. A simple and quick review of the web offers many examples. The International Ecotourism Society promotes a sea turtle conservation project in Costa Rica (TIES 2012). Oprah Winfrey supports going to the Grand Canyon for the purpose of documenting the behaviour of beavers, reseeding grasses and shrubs, and restoring deer habitat. There are even practical books on the topic. Brodowsky (2010) in association with the National Wildlife Federation lists over 300 different organisations that have been developed to place ecotourists into situations where they can, in the words of Oprah.com, 'do good and feel good'. The academic literature is robust in this area as well (see for example Gray and Campbell 2007; Coghlan and Fennell 2009).

The UK-based Operation Wallacea is an organisation that runs scientific studies and community-based projects all over the world. Multiple programmes can be found in Indonesia, Honduras, South Africa and Mozambique, Peru, Cuba, Egypt, Madagascar, Guyana, and Mexico and Guatemala. In a study that profiled volunteer ecotourists in southwest Sulawesi, Indonesia, Galley and Clifton (2004) found that these tourists were young (in low twenties), mostly female, single, well educated, driven by research-related motives, and maintained a high level of understanding of the basic tenets of ecotourism (see also Weiler and Richins 1995).

Rattan *et al.* (2012) studied 200 participants at the Elephant Nature Park (ENP) in Chiang Mai, Thailand – a sanctuary that aids sick, injured and abused elephants that have been used in tourism, logging and other enterprises – for the purpose of understanding how volunteer tourists at the ENP affect other tourists. Results of the study indicated that participants (non-volunteer tourists) increased their awareness of conservation issues on a number of measures based on pre-survey and post-survey scores. For example, on questions like 'Volunteer tourism is an important aspect of conservation', 'I would like to volunteer at the ENP', and 'I would like to volunteer with organisations back home that advocate and protect animal rights', participants indicated a higher level of agreement in their post-survey responses. As such, the ENP volunteer tourism model was said to be effective at raising the awareness of conservation as well as the importance of volunteerism in the minds of non-volunteers. (See also Nolan and Rotherham 2012 who examine aspects of altruism and self-interest in the context of volunteer consumer perceptions, and the relationship between volunteer, producer and experience.)

Conclusion

Learning, and this includes environmental education, is a key criterion that helps to distinguish ecotourists from other types of tourists. Simply stated, ecotourists are those travellers who are interested in learning about the nature and natural resources of a destination. This is important because it opens the door for an appreciation of what is required to safeguard the natural and cultural assets of these places. The problem is the extent to which learning translates into pro-environmental behaviours both at the destination and longer term; that is, to changing behaviours and lifestyles at home. Research shows that these changes are not long lasting, and that there are concerns that at least some types of ecotourists – and this is likely to be the majority – are changing their attitudes and behaviours in only a situational capacity. The next chapter explores an important dimension of these attitudes and behaviours in the form of ethics. It is arguable that this last core criterion is perhaps the most important in helping to define the nature of the ecotourism industry.

Summary questions

1 How does learning differ from education?
2 Which country has been successful at initiating an EcoGuide programme? What are the dimensions of this programme?
3 What are the characteristics of good interpretation?
4 Explain the concept of learning drop-off.
5 Ecolodges come in many shapes and sizes. How have some ecolodges crossed the line in reference to what is a legitimate or authentic ecolodge, and what is not?
6 Explain how a volunteer ecotourist might gain a much different experience overall compared to a more conventional ecotourist.

7 The moral imperative

Grub first: then ethics. – Brecht

In the words above, Brecht strikes to the heart of the dilemma facing ecotourism in many regions of the world today. One cannot help but relate the sentiments of Brecht to Maslow's (1954) hierarchy of needs: simply that people will be moved to satisfy lower-order needs first (those which are physiological, including food, shelter and safety) before upper-level, or psychological, ones (leisure, self-respect and self-actualisation). The implications of Maslow's model to human behaviour are suggestive of the fact that humans are often caught in a duality between what is universally right and personally satisfying or rewarding. It is perhaps why people continue to practise slash-and-burn agriculture; why governments are reluctant to explore other forms of energy use beyond fossil fuels; and why tourist operators are prone to placing more people on sensitive environments than they are supposed to. There appear to be two main motivations for this action along a broad continuum. At one end of the continuum is the need to survive and support a family, while at the other is the need to prosper economically at all costs.

In this chapter ecotourism is discussed in the context of an ethical imperative – an essential ingredient needed to better conceptualise ecotourism both in theory and practice. This is not just far-fetched idealism, but comes after careful consideration of the foundations of human nature – a topic not at all familiar to tourism theorists. The theory of reciprocal altruism explains why we are both self-interested and cooperative, and the application of this theory to tourism puts into perspective why we are faced with a variety of ethical issues. The chapter also focuses on defining ethics as well as summarising the extent of tourism research and ethics to date.

Ethical by nature

It comes as no surprise that ethics has been a central theme in the writings and teachings of civilisations for over 2,500 years, because we are in fact ethical by nature. Biologists argue that *Homo sapiens* crossed an evolutionary threshold separating us from the so-called 'lower' animals because of our unique ability to: (1) anticipate the consequences of our actions; (2) make value judgements; and (3) choose between different courses of action (Ehrlich 2000). Many thousands of years of living in small, stable, dependent communities, where we knew each other very well and could readily identify cooperation and cheating within the group, endowed us with sensitivities towards sympathy, trust, gratitude, loyalty, guilt, shame, anger and contempt, among other survival traits. Over time – evolutionary time – the resultant psychological system which developed to regulate altruism and cheating in individuals (as theorised by Trivers 1971) called 'reciprocal altruism' (see Fennell 2006a, 2006b) is said to be the foundation for the human capacity

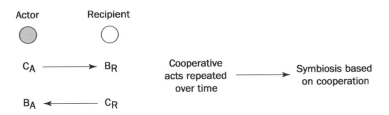

Figure 7.1 Reciprocal altruism. The actor at a cost, C_A, provides a benefit to the recipient, B_R. The recipient later chooses to act altruistically towards the actor at a cost, C_R, in order that the actor may benefit, B_A. Repeated interactions decrease the motivation to cheat and increase cooperation

Source: Fennell (2006a)

to be ethical (Mayr 1988). And while the *capacity* for ethics is a function of biology, the *products* of this capacity are a function of culture.

Reciprocal altruism is founded on a system of interactions between individuals based on costs and benefits, whereby individuals who are altruistic towards another (altruism defined by Barash 1982: 389 as 'behaviour that reduces the fitness of the performing individual while increasing that of the recipient') expect return favours (reciprocity) down the road which will ultimately increase their fitness or well-being (Figure 7.1). The dynamics of group interaction in the past were very much about small, stable, dependent communities, as noted above, where the universe was largely predictable and unchanging. We knew each other well and because of this were able to keep in check those who cooperated and those who cheated. This system has worked well for millennia. But the problem is that we have hurtled from Stone Age to Space Age within the last 5 seconds of a 24-hour evolutionary clock, facing 'modern challenges with hard-wired instincts not much changed since the days we were roaming the savannah naked' (Fulgosi 2006: A22). As such, we may live in a world of technology, innovation and change, but our brains are better designed to live in the other world of solidarity and altruism (Hayek 1944/1994).

While reciprocal altruism explains our evolutionary past, how does it relate to our tourism present? Fennell (2006a) argued that because tourism is based on short-term interactions between hosts and guests (e.g. a 5-day holiday package), cooperative relationships cannot be developed in the absence of a sufficient amount of time to exchange needed benefits and costs. In the absence of the time needed to build cooperation, cheating may take place (e.g. operators ripping off tourists), based on the belief that service providers might never have to interact with a particular tourist again. Fennell argued that this type of problem might also occur in the resort or destination on the whole based on the cumulative effects of cheating (i.e. the development of a cheating norm among service providers in getting ahead).

The foregoing illustrates that there is a basic human element, beyond the broad social, economic and political explanations, that is missing in our attempts to come to grips with the many problems that plague the tourism industry. Simply stated, the problems that continue to constrain the industry are explainable, to a large degree, from the perspective of human nature. This has been touched on by Przeclawski (1996: 236) who said that, 'Tourism can not be explained unless we understand man, the human being'; as well as Wheeller (2004) who observed:

> If, as we are led to believe, tourism is the world's largest industry, then we should remember that it is a world driven largely by avarice, greed, self-interest. 'A much wants more, what's in it for me, now mentality'. These traits are in us. And these forces

drive tourism just as they drive everything else . . . We need, therefore, to look first at ourselves and then at society in general when we address tourism. But do we?

Tourism is an expression of the self as a reflection of who we have become over thousands of years of evolution. It may come as a surprise that before law, and even before politics, commerce prevailed in primitive societies through the trade of scarce commodities for mutual benefit but also to forge lasting (hopefully!) alliances to buffer against warfare. Cost–benefit calculations in time all lie at the heart of human behaviour – and at the heart of tourism enterprises and interactions. Important in this however is that although we often take the self-interest path instead of the altruism one, we are just as prone to acts of morality as we are to the alternative – indeed, the balance is what makes us human, as described by Midgley (1994: 9) in the context of freedom and constraint:

> If freedom and morality are indeed closely linked . . . it is perhaps a rather paradoxical fact that the first effect of freedom should be to put us under these new constraints [impartiality, truthfulness, parsimony, and so on]. Our freedom is exactly what gives us these headaches, what makes possible this moral thinking, this troublesome kind of search for priority among conflicting aims. By becoming aware of conflict – by ceasing to roll passively from one impulse to another, like floods of lava through a volcano – we certainly do acquire a load of trouble. But we also become capable of larger enterprises, of standing back and deciding to make lesser projects give way to more important ones. That, it seems, may be why moralities are needed.

On the basis of this body of knowledge, Fennell and Malloy (2007) argue that ethics (and in particular codes of ethics in their work) is simply an extension of who we are at the core. Eliminating ethics or at least suppressing it in tourism for the sake of profit is to exercise freedom and to deny constraint – both of which are two sides of the same coin.

Ethics

Although ethics has been defined in many different ways, the basis of its meaning relates to the study of what is good or right for human beings (Hoffman et al. 2001); or more fully as, 'rules, standards, and principles that dictate right conduct among members of a society or profession. Ethics are based on moral values' (Ray 2000: 241). In the case of professions, ethics are based on moral values that the organisation holds to be true, which should be a reflection of the broader societal values in which the profession is placed. While ethics has historically been considered a discipline with theoretical applications, the move towards a more practical use of this type of philosophy began to surface in the business literature during the mid-1970s when the first conference on business ethics was organised at the University of Kansas (Bowie 1986). Studies in this area intensified as a result of a number of business-related ethical transgressions, including the abuse of power and business scandals (Sims 1991) at local, regional, national and international scales. Consequently, applied ethics has evolved both in business and in society as a whole to include a number of key areas related to human well-being and development, including business, the legal and medical professions, the biosphere and environment, and, accordingly, tourism (see Figure 7.2). Academics have long been involved in philosophical debate over the ethical nature of humankind, and it is from here that the applied side has been able to act on these transgressions. It appears that in ethics, like other disciplines, there is some polarisation with respect to the utility of applied ethics within society, which

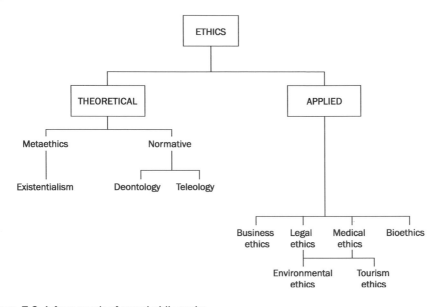

Figure 7.2 A framework of moral philosophy

Source: Adapted from Honderich (1995)

relates to the perceived role that theory can play in helping to shape application, and also how its application can help drive theory. This dilemma is reinforced by Fox and DeMarco (1986) who write that:

> Work in applied fields cannot wait for theory to advance to the point of providing clear guidance. Indeed, theory will probably not advance significantly unless current investigations into practical issues reveal areas of agreement and principles on which philosophers can rely in reconstructing their theories.
>
> (Fox and DeMarco 1986: 17)

In the same edited compendium, Singer (1986: 284) reinforces the relevance of applied ethics as having more of an impact, with theoretical ethics increasingly viewed as having value only insofar as it can throw light on the problems of applied ethics.

From the theoretical side, research has tended to emphasise both meta-ethics (existentialism) and normative ethics (deontology and teleology), with the latter being more prominent by virtue of the fact that it deals with the behaviour of businesses, institutions and so on. By contrast, existentialism advocates a distinctly individualistic mode of thinking, focusing on the subjective and personal dimensions of human life. Existentialists concern themselves with the question of the nature of being in understanding human existence (see Fennell and Malloy 2007). Using the example of geography, Tuan (1971) wrote that geographers establish contact with the world in two ways: nomothetically (environmentalism) and ideographically (existentialism). Environmentalism operates in a world of objects, existentialism in a world of purposeful beings. Whereas it is the goal of the environmentalist to search for general laws to discover meaning and order, the existentialist inherently strives for meaning in the landscape, 'by tracing man-nature to its basis in man's bodily relation to the world' (1971: 185). (More on the ethical theories of deontology, teleology and existentialism, as they relate to tourism, follows.)

Ethics and tourism: too little research . . . still

Lea (1993) argues that the origins of ethical concern in tourism studies are traceable to the sociology of development literature (human–human interactions), and more recently alongside an ethic that deals with the natural world (human–environment interactions). The significance of development issues in tourism is seen through statements like the following by a Hawaiian delegate of a tourism conference: 'We don't want tourism. We don't want you. We don't want to be degraded as servants and dancers. That is cultural prostitution. I don't want to see a single one of you in Hawaii. There are no innocent tourists' (Pfafflin 1987: 578). Although this is an example of a social impact, more fundamentally it is an ethical issue. We call attention to issues of this nature because we understand that there has been an injustice to either someone or a group of people. In the case of Hawaii, tourism is an economic cornerstone of the Hawaiian economy. To some people, its presence has contributed significantly to cultural dislocation within the region. Clearly, though, this is not the only sentiment that exists within Hawaii, especially from the perspective of those who stand to gain most from the industry. It is those who are realising the economic benefits of the industry who are in most cases supportive of its existence.

These relationships have been discussed by D'Sa (1999), who says that northern investors dictate tourism development in Third World destinations, who take on local elites as junior investors in procuring land and resources, with local people left to pick up the many pieces. Tourism's linear, instrumental manner of thinking has ensured that the interests of the marginalised are deemed morally irrelevant in the face of industry productivity (Stark 2002). The presence or absence of acceptable ethical behaviour in tourism settings is, therefore, very much a function of the myriad interactions between decision-makers, tourists, operators and local people in space and time (Fennell and Przeclawski 2003). There is thus a sense of urgency among theorists that a balance must be struck between the various stakeholders of the tourism industry in ensuring that the goodwill of some stakeholders (e.g. tourists and local people) is not overridden by the misgivings of other stakeholder groups (e.g. government or industry). Obviously the sheer number of interactions between different groups creates a myriad of cases where positive or negative impacts are generated (the two examples in Case Study 7.1 serve to illustrate this point).

CASE STUDY 7.1

Ethics in question: operators, local people and tourists

On a trip to the jungles of the Yucatan Peninsula in Mexico, our group was taken to a farmer's residence before venturing off to see some of the famous Mayan ruins. At this farm the operator took great joy in showing us three jaguars that were secured to a fence using four-foot chains. The operator's rationale for showing us the jaguars was that he was certain that we would be unable to see jaguars in the forest over the course of our tour. He knew that such a sighting would be highly valued by ecotourists. This was his way of guaranteeing that we saw these magnificent beasts. Clearly, the animals did not mean anything to the operator, and he was taken aback by the suggestion of one of our party that he would have rather seen a picture of a jaguar in a book than a live animal under these circumstances.

The second example involves the interference of local people in an ecotourist setting. While in Costa Rica on an ecotour, we had the pleasure, at a comfortable distance, of

continued

witnessing a green sea-turtle laying her eggs one night on a beach. After finishing her task she began to return to the sea, when suddenly a local man appeared and jumped on her back as she made her way. The utilitarian value that he placed on the turtle was clearly one that reflected the sentiment of many Costa Rican people at the time, one that is not at all consistent with that of ecotourists. Turtles and turtle eggs are highly prized as a source of food by local people, even though many signs had been put up clearly stating that the personal consumption of turtle eggs was against the law (owing to declining stock) (see Plates 7.1 and 7.2).

In regard to the environmental ethic component identified by Lea, above, there is a huge volume of work that has proved useful to tourism scholars. Myers (1980) illustrates that most people are concerned about issues like the disappearance of species, but not necessarily at the expense of other, more pressing matters such as population and pollution. Similarly, Elliot (1991) writes that there are degrees of environmental ethics to which people submit. These range in philosophy from human-centred ethics, which advocate a stance whereby ethics are evaluated on the basis of how they affect humans; to ecological holism, where the biosphere and the major ecosystems of the planet are morally considerable. Plants, animals, rocks and so on matter only inasmuch as they maintain the significant whole. Other intermediary ethical stances include animal-centred ethics, where individual members of a species are the focus; life-centred ethics, where all living things are valued (but not necessarily equally); and ethics which imply that all living and non-living things have intrinsic value (see Figure 7.3).

THE FAR SIDE® By GARY LARSON

"On three, Vince. Ready?"

Figure 7.3 Ethics and African game reserves

Plate 7.1 This jaguar – chained up for 24 hours a day by a local farmer – was promoted as an ecotourism attraction by the tour operator in Mexico

Plate 7.2 In the turtle egg-laying season both tourists and local people must be sensitive to the needs of the female, which include space and the absence of harassment

A human-centred focus on ethics, or anthropocentrism, is more of a dominant 'Western' worldview, while its antithesis, biocentrism, is more of a minority, harmonious world-view. (See Chapter 5 for a related discussion on conservation and environmentalism). The anthropocentric paradigm posits that nature can be conceived only from the perspective of human values. Humans, therefore, determine the form and function of nature within human society. Conversely, the biocentric philosophy considers that all things in the biosphere have an equal right to exist; that is, they are all equal in intrinsic value (Devall and Sessions 1985). In the realm of parks and protected areas, Swinnerton, in Philipsen (1995), writes that resource managers and other decision-makers around the world have taken competing views on the place of environmental management in parks. Such views range from preservation to conservation to exploitation, and demonstrate the inconsist-ency in use of these terms (Figure 7.4). Those managers advocating the preservationist viewpoint adhere to more of a biocentric philosophy through the practice of little inter-vention, placing high value on natural resources, responsible use and very small numbers of tourists.

There does appear to be a swaying of public sentiment in favour of a more environmen-tally oriented ethic according to Philipsen (1995). In citing the work of many authors (Dunlap and Heffernan 1975; Pinhey and Grimes 1979; van Liere and Noe 1981), Philipsen feels that the choices people make with respect to tourism reflect the importance of environmental values in their lives. Urry (1992), for example, writes that the collective tourist 'gaze' of the mass tourism phenomenon seems to be declining in the face of a more romantic gaze exemplified through green tourism; while Poon (1993) has introduced a 'new tourism' perspective based on these changing values. In support of this point, Yeoman *et al.* (2006) write that there are few markets anywhere in the world not affected by what has come to be referred to as 'ethical consumption'. These authors define it as,

	Resource protection	Resource development	
	Preservation ←→	Conservation ←→	Exploitation
View of resource	Biocentric/ anthropocentric	Anthropocentric	Anthropocentric
Level of intervention	No intervention	Limited intervention	Unlimited intervention
Measures of natural value	Undisturbedness Naturalness Completeness	Biodiversity Rareness	
Land use strategy	Segregation	Combination	Segregation/ combination
Access regulations	No use Responsible use	Controlled use Responsible use	Unlimited use Abusive use
	Very small numbers	Small numbers	Big numbers Mass tourism

Figure 7.4 Characteristics of resource protection and development

Source: Philipsen (1985)

'the motivation to purchase which lies beyond the stimulus of price, quality and opportunity and which invokes philosophical concerns which may be pre-existent in the mind of the consumer' (Yeoman *et al.* 2006: 184). This has been reported by Weeden (2001), who found in a study of specialist tourism operators that customers were interested in taking ethical holidays. However, how this translates from tourist need to ethical marketing and service delivery is still subject to debate. Because ethics has rarely been emphasised in tourism in the past, standard practices continue to be adopted and accepted whether they are ethical or not (Fleckenstein and Huebsch 1999; Burns 1999).

Fundamentally, there is a very weak foundation of research in tourism ethics studies to date (D'Amore 1993; Payne and Dimanche 1996). There does, however, appear to be growing recognition within tourism's public, private and not-for-profit sectors that, from the perspective of sustainable development, ethical concerns have not fallen upon deaf ears. According to Hughes (1995), it is only the technical, rational and scientific province of sustainability that has predominated to date, rather than the ethical province which was most responsible for the initial drive for a newer and more holistic development paradigm.

In the past, tourism ethics has been relegated to the area of hospitality management owing to the emphasis of hospitality's relationship to service and business (Wheeler 1994). This research provided the foundation for the move to establish the International Institute for Quality and Ethics in Service and Tourism (IIQUEST), which was designed to bridge the gap between ethics and issues related to community relations, sexual harassment, the rights of guests and so on (see Hall 1993). However, the Rio Earth Summit of 1992 was a principal catalyst in generating more interest in the utility of ethics in the realm of tourism research where those in attendance committed themselves to Agenda 21. Genot (1995) outlines chapter 30 of this plan:

> Business and Industry, including transnational corporations, should be encouraged to adopt and report on the implementation of codes of conduct promoting best environmental practice, such as the International Chamber of Commerce's Business Charter on Sustainable Development and the chemical industry's responsible care initiative.
>
> (Genot 1995: 166)

Recently, a number of articles have surfaced either in response to the dearth of literature on tourism and ethics, and/or as a logical progression in the evolution of tourism research. This recent intensification is indeed timely as tourism researchers grapple with more philosophical issues that relate to business, society and the environment. These additions to the literature include general discussions on ecotourism and ethics (Duenkel and Scott 1994; Kutay 1989), more specific commentary on ecotourism and ethics (Karwacki and Boyd 1995; Wight 1993a), tourism ethical decision-making on the quality of life in Third World countries, and ethics and marketing in the ecotourism industry (Wight 1993b).

Hultsman (1995) proposes the use of a framework of ethics and tourism based on the notion of 'just' tourism, which refers not only to acting in a fair, honourable and proper manner, but also to the fact that tourism is 'merely' or 'only' a 'small thing'. With reference to the latter aspect of the framework, Hultsman emphasises that tourism has become more important as an economy than as an experience and, as such, has lost some of its intrinsic qualities that are derived through the pleasure of experience. He notes that, 'Should tourism reach the point of being considered by service providers as first a business and second an experience, it is no longer "just tourism"; it is industry' (1995: 561). He concludes his paper by suggesting that ethical issues need to be included in the textbooks used in tourism curricula, and further that, as noted by many other authors, we ought to be concerned that there is a real lack of professionalism in the delivery of tourism services. This is echoed by theorists such as Tribe (2002), Yaman (2003) and Jamal (2004), all of whom argue strongly for a more substantive ethical agenda in tourism research.

In other related research, Upchurch and Ruhland (1995: 37) focused on advancing the hospitality industry's understanding of ethical work-climate types through an analysis of the normative ethical theories of egoism ('an individual should follow the greatest good for oneself'), benevolence ('actions that are delivered on a fair and impartial basis, are based on maximizing the good, and follow impartial distribution rules'), and principle (which 'is a theory suggesting that decisions are based on rules. The outcome or actions should be based on the merit of the rule'). Respondents were Missouri lodging operators who filled out the Likert-style ethical climate questionnaire designed to measure the three ethical theories. These authors found that benevolence was the most frequently perceived ethical climate type present in the lodging respondents (which, they say, is consistent with the literature). Benevolence, therefore, indicated certain responsiveness to the clientele in a socially oriented manner.

This element of social responsibility is also discussed in a paper on business ethics and tourism by Walle (1995). He discusses the dichotomy that exists in business ethics theory as established by the work of Friedman and Davis. He illustrates that for the Friedman school, the *modus operandi* of business is to generate profits, and that it is not the responsibility of business leaders to dwell upon social policies and strategies (within the law). On the other hand, the Davis school advocates a more socially responsible behaviour for business, which will in turn lead to greater profits and prevent government intervention, through the generation of positive publicity (Table 7.1). Through these extremes, Walle argues that tourism, because of its uniqueness, cannot follow the universal or generic strategies of mainstream business, which focus on the organisation and its customers (e.g. manufacturing). Instead, he makes the point that:

> Tourism is not a generic industry since it uniquely impacts on the environment, society and cultural systems in ways which require a holistic orientation within a broad and multidimensional context. Contemporary business ethics, however, has been slow to embrace such a holistic perspective. Historically, the focus has been on the organization and its customers. Impacts on third parties (externality issues) have often been ignored.
>
> (Walle 1995: 226)

Walle's (1995) initial conceptualisation on social obligation, responsibility, and responsiveness is reworked to illustrate how such orientations might be applied to tourism's uniqueness (Table 7.2).

In work related more specifically to ecotourism, Fennell and Malloy (1995) drew a relationship between meta-ethics (existentialism) and normative ethics (deontology and teleology) in discussing how each can contribute to comprehensive ethical decision-making in the ecotourism industry. This, they reason, is a deviation away from the traditional manner by which to resolve ethical decisions; that is, mediating between the two dichotomous approaches of meta-ethics and normative ethics.

Existentialism implies that an act is right or wrong according to the actor's free will, responsibility, and authenticity (Guignon 1986). Existentialism or authentic behaviour favours neither ends nor means, but rather the premise that individuals need to be self-aware and prepared to take full responsibility for their actions. In essence, this provides the opportunity for individuals to choose freely their behavioural patterns within the confines of their own acceptance of responsibility for all consequences on all things. Deontology is an ethical approach which suggests that an act is right or wrong on the basis of rules or principles of action or duties or rights or virtues (Mackie 1977). This approach

Table 7.1 *Ethical orientations: a comparison*

	Social obligation no. 1 (Friedman)	Social responsibility no. 2 (Davis)	Social responsiveness (extension of No. 2)
General overview	Legal and profitable	Current social problems are responded to	Future social and/or environmental problems are anticipated and addressed
Choosing options	The sole consideration aside from profit is legality	Decisions respond to social issues which overtly need to be addressed	Decisions based on anticipation of future needs and/or social problems even if they do not impact or are caused by the firm
Strategies are evaluated with reference to:	Is the strategy legal? Is the strategy profitable enough?	Has the organisation responded to problems and issues which have emerged as significant?	Future problems are addressed even if the organisation is not directly causing them

Source: Walle (1995)

Table 7.2 *Special ethical considerations of tourism*

Tourism's perspective	Social obligation no. 1 (Friedman)	Social responsibility no. 2 (Davis)	Social responsiveness (extension of no. 2)
Progress is not inevitable or inherently beneficial	Since the concept of 'progress' is not universal or inevitable, we should not place an over-reliance upon it in our strategies/ tactics	Tourism has a responsibility to encourage development which meshes with the local environment and culture, not in *accordance* with a universal concept of 'progress'	Since 'progress' leads to concomitant changes in culture and the environment, tourism strategy should be appropriate and mitigate its impact
Tourism can be undermined by pressures of the industry	Change wrought by tourism might undercut the industry. Such potential should be prevented and mitigated when doing so is a good tactic	Tourism causes negative impacts and pressures on people and the environment which should be mitigated	The industry has both practical and ethical reasons to respond to impacts on the environment and local people
All relevant stakeholders need to be considered when strategies are forged	Government regulation and loan conditions might demand responding to the needs of all relevant stakeholders	Tourism should respond to the needs of various stakeholders which are impacted on by the industry	The industry should anticipate future impacts from various sources and respond in proactive ways

Source: Walle (1995)

advocates behaviour that is means-based or intentions-based in its orientation. Deontology, or right behaviour, provides us with guidance through the provision of rules and regulations to follow; in essence, our 'duty' is laid out for us and we 'ought' to adhere to it. Conversely, teleology, or good behaviour, is an ethical approach which suggests that an act is right or wrong solely on the basis of the consequences of its performance (Brody 1983). Because of its orientation towards the consequences of one's action, it is ends-based. In this sense, the stakeholder is released from following the tradition or dogma of the past and is able to choose in a manner, which is consistent with the changing circumstances of societies and cultures.

Fennell and Malloy (1995) go on to suggest that although each of the three (existentialism, deontology and teleology) advocates radically different perspectives, stakeholders or decision-makers in search of comprehensive ethical decisions may employ all of these theories in arriving at ethically good, right and authentic solutions. Also, according to these authors, it is not the case that people rely exclusively on one pure form of ethics as a means by which to make decisions. They feel that what might be termed a triangulated approach would correspond to the many demands, both organisational and moral, that exist within the tourism industry (Figure 7.5). For the ecotourist, this would accomplish two things:

> First the individual is provided with an ethical criteria [*sic*] with which to assess and resolve issues in a comprehensive fashion that is normative as well as introspective. As a result, it is hoped that what for many may be a latent sense of ethics, may become more apparent and considered as a result of exploring teleological, deontological, and existential perspectives. A second consequence of this model is that it is used as a point of departure for not only empirical research of the behaviour of ecotourists, but also for the development of more comprehensive ethical conduct for the ecotourism industry in general (e.g. organisational culture, climate).
>
> (Fennell and Malloy 1995: 178)

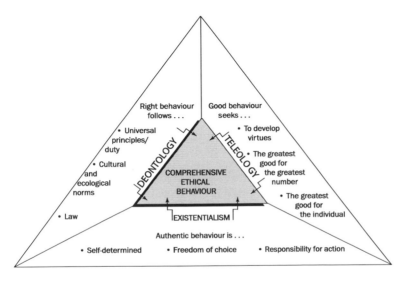

Figure 7.5 A model of ethical triangulation

Source: Fennell and Malloy (1995)

In other research, Malloy and Fennell (1998a) examined the organisational culture literature in attempting to differentiate between ethical and non-ethical work climates in the ecotourism industry (Figure 7.6), on the basis of work by Schein (1985) and Kohlberg (1981,1984). Schein (1985) wrote that culture sets certain standards of acceptable behaviour within an organisation. Individuals are thereby subject to a process of socialisation based on normative and value-based behaviours of the organisation. On the other hand, Kohlberg's work is fundamental to our understanding of moral development within society from the perspective of three distinct phases: pre-conventional (people who act to avoid punishment, receive rewards from external sources, seek pleasure and give little consideration to social norms and ecological principles); conventional (individuals act to gain approval within society by adhering to social sanctions); and post-conventional (defined by the move to be influenced not by external forces but rather from within for the good of the community, but also, more broadly, for the good of humanity and the planet). Malloy and Fennell (1998a) termed these three phases, applied to ecotourism, respectively: (1) the market ecotourism culture, (2) the sociobureaucratic ecotourism culture, and (3) principled ecotourism culture. According to these authors, it is this last stage that the ecotourism industry must strive to reach, which demands not solely an economic and/or sociological agenda, but rather a socio-ecological change that reflects the general goals of ecological and social holism. The transitions between stages are influenced by broad political processes (e.g. the development of policies to control tourism impacts); cultural processes (e.g. promotion within firms based on socially and ecologically responsible behaviour); and technical processes (e.g. certification of ecotourism guides). Two other studies have been conducted more recently to place the work of Malloy and Fennell into context. Work by Ross (2003) suggests that organisational cultures and ethical behaviour can be advanced through trust and communication. Finally, the market culture mentality in ecotourism has

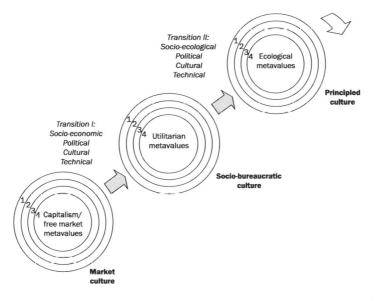

Figure 7.6 Moral development in ecotourism organisational cultures. 1, artefacts of the ecotourism organisation; 2, patterns of individual behaviour in the ecotourism organisation; 3, values and beliefs of the ecotourism organisation; 4, basic assumptions of the ecotourism organisation

Source: Malloy and Fennell (1998a)

been addressed by Holden (2003), who argues that service providers are often more interested in short-term benefits based on the instrumental (rather than intrinsic) use of nature.

What is encouraging in efforts to further develop links between ecotourism and ethics
are recent studies that have taken an empirical approach. Following the lead of others who
suggest that ecotourism ought to rely on principles that are grounded in ethical theory
(Blamey 2001; Orams 1995), Nowaczek and Smale (2005) adapted Fennell's (2001) indicators of ecotour operator ethics in their study of the Peruvian Amazon. These indicators
are designed to initiate further research on programme aspects such as general ethics (e.g.
respect shown to plants and animals), local people (e.g. ownership or participation in the
ecotourism business by local people), environmental education (e.g. level of on-site
education for tourists), operator professionalism (e.g. knowledge of the site itself),
contribution to conservation (e.g. physical contributions such as the removal of garbage)
and accommodation and transportation (e.g. use of hotels that are locally owned). Based
on a 5-point scale ranging from 'inferior' to 'superior', ecotourists felt that the service
provider was exceedingly ethical, with all responses higher than 4.0 on the scale (general
ethics ($M = 4.67$, $SD = 0.44$) was most highly rated of the six indicator sets). More specifically, the authors found that older visitors rated the ethical behaviour of the operator more
favourably than younger ecotourists, and that as the size of the ecotour group increased,
ratings of ecotour operator ethics decreased (as discussed at the end of Chapter 2).

The ethical imperative in tourism has also been discussed through the notion of responsible tourism, where research activity and practice has been steady in the past (see for
example Miller 2001; and Hudson and Miller 2005). One version of responsibility in
tourism is the ambassadorship idea that has been employed in Antarctica for several years,
despite the fact that very little empirical evidence exists to place it into context. Cited in
Maher *et al.* (2003), Burton (2000: 6) writes that, ' "ambassadorship" ' is the process of
advocating the 'preservation of the continent [by] those who have been to the "ICE" and
so have a first hand experience of the values [being sought] to protect'. Furthermore, the
Ecotourism/Heritage Tourism Advisory Committee expeditions (1997: 7) also cited in
Maher *et al.* suggest that tourism creates:

> 'ambassadors' by raising awareness ... through sharing with them the unique natural
> history of Antarctica and the Sub-Antarctic, allowing Expedition members to visit historic
> sites and discussing with them the conservation issues confronting the Antarctic Continent.

Ambassadorship is said to be loosely synonymous with 'stewardship' and 'advocacy', but
it is not certain what it means from anticipation, on-site experience, behaviour and recol-
lection contexts in Antarctica, according to Maher *et al.* The authors suggest, like Burton
above, that it is a richly value laden concept but uncertainty exists how ambassadorship
translates into the preservation of the continent. Is it just a word that resonates with people
while on-site, or can visitors be activated to raise awareness about the special problems in
Antarctica in a sustained fashion down the road?

Researchers have also started to examine equity issues in tourism and ecotourism.
Gender inequalities become ethical issues when patriarchal societies establish divisions of
labour. Reimer and Walter (2013) found a clear gender division in terms of the jobs
deemed to be suitable for men and women in southwest Cambodia. Men were typically
involved as guides, accountants, administrators, patrollers, drivers, garbage collectors,
construction workers and porters. By contrast, women were involved as cleaners, washers
and cooks. In this way, ecotourism tends to intensify traditional gender roles with women
taking a more subordinate role in the ecotourism industry than men.

Equity studies have also examined the shifting of the burden of ecological problems in
the north to people living in the south (Pellow *et al.* 2001). Pollution is a case in point,

Table 7.3 *Key concepts of environmental justice*

Key concepts	Definition
Environmental justice	'The fair treatment and meaningful involvement of all people regardless of race, colour, national origin, or income with respect to the development, implementation, and enforcement of environmental laws, regulations, and policies' (Liu 2001: 11; US EPA 2007).
Environmental equity	'The distribution of environmental risks across population groups and our policy responses to these distributions' (US EPA 1992: 2).
	An elusive and politically charged idea that defies a simple definition.
	Assumes 'no specific outcome and causes and leaves it for an analyst to determine the relationship between environmental risk distribution and population distribution' (Liu 2001: 13).
Environmental racism	'Any policy, practice, or directive that differentially affects or disadvantages (whether intended or unintended) individuals, groups, or communities based on race or color' (Bullard 1996: 497).
Environmental discrimination	The results of, and process by which, environmental policies create intended or unintended consequences, especially those which have disproportionate impacts on individuals, populations, or communities, minority populations or races, women and lower income groups.

Source: Adapted from Lee and Jamal (2008)

where northern countries like Canada and the USA sell their waste to the south or have their industrial wastes filter to the southern countries via air and water. Issues of inequality, racism, injustice and discrimination are therefore emerging as significant problems in the relationship between the north and south. Even within countries of the north there are equality issues when it comes to the handling and wastes. These issues are seldom discussed in the context of tourism, and they boil down to a sense of powerlessness of those who are most affected by the actions of dominant others (Fennell 2006b).

Lee and Jamal (2008) have provided a useful breakdown of key concepts in the broad environmental justice literature, and follow this up with distributive justice issues that are found in tourism (Table 7.3). These authors have also identified a number of environmental justice issues tied specifically to tourism. Some of the issues include the over-exploitation of water resources, air pollution, deforestation, threats to public health, the number and distribution of recreation sites, privatisation and accessibility to natural goods, displacement of local people, and the distribution of economic benefits from environmental resources among stakeholders. Some of these ethical issues are further examined in Chapter 8 on tourism impacts.

Codes of ethics

A recurrent theme in many of the aforementioned publications is the advocacy of codes of conduct/ethics for the tourism industry. The British Columbia Ministry of Development, Industry and Trade (1991: 21) defines a code of ethics as, 'a set of guiding principles which govern the behaviour of the target group in pursuing their activity of interest'. From an industry perspective, codes function as, 'messages through which corporations hope to shape employee behaviour and affect change through explicit statements of desired behaviour' (Stevens 1994: 64). In tourism, such codes have gone beyond the realm of business by ensuring that local people, government and tourists abide by predetermined guidelines. There has been a proliferation of such codes over the past few years by a variety of organisations, including government, non-governmental organisations (NGOs) and industry, many of which can be found in the work by Mason and Mowforth (1995, 1996) and the United Nations Environment Programme Industry and Environment (1995). An example of such a code is the Country Code, which appears to be the first of its kind (Figure 7.7), and which has more recently evolved into the Countryside Code, launched in England in 2004.

According to Scace *et al.* (1992) codes fall into two main categories: codes of ethics and codes of practice/conduct. The former are philosophical and value-based, whereas the latter are more applicable and specific to actual practice in local situations. An example of a guideline falling within a code of ethics would be 'respect the frailty of the earth'; while an example of a code of practice guideline would probably be oriented more towards acceptable business practice with reference to the organisation's 'commitment' to the customer. This dichotomy, though, is somewhat misleading as it implies that codes of practice/conduct are not concerned with values. Although codes of conduct ought to be specific in actual situations, as suggested, they should also be founded upon a sound ethical principle as outlined in the previous discussion on ethics and business (Fennell and Malloy 2007).

The link between the tourism industry and ethics has been discussed by Payne and Dimanche (1996: 997) who suggest that codes of ethics ought to project a number of key

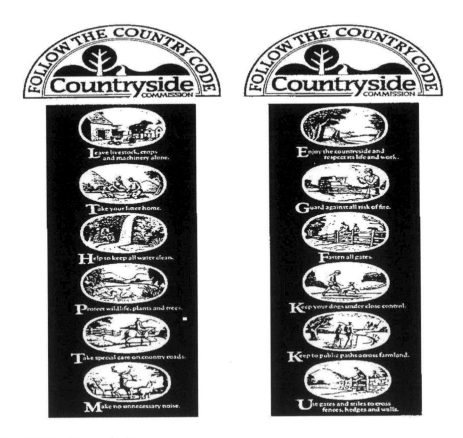

Figure 7.7 The Country Code

Source: Mason and Mowforth (1995)

values, including justice, integrity, competency and utility, in illustrating that: (1) the tourism industry must recognise that its basis is a limited resource, the environment, and that sustainable economic development requires limits to growth; (2) the tourism industry must realise that it is community-based, and that greater consideration must be given to the socio-cultural costs of tourism development; and (3) the tourism industry must also recognise that it is service oriented, and that it must treat employees as well as customers ethically.

Genot (1995) further underscores the practical reasons for industry to adopt codes of conduct, stating that, among other factors, a sound environment means good business, meeting consumer demand, unifying industry efforts and image, and assuring product quality. Genot feels that the following principles are at the core of any code of ethics: environmental commitment, responsibility, integrated planning, environmentally sound management, cooperation between decision-makers and public awareness. Many of these principles are elaborated upon in other publications devoted specifically to the ethical conduct of operators and other members of the tourism industry (see Dowling 1992; Ecotourism Society 1993), in specific regions like the Arctic (Mason 1994, 1997b), with specific activities like whale watching (Garrod and Fennell 2004; Gjerdalen and Williams 2000), and in an effort to contrast eastern and western tourism development paradigms (Theerapappisit 2003).

Stronger consideration needs to be given to the formation and understanding of these codes, according to Malloy and Fennell (1998b). These authors subjected 40 codes (414 individual guidelines) to content analysis on the basis of who guidelines were developed for, who guidelines were developed by, the type of tourism they were directed towards, the orientation of the code, the mood of the message and the main focus of the guideline. These variables were juxtaposed with two philosophies of ethics (deontology and teleology), in addition to determining whether the codes were relevant to a local ('local' included local, regional and national scales) or cosmopolitan (meaning a universal code) condition. Deontological guidelines were those which, for example, advised tour operators to abide by organisational policies developed for the purpose of honouring established environmental or cultural norms. An example of a deontological code is as follows: 'Be aware of the periphery of a rookery or seal colony, and remain outside it. Follow the instructions given by your leaders'. Teleological guidelines included those that explained some form of consequence from action or inaction. An example of a teleological code is as follows: 'Do not enter buildings at the research stations unless invited to do so. Remember that scientific research is going on, and *any intrusion could affect the scientists' data.*' (The consequence of the action is highlighted in italics.) Figure 7.8 illustrates the methodology employed in the study for each of the 414 guidelines.

In this research, approximately 77 per cent of all guidelines in the study were found to be deontological in nature. As such, these guidelines, according to Malloy and Fennell, fail to provide the decision-maker (e.g. tourist) with the rationale for abiding by a particular code, with the assumption that an explanation of consequence (i.e. the teleological premise) is unnecessary to the tourist or other stakeholder. More specifically, it was found that most of the codes were developed by associations (NGOs), and for tourists; many of the codes were ecologically based, rather than socially or economically based; about 85 per cent were positively stated; and the focus of such codes was on people or the resource base. Malloy and Fennell recommend that future research in the area of codes of ethics should explore the deeper philosophical meaning of codes, advocate more of the teleological perspective in the development of codes (i.e. the consequences of one's action if the code is not adhered to), and aim at a better understanding of the extent to which codes actually change behaviour.

This latter component, behavioural change, has more recently been investigated by Bhattacharya *et al.* (2006), in a study of ecotourists to Van Vihar National Park, Bhopal,

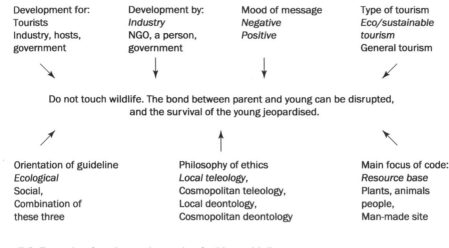

Development for:	Development by:	Mood of message	Type of tourism
Tourists	*Industry*	*Negative*	*Eco/sustainable*
Industry, hosts,	NGO, a person,	*Positive*	*tourism*
government	government		General tourism

Do not touch wildlife. The bond between parent and young can be disrupted, and the survival of the young jeopardised.

Orientation of guideline	Philosophy of ethics	Main focus of code:
Ecological	*Local teleology,*	*Resource base*
Social,	Cosmopolitan teleology,	Plants, animals
Combination of	Local deontology,	people,
these three	Cosmopolitan deontology	Man-made site

Figure 7.8 Example of an Antarctica code of ethics guideline

Source: Malloy and Fennell (1998b)

India (see also Chapter 6 on learning and behavioural change). Bhattacharya *et al.* found that 60 per cent of ecotourists had no awareness about the park's code of conduct, with only 10 per cent of visitors having any level of awareness about codes in general. Consistent with this low level of awareness was the fact that 26 per cent of visitors were opposed to receiving any guidelines or instructions regarding tourism in the park. What this study points to is that visitors, who actively include nature and wildlife as part of their recreational regime, have had little exposure to codes of ethics as behavioural modifiers and, when presented with these, choose not to follow them. (See Adams *et al.* 2001, who found that the process of developing and implementing a code may be more important in increasing ethical awareness than the content of the code itself.) Code of ethics compliance behaviour of ecotourists has been investigated by Sirakaya and Uysal (1997), who found that sanctions and deterrents do not play a strong role in predicting conformance of the sample. By contrast, compliance was increased through the use of education about the benefits of compliance and the expected practices within the destination. The authors felt that policymakers need to be aware of these differences when designing guidelines, and that guidelines need to be distributed to all operators if their tenets are expected to be met. Furthermore, Waayers *et al.* (2006) found that ecotourists are not always compliant with codes of ethics when it comes to their interactions with wildlife. The authors found that even though tourists were made aware of the rules of engagement with nesting marine turtles, 77 per cent of the sample breached the code, with 51 per cent of these breaches resulting in disturbance to turtles trying to nest. These breaches included shining a torch on the turtles, moving closer than three metres to the turtles, and not staying behind the turtles at all times.

Strong and Morris (2010) found that in the case of the Grey seal around Ramsey Island, UK, there is a need to take a closer look at the marine code of ethics. Disturbance to the seals is taking place even though operators are staying true to the guidelines within the code. Recommendations by the authors are for a review of the current guidelines, with particular reference to boat distance away from the seals and boat speed, as well as questions about the ability of the industry to manage common pool resources through voluntary measures like codes.

What appears to be helpful in getting groups like service providers or tourists to follow codes of ethics is personal involvement: groups will support the code of ethics (instead of a code developed by an external group like government) if they have been involved in its creation and implementation (see the case study on Antarctica in this chapter). Such is the case in northern Florida concerning the swim with manatees tourism, where Sorice *et al.* (2006) identified a series of issues in regard to enforcement, water quality, harassment, density and crowding and education that constrain the efforts of government. This corroborates the work of other researchers who feel that the non-regulatory approach is better suited to tourism over the regulatory or big stick method (See Fennell and Malloy 2007, for an extended discussion on tourism codes of ethics; see also Rivera 2004, who found that other measures beyond the programme itself were required to stimulate proactive voluntary environmental behaviour in Costa Rica's Certification for Sustainable Tourism programme, where coercive institutional pressures through government monitoring as well as normative tourism industry association forces induced higher environmental performance.)

Despite what appears to be normative acceptance of the utility of codes of ethics in the tourism industry, it is clear that they have not been universally embraced by tourism researchers. Wheeller, for example, is not convinced of their utility, as demonstrated in a stinging review of ecotourism. He feels that there are no answers to the confusion surrounding both ecotourism and sustainable tourism, but simply

> a never-ending series of laughable codes of ethics: codes of ethics for travellers; codes of ethics for tourists, for government, and for tourism businesses. Codes for all – or, more likely, codeine for all . . . But who really believes these codes are effective? I am pretty wary of platitudinous phrases like 'we are monitoring progress'. Has there been any progress – indeed, has there been any monitoring? Perhaps I am missing it and the answer itself is actually in code.
>
> (Wheeller 1994: 651)

CASE STUDY 7.2

Ecotourism in Antarctica

At 13.6 million km², Antarctica is the largest remaining wilderness on earth. It is covered by a sheet of ice which is on average 2.1 km thick. While no humans live there permanently, the region is visited by hundreds of scientists each summer at one of 37 research stations operated by 18 different countries. Because of the unique location of Antarctica and its international significance, no one nation has sovereignty over any part of Antarctica. The region is governed by the Antarctic Treaty System of 1959 (came into effect in 1961), which provides for the conservation, research and management of Antarctic resources, and how these are to be used. As regards tourism, the Treaty in Recommendation VIII–9/1975 acknowledges that tourism is a natural development in this area and that it requires regulation (Heap 1990, cited in Bauer and Dowling 2003).

In a study of ship-based polar tourism, Grenier (1998) notes that the Argentinian ship, *Les Eclaireurs*, was the first to bring tourists to the Antarctic Peninsula (the 1,600 km spine that juts from the Antarctic continent towards South America) in 1957–58. In total, four cruises from Argentina and Chile brought about 500 tourists in that season. Lars-Eric Lindblad, however, is generally acknowledged to be the first to run international ecotours to Antarctica from New York (he chartered an Argentinian naval ship), starting in 1966. By the end of the 1990s, 84,173 ship-based passengers had visited Antarctica, with 14,623 alone visiting the region in 1999/2000 (Bauer and Dowling 2003). In the 2003–4 season, approximately 27,000 passengers visited Antarctica with a projected growth to 32,500 by

continued

2009 (Bastmeijer and Roura 2004). The importance of ship-based tourism remains today, with up to 90 per cent of all tourists arriving this way.

Most trips to the Antarctic Peninsula depart from Ushuaia, Argentina, Punta Arenas, Chile or Port Stanley in the Falkland Islands, with the trip from the South American continent to Antarctica taking approximately two days. Ships accommodate anywhere from 35 to 400 passengers. Itineraries to this region of Antarctica vary according to the type of vessel used. Trips beyond the Antarctic Peninsula require an ice-breaker. Those travelling to the Ross Sea section of Antarctica depart from Hobart, Tasmania, Invercargill, New Zealand or Christchurch, New Zealand, from late December to late February, with mid-January to mid-February reported to be the best time to view wildlife. Ecotour activities on such trips normally include activities such as viewing penguins (some ships have helicopters with which to visit certain penguin colonies), whales, seals and sea birds; visiting historic sites, scientific and whaling stations; and appreciating the grand scenery of the region. Tourists typically spend about two hours on shore at a time walking around and photographing wildlife.

Due to the absence of legislation regarding tourism in the Antarctic treaty, tourism operators, through the efforts of Lars-Eric Lindblad, took it upon themselves to develop codes of ethics for operators as well as for tourists. Lindblad also inspired a coalition of cruise operators, called the International Association of Antarctica Tour Operators (IAATO), to help administer these guidelines (Stonehouse 2001). The Rules for Antarctic Visitors, as distributed by Marine Expeditions, are as follows:

- Do not leave footprints in fragile mosses, lichens or grasses.
- Do not dump plastic or other non-biodegradable garbage overboard or on the continent.
- Do not violate the seals', penguins' or seabirds' personal space.

Start with a baseline distance of 5 m from penguins, seabirds and true seals, and 18 m from fur seals.

- Give animals the right of way.
- Stay on the edge of, and don't walk through, animal groups.
- Back off if necessary.
- Never touch the animals.
- Do not interfere with protected areas or scientific research.
- Do not take souvenirs.

Despite these guidelines, Grenier (1998: 186) has put together a list of some ethical transgressions that he has observed in his polar travels over the course of many years. These include:

- A musician scaring away a group of penguins by giving his 'first Antarctic flute concert'.
- Passengers collecting stones, feathers and bones for souvenirs.
- A passenger tossing stones at the foot of a penguin to improve a photographic opportunity.
- Visitors touching penguins.
- A tourist walking straight into a penguin rookery, despite warning, stating that he had 'paid for the right to go as he pleased'. The penguins were scared off, exposing chicks to predators and freezing temperatures.

- A captain positioning himself next to a seal for a better photo opportunity.
- A zodiac driver consuming alcohol before an excursion conducted after sunset.
- A crew member harassing juvenile penguins dying after having been abandoned by their parents.

Davis (1998) notes that the management of people in this type of wilderness is difficult because of the lack of sovereignty by any one country, and because of the differing cultural views on wilderness management philosophies. Tourists have consistently shown interest in the development of more facilities on the continent – like toilets, accommodation, post office, gift shop and other activities such as diving, skiing and camping. Davis's preference is to stipulate which activities are acceptable as opposed to regulating all activities equally.

Hall and Wouters (1995) observe that because of the fact that only 2 per cent of the land is ice free, there is a tremendous amount of competition between humans and wildlife, the latter using this area for breeding. This is compounded by the fact that the tourism season coincides with the peak breeding season. The constant human intervention is thought to cause behavioural changes in animals, denudation of habitat, and the spread of new organisms which might have the potential to spread animal and plant diseases. In 1997, for example, a poultry virus brought to the continent either by scientists or tourists was found to have infected Antarctic penguins. The concern is not only the type of impact occurring in Antarctica, but also who causes these shocks. Tangley (1988) writes that while scientists worry about the impact of tourism, conservationists say it is science that threatens the fragile Antarctic environment. This issue has touched off a debate on the responsibilities and effects of both parties, with scientists most vociferous about the dangers of an ever-increasing presence of tourists on the continent.

Because of the many problems and issues brought on by tourism in Antarctica, Bastmeijer and Roura (2004) suggest that the precautionary principle should be applied explicitly in Antarctica under certain conditions. Precaution would be useful in: (1) improving the applicability of EIA to tourism; (2) improving the assessment of cumulative impacts; (3) prohibiting tourism in locations that lack proper monitoring; (4) establishing temporal and spatial limitations for certain sites; (5) concentrating tourists in certain areas and managing them sustainably; (6) preventing access to sites not visited previously; and (7) implementing restrictions on tourism activities currently permissible (see also Fennell and Ebert 2004).

Like the Galapagos, Antarctica is truly one of the world's prime ecotourism hotspots. How it comes to be managed is absolutely critical as a case study which might provide meaningful leadership in the management of other tourism regions around the world. In considering this region, the question of humankind's obsession with visiting every corner of the planet comes to mind. Some argue, as scientists in Antarctica might support, that tourists have no reason or purpose to be in Antarctica. Others suggest that it is our right as humans to go where there is opportunity to see and experience new lands. The question might become more important over time as we see tourism continually expand into the world's most pristine areas. The argument for tourism in these areas is that it helps to support economically depressed communities. Antarctica is perhaps the only place in the world where this position does not hold true.

Websites:
  http://sedac.ciesin.org/pidb/texts/antarctic.treaty.protocol.1991.html
  http://astro.uchicago.edu/cara/vtour/
  http://www.lonelyplanet.com/destinations/antarctica/antarctica/facts.htm

Accreditation and certification

One of the most intensely examined areas in ecotourism over the past decade is accreditation and certification of operators and their programmes (the two terms often viewed as synonymous under the label of certification) as the basis for the education and the provision of skills of those working in the industry. The heightened level of attention afforded to these two concepts has been discussed by Morrison *et al.* (1992), who identify a number of programmes in North America, including Certified Travel Counsellor, Certified Hotel Administrator (the two longest-standing programmes), Certified Tour Professional, Certified Meeting Professional, Certified Hotel Sales Executive, Certified Festival Executive, Certified Incentive Travel Executive, Certified Travel Marketing Executive and Certified Exhibit Manager. The World Trade Organization (Dankers 2003) defines certification as a procedure by which a third party gives written assurance that a product, process or service is in conformity with certain standards; and accreditation as the evaluation and formal recognition of a certification programme by an authoritative body.

Morrison *et al.* (1992) point out that certification in the past has dealt with the individual professional, while accreditation is concerned with programmes and institutions. These authors observe that certification programmes have grown to replace formal institutions like universities in gaining expertise and professionalism in the tourism industry. They caution, however, that although certification will indeed increase in the future, it should not be a substitute for travel and tourism degrees. In addition, whereas universities are able to offer a good grounding in the theoretical aspects of tourism, they may also diversify into offering certification options to graduates through programme offerings. Outside tourism, the relevance of accreditation has been debated in several other disciplines. Scalet and Adelman (1995), for example, discuss the pros and cons of accreditation in fisheries and wildlife programmes, while Silverman (1992) examines accreditation of undergraduate programmes in environmental health science.

In the realm of adventure recreation, there has been significant debate on how best to certify the adventure programme experience, including standards of programme quality, professional behaviour, professional conduct and risk management (Gass and Williamson 1995). Similarly, these have revolved around the certification of leaders and guides through organisations such as the American Mountain Guides Association and the Wilderness Education Association, while the American Camping Association and the National Recreation and Parks Association have implemented accreditation programmes to meet their own specific needs. Two studies found overwhelming support for accreditation over individual certification as it (accreditation) provided an umbrella to ensure a higher degree of professional credibility and less reliance on individual operators (see Cockrell and Detzel 1985; Bassin *et al.* 1992). These findings were considered in the development of an American national accreditation process that focused on programming in the adventure recreation field by the AEE (Association for Experiential Education 1993). This document systematically evaluates operations in areas related to ethics, risk management, staff qualifications, transportation and technical skills. For example, with respect to the environmental aspects of river rafting, code 43.B.01 suggests that:

> Staff are familiar with the operating sites for rafting. *Explanation:* A pre-site investigation is conducted to understand the conditions as well as the education/therapeutic possibilities. This includes locating appropriate takeout areas and access to safe transportation sites.
>
> (AEE 1993: 82)

According to Gass and Williamson (1995: 25), this national programme maintains the following accreditation procedure:

1 *Self-assessment and documentation stage.* Operators must compare their programme against the AEE programme guidelines.
2 *Verification stage.* A team of between two and four reviewers visit the operator and conduct interviews, inspect equipment and observe activities as necessary. A standards report card is used to determine whether a component of the programme 'passes', 'fails' or is 'conditional' based on the need to remedy the situation, or does not apply in the case where an activity is not associated with a programme.
3 *Follow-up stage.* Findings from the verification stage are forwarded to the Program Accreditation Services committee, which sends a recommendation to the AEE Executive Board, which in turn decides whether the programme should receive accreditation.

In further support for the idea of accreditation, Gass and Williamson (1995: 23) offer the following: (1) accreditation provides adventure programmes with the ability to achieve standards without losing the flexibility to decide and design how these standards are met; (2) accreditation takes a systematic view of the process of adventure programming rather than dividing it into individualised categories; (3) accreditation encourages ongoing improvement through internal and external review; and (4) it assures clients, agencies and resource managers that a programme has clearly defined objectives which can be reasonably met.

As suggested in Chapter 3, ecotourism shares some aspects of adventure recreation/tourism, more from the context of setting than in programming. This element of setting (e.g. tropical rainforests) may provide a component of risk that is absent from other, more conventional forms of tourism. To a lesser extent than adventure travel, ecotourism may be seen as bridging a gap between various forms of adventure recreation and tourism. In comparison to outdoor recreation, however – in particular, outdoor guides and leaders – tourism lags far behind, and as in adventure recreation, programming in ecotourism will be a key issue in the future (see also Chapter 11).

Recently, the topic of accreditation and certification has been debated in ecotourism, consistent with this trend in the adventure pursuits industry. For example, Wearing (1995: 34) suggests that professionalisation and accreditation in ecotourism will continue to be at the forefront of discussions related to regulation and control, as a means by which to provide focus to an industry that is almost impossible to limit (in terms of expansion). He feels that accreditation affords the opportunity to improve tourism industry standards while at the same time ensuring high-quality services and programmes in a very competitive marketplace. Wearing identifies a series of advantages and disadvantages of accreditation as they relate to ecotourism, as follows:

Advantages of accreditation

1 The ecotourist has the knowledge of what will be taught, and the approaches used will be the best and the safest.
2 The ecotour operator has a recognised and accepted methodology and thus increased social status.
3 The employer of ecotourism guides has knowledge of an accepted industry standard.
4 The government has the knowledge that new operators have been taught to a minimum standard.

Disadvantages of accreditation

1 The idea that one needs a service which is best provided by an expert creates a relationship of mutual dependency and a social distance between the ecotourist, the operator, the host community and the natural environment.

2 Ecotour operators are assessed on their accomplishments rather than their human characteristics to the effect of depersonalising the human–natural environment relationship.

3 Professionals can tend to rationalise and focus on facts, objective data, and procedures, thus potentially losing the intrinsic, intuitive association so often formed through ecotourism.

4 Through commodification, the operator may come to view the natural and cultural environments as a means to an end and so become alienated from the natural world and hence unable to encourage the ecotourist or encourage environmental awareness.

5 Loss of perceived freedom by the individual ecotourist, who feels he or she must do as the operator says; this restricts the ecotourism experience.

6 The imposed structure that is required to assess or measure the ecotour operator entails limitations.

The message put forward by Wearing is that although accreditation would certainly reduce risk, increase standards and increase status, there is a danger of restricting innovation and accessibility in the ecotourism industry. What emerges as a significant dilemma, therefore, is the degree to which the industry can afford to emphasise innovation and accessibility at the expense of ensuring that proper standards are being upheld. The issue is one that clearly requires more time and education for operators in the field. In addition, it is a normative concern as the likelihood for compliance is directly related to how operators feel individually, but also collectively, about the process of accreditation.

A good example of the accreditation process in action can be found in the Nature and Ecotourism Accreditation Program of Australia, where those working in the field (i.e. operators) submit to a standardised procedure. At present, operators are given a core level of accreditation if they satisfy the programme's basic criteria; however, the system encourages operators to implement measures beyond the standards of the core criteria in earning advanced standing. Eligibility for accreditation is based on eight criteria (Chester 1997: 9). According to these criteria, ecotourism:

1 focuses on personally experiencing natural areas in ways that lead to greater understanding and appreciation;

2 integrates into each experience opportunities to understand natural areas;

3 represents best practice for ecologically sustainable tourism;

4 contributes to the conservation of natural areas;

5 provides constructive, ongoing contributions to local communities;

6 interprets, involves and is sensitive to different cultures (particularly indigenous ones);

7 consistently meets client expectations; and

8 marketing is accurate and leads to realistic expectations.

There was resounding support for the idea of ecotourism accreditation in Australia during the mid-1990s (McArthur 1997), which culminated in the launching of the National Ecotourism Accreditation Program (NEAP) at Australia's 1996 national meeting. Accreditation is offered to those operators working in the accommodation, tour operations and attractions sectors, and operators are accredited by first completing an application

Table 7.4 *Ecotourism accreditation fees in AUS$ (from November 1996)*

Fees – based on annual business turnover ($)	0–100,000 turnover p.a.	100,000– 250,000	250,000– 1,000,000	1,000,000– 3,000,000	3,000,000 and over
Ecotourism accreditation document fee (per. doc):					
Members	75	75	75	75	75
Non-members	175	175	175	175	175
Application fee (only in year 1)	75	100	175	250	400
Annual renewal fee	100	150	300	500	750
Total fee in year 1:					
Members	250	325	550	825	1225
Non-members	350	425	650	925	1325

Source: McArthur (1997)

(through a nomination process involving three referees), and by paying a one-off application fee and an annual fee. Table 7.4 provides an overview of this fee structure. One of the key features of the Australian accreditation initiative is the level and style of enforcement employed in the programme. Each of the applicants is monitored in four ways: (1) the applicant's honesty in completing application documents; (2) feedback from clients; (3) feedback from referees (random selection of operators); and (4) through random audits.

The most recent change made to NEAP is the inclusion of a nature tourism category allowing the vast number of nature tour operators in Australia to receive accreditation. These nature-based and sustainable-oriented operators, who do not fit the stricter definition of ecotourism, may still be recognised according to the guidelines of the NEAP initiative. The other positive change to NEAP is the recognition that programmes are accredited rather than operators. This means that although operators may run a number of programmes, it is only perhaps one or two, for example, of these programmes which are worthy of accreditation status. This gets away from a more broadly based operator-specific method which was not necessarily representative of the philosophy of ecotourism within all programmes. Rainforest Alliance is spearheading a global accreditation body, known as the Sustainable Tourism Stewardship Council (STSC). This body includes input from numerous bodies at all levels for the purpose of improving the credibility of sustainable tourism certification programmes. In general the STSC will be the main accrediting body that will approve sustainable tourism certification programmes (see Font *et al.* 2003; Rainforest Alliance 2008, http://www.certificationnetwork.org/?id=council).

Honey and Rome (2001) refer to two main types of certification that are found in practice: (1) process-based environmental management systems, which are driven by environmental management systems such as environmental auditing and environmental impacts assessments; and (2) performance-based methods which use environmental, sociocultural, and economic criteria, standards or benchmarks. As many of these concepts are relatively new to tourism, it is worth while to provide some basic definitions of terms, which may be little understood (see also Font and Buckley 2001; Issaverdis 2001), following from Honey and Rome (2001):

• *Environmental Management System (EMS).* The EMS is part of the overall management system that includes the organisational structure, responsibilities, practices, procedures, processes and resources for determining and implementing the environmental policy. An environmental management system includes tools such as environmental impact assessment, environmental auditing and strategic environmental assessment.

- *Audit.* A systematic, documented, periodic and objective evaluation and verification of how well a particular entity (company, product, programme, individual, destination etc.) is doing compared to a set of standards.
- *Benchmarking.* The process of comparing performance and processes within an industry to assess relative position against either a set industry standard or against those who are 'best in class'.
- *Best practice.* Used to designate highest quality, excellence or superior practices in a particular field by a tourism operator. It is widely used in many award and certification programmes, as well as in academic studies, to designate best in a particular class or a leader in the field. 'Best', however, is a contextual term. There is no set standard of measurement and the term is often loosely or ill-defined. (As suggested by Issaverdis 2001, there is little difference between best practice and benchmarking.)
- *Ecolabelling.* Describes a scheme in which a product or service may be awarded an ecological label on the basis of its 'acceptable' level of environmental impact. The acceptable level of environmental impact may be determined by consideration of a single environmental hurdle or after undertaking an assessment of its overall impacts. (See also Font and Buckley 2001.)
- *Ecotourism certification.* Programmes that cover businesses, services and products that describe themselves as involved in ecotourism. They focus on individual or site-specific businesses, have standards that are tailored to local conditions, and are largely or totally performance-based.

The importance of these initiatives to ecotourism was noticeable during the World Ecotourism Summit in Quebec City, Canada, in May 2002, where a number of researchers and organisations gathered to discuss the fate of such programmes. As Buckley (2002) notes, certification has become politicised because of the sheer number of such schemes in operation (upwards of a hundred), and the associated control over those which are most prominent. Buckley suggests that the World Ecotourism Summit was used as a platform by Green Globe 21 and the Ecotourism Association of Australia to launch a draft International Ecotourism Standard for certification, which he observes is the Australian NEAP process offered in other countries through Green Globe 21. Where this leaves other schemes is an immediate point of interest, along with the associated implications such a monolithic unit might create. Just as important in the debate over the significance of such schemes is the basic question of whether accreditation schemes are worth all of the energy put into them. Buckley notes that they may be less effective than simple government regulations such as codes and laws. At present, it remains that these schemes are politically charged and have an uncertain future, and the debate over their utility promises to be an interesting one.

Pessimism over ecotourism certification has also been expressed by Jamal *et al.* (2006), who argue that such schemes cannot do justice to the experiential aspects of ecotourism because there is simply too much of an emphasis on the institutionalisation of modernity through objective forms of evaluation which, 'strengthens power structures rather than overturns them' (2006: 168; refer back to Chapter 6). Certification schemes have turned more towards commodification than well-being and participatory democracy, which is really what ecotourism ought to be about. The instrumental reason inherent in these broad schemes has been discussed by Fennell (2006b: 212–213) in the context of the 2002 Ecotourism Summit in Quebec City, as follows:

The apparent need to have such meetings controlled by governing bodies and NGOs, those entities which are the biggest and have the most prestige and clout, at the expense of independent free speech and scholarship, emphasises the policy fetish. These groups have in common the same specific organisational agenda that becomes threatened by thought which takes place outside the organisational vacuum. But the nature of this vacuum creates an inescapable capsule which prevents and compromises their ability to strike a balance between what they want and what is really needed . . . The Quebec City meeting which was quite restrictive in its programme and organisation, provides ample evidence of how instrumental reason can both flatten and narrow our options.

These schemes have also been discussed by Sasidharan *et al.* (2002), who observe that less-developed countries (LDCs) could potentially develop an internationally recognised tourism ecolabelling programme that would help place them on the path of environmentally conscious development and management. The impetus behind such a move would be to better position ecotourism enterprises in the LDCs to capture the market of high-spending Western tourists. After careful examination of a variety of existing schemes, and through consideration of related knowledge, the authors find such an enterprise excessively challenging for the following reasons: (1) ecolabelling issues between countries would vary depending on ecological, social and economic resource conditions; (2) there is no conclusive evidence that ecolabels actually improve the environment; (3) environmental education has been found not to stimulate environmentally responsible purchasing; (4) ecotourists may not respond positively to ecolabel programmes and those that market them; (5) the high cost incurred in running an environmentally sensitive operation in addition to the cost of being part of an ecolabel programme may drive service provider prices up; (6) the higher costs may dissuade ecotourists from supporting ecolabel programmes and services, allowing the non-ecolabelled service providers to gain at the others' expense; (7) there may be conflicts of interest between the various stakeholder groups involved in facilitating and running the ecolabel programme, especially as it contrasts with the profit motive of operators; and (8) the investment in technology needed to be a part of the programme for service providers, while maintaining profit margins, would make it difficult for the firms to meet the high standards of the programme. The foregoing, according to the authors, would furnish, 'the emergence of large, multi-national tourism enterprises as "environmental market leaders", thereby providing them with a marketing edge over small-scale enterprises of developing countries' (Sasidharan *et al.* 2002: 171–172). Thus ecolabelling schemes appear to be more akin to strengthening power structures (Jamal *et al.* 2006) and instrumental reason (Fennell 2006b) than what is really needed for LDCs.

Ecolabels provide reliable information to consumers about the environmental impact of their products and services, which may be used to influence the decisions of consumers. In this way, service providers hope to win the confidence of potential buyers through their expressed environmental values (Piper and Yeo 2011). Important in all of this is that the actual products and services provided are parallel to the marketing strategies of the company or organisation. Piper and Yeo (2011) write that there are approximately 340 international ecolabel programmes, with 40 of these dealing specifically with tourism.

Even though ecolabels and certification programmes have been in existence for some time, awareness of these initiatives amongst tourists is often low. Puhakka and Siikamäki (2012) surveyed 273 visitors to Finland's Oulanka PAN Park and, consistent with other studies (see Fairweather *et al.* 2005), found a low level of awareness of ecolabels. However, even though there was not much knowledge of these programmes among the sample, there were generally positive attitudes expressed for these programmes. Tourists wanted more information on ecolabels and certification.

Animal ethics

Previously in the chapter it was mentioned that animal-centered ethics was one of a few forms of environmental ethics (Elliot 1991). This is an area of research that has received little attention – until recently. The reasons for this lack of emphasis are not at all well defined, especially when issues arise that are clearly contentious. For example, some theorists believe that zoos are acceptable as a form of ecotourism (see Mason 2000; Curtin 2004), even though research points to the fact that zoos are more about entertainment than education and conservation (Ryan and Saward 2005; see also Cater 2010). Furthermore, there are questions over whether activities like hunting and fishing are ecotourism (see Part I), and sled dog tourism has come under scrutiny as a form of ecotourism. There are many cases where the consumptive use of animals simply do not fit with ecotourism's philosophy (refer to Chapter 3 for the example of billfishing as ecotourism).

In fact, there are many animal ethics theories at hand that may be useful in better framing these issues and debates. For example, Fennell and Sheppard (2011) used utilitarian theory as a lens to deconstruct the sled dog cull after the 2010 Vancouver Winter Olympics. At the end of January 2011, a story broke over the mass slaughter of 56 sled dogs owned by Howling Dog Tours. Robert Fawcett, the man responsible for killing the dogs (shooting and stabbing), was charged and convicted of causing unnecessary pain and suffering. He was fined $1500, with no prison sentence. Fennell and Sheppard concluded that the killing of dogs was morally wrong because: (1) the act achieved more harm than good; and (2) dogs showed signs of fear and distress demonstrating self-consciousness.

Animal welfare has been used in regard to the same issue. Soon after the cull, the British Columbia (BC) government assembled a task force for the purpose of developing new standards of care for the sled dog industry (Government of British Columbia 2011a). The task force included members of government, veterinarians, animal welfare scientists, sled dog tour owners, a member of the sled dog mushers association and a member of the BC SPCA (Society for the Prevention of Cruelty to Animals). By mid-2011, the Prevention of Cruelty to Animals Act was updated, with fines increasing to $75,000 and prison sentences of 24 months for the worst offenders (Government of British Columbia 2011b). Canada's first Sled Dog Code of Practice was unveiled on 30 January 2012 (Government of British Columbia 2012), and included detailed sections on best practice related to Health and Welfare, Nutrition, Housing, Husbandry, Transportation and Euthanasia. In the section pertaining to Euthanasia, the code requires that:

- Euthanasia must not be used as a means of population control for healthy, re-homable sled dogs, unless it is demonstrable that all options of re-homing have been exhausted.
- Euthanasia must be carried out by a competent person to ensure death occurs quickly and without unnecessary pain, suffering or distress.
- Euthanasia by firearm must be performed according to *Guidelines for euthanasia of domestic animals by firearms*, unless an alternative method can successfully be used to ensure death occurs quickly and without unnecessary pain, suffering or distress.

Ecocentric theory has also been used in the case of Fraser Island, Australia, where the death of a nine-year-old boy from a dingo attack provoked re-evaluation of policies on how to manage the interactions between tourists and dingoes. Burns *et al.* (2011) argue that the way forward on Fraser Island is to incorporate an ecocentric ethic in the management of dingoes and non-consumptive wildlife tourism. The implementation of an ecocentric perspective 'does not place the rights of humans above those of non-humans, but instead argues that non-human species have the same rights to live and prosper as humans' (2011: 182). Furthermore, Tremblay (2008) argues for a switch in the management of wildlife drawn into ecotourism, from a focus on motivations and impacts on animal welfare, to one where habitat sustainability, large-scale impacts and multiple uses is the norm.

The indecisiveness around zoos, hunting and fishing as ecotourism (or not) prompted Fennell (2013b) to argue that zoos and other forms of consumptive tourism should not be classified as ecotourism. Despite the fact that tourists are able to get close to animals of interest for a more embodied experience, zoos are captive environments that prevent animals from living the lives they are supposed to live. As such, what has made the zoo acceptable – as well as hunting and fishing and other forms of animals use – is the absence of guidance determining how the use of animals in tourism should correspond with ecotourism in theory. Fennell offers the following first principle in an effort to move the debate forward. He argues that the ecotourism industry should:

> Reject as ecotourism all practices that are based on or support animal capture and confinement, or other forms of animal use that cause suffering, for human pleasure and entertainment.

> Embrace as ecotourism interactions that place the interests of animals over the interests of humans. This would include encounters with free-living animals that would have the liberty to engage or terminate interactions independent of human influence.
>
> (Fennell 2013b)

Conclusion

One of the concerns that people in the field of ecotourism share, at least from the research standpoint, is the lack of empirical data to substantiate many of the claims being made about ecotourism as a more responsible form of tourism over other types; that is, are ecotourists more ethically based than other types of tourists? What about their values and attitudes? Some of these questions have been at least partially addressed by Dolnicar (2010). Dolnicar found that there are two measures that are most important in predicting the pro-environmental behaviour of tourists. These are feeling morally obliged to behave in an environmentally friendly manner, and income. In a related study, Dolnicar et al. (2008) provided a summary of empirical studies on ecotourists, with the conclusion that, generally speaking, ecotourists have higher incomes and demonstrate more concern for the natural environment than other types of tourists. It follows that individuals are perhaps drawn to ecotourism not only because they can afford to travel as such, but also because of this moral predisposition for behaving in an environmental manner. Dolnicar (2010) argues that the implications for the tourism industry are that decision-makers ought to take more of a demand-side approach to sustainable tourism, especially in parks and protected areas, through the identification of those market segments that have a high intrinsic inclination to protect the natural world.

A potentially fruitful area in differentiating ecotourists from other types of tourists is through an analysis of values. Conventional measures used in other social science fields include Rokeach's Value Survey, the Values and Lifestyle Scale, and the List of Values Scale (see Madrigal and Kahle 1994; Kahle et al. 1986). Madrigal (1995) writes that owing to their centrality to a person's cognitive structure, personal values, and therefore scales of this nature, are effective predictors of human behaviour. In relation to attitudes, the New Environmental Paradigm (NEP) scale has been moderately successful at measuring beliefs about humankind's dominance and harmony with the natural world (Jurowski et al. 1995), and may also lend some support in differentiating between, for example, ecotourists and mass tourists within a destination. Finally, in the context of the business literature, Reidenbach and Robin (1988, 1990) have developed a Multidimensional Ethics Scale (MES) which is designed to determine the ethical differences between subjects on the basis of deontological, teleological, justice and relativistic theories of ethics. Using this scale, Fennell and Malloy (1997) found that ecotour operators were moderately more ethical than other types of operators (fishing, cruise line, adventure and golf) on the basis of their responses to three tourism scenarios, providing some empirical evidence to suggest that, at least from the perspective of operators, ecotourism is in fact a more ethically based sector of the tourism industry.

Summary questions

1 What is the difference between theoretical and applied ethics, and why are they so important together?
2 Speculate as to why there has been such reluctance in taking a more proactive role in understanding how ethics might benefit the ecotourism industry.
3 How do deontological and teleological codes of ethics differ? Which is better, and why?
4 What is ethical triangulation?
5 Which code of ethics is reported to be one of the oldest, as regards tourism and outdoor recreation?
6 How do accreditation and certification differ, and how might they contribute to a heightened level of professionalism in the field?
7 How has the Australian accreditation programme been changed most recently to take into consideration nature tourism? What might the positive and negative implications of this move be to the ecotourism industry in that country?

Part III

Topics and issues important to ecotourism

This final part includes a number of main topics and issues that are prevalent in the ecotourism literature. It begins with a discussion of the socio-cultural and ecological impacts of ecotourism. It may be argued that any type of tourism, including ecotourism, has impacts. It then becomes important to manage these impacts to be as minor as possible. The issue of impact is compounded by the fact that ecotourism involves long-haul travel and associated high usage of fossil fuels to satisfy hedonistic ends. The more people who travel, the bigger the problem, and this fact alone calls into question whether ecotourism can be truly ethical or responsible.

Continuing forward from the discussion on local participation and benefits in Part II, Chapter 9 takes a broader scale look at economics and marketing in ecotourism. Leakages, multiplier effect and specific examples of how revenue is used in parks and the value of ecotourism as compared to other land uses are central components of this chapter. This new edition also includes a section on demarketing as a technique that actively dissuades people from purchasing a product (as in visitation to parks and protected areas), for the purpose of maintaining socio-cultural and ecological integrity.

Chapter 10 focuses on development, governance and politics in ecotourism. As there is often an uneasy relationship that exists between the various stakeholders in tourism, including local people, tourism, government and protected areas, careful management through cooperative endeavours proves beneficial. A number of environmental governance models are discussed and these are matched with different case studies from the ecotourism literature. These different models on governance emphasise different roles and relationships in attempts to satisfy issues around shared involvement and inclusivity.

Chapter 11 provides guidance for practitioners (and academics) on how to build effective ecotourism programmes. Good definitions, lead to good policies, which in turn lead to good programmes. As these programmes are most often the face of ecotourism in the eyes of ecotourists, it is important to get these programmes right. It is only when there is consistency amongst the various agents involved in ecotourism – policy-makers, academics, practitioners and tourists – that ecotourism will work both in theory and practice. Aspects of planning, implementation and evaluation in programme planning are emphasised in a model that will hopefully help achieve the social, ecological and economic goals of ecotourism.

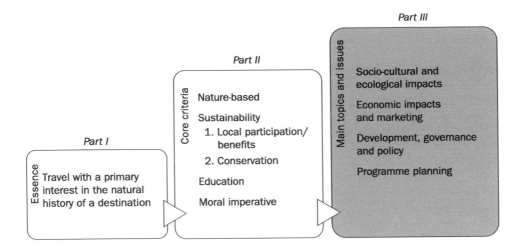

Figure P3 The structure of ecotourism

8 ▸ Socio-cultural and ecological impacts of ecotourism

Tourism research typically centres on topics related to the social, ecological and economic impacts of the tourism industry. Social impact studies involve an analysis of how the industry has affected local people and their lifestyles, whereas ecological studies have emphasised how the industry has transformed the physical nature of local and regional landscapes. Such studies seem to be in contrast to tourism economic research, which in most cases illustrate the income-generating power of the industry at many scales. Given that impact research is quite voluminous, it is not the purpose of the following discussion to provide a complete overview of research in these areas. Instead, this chapter focuses most extensively on issues related to ecological impacts, carrying capacity and, less specifically, on social impacts. Economics and marketing in ecotourism are the topic of the next chapter.

Social impacts of tourism

One of the most influential frameworks developed to analyse tourism's impact on local people is based on the work of Doxey (1975), who, in a general context, was able to encapsulate the evolving sentiment that local people express as tourism expands and occupies greater proportions of a local economy over time. Doxey wrote that there are essentially four main stages to consider in the assessment of local feelings towards the tourism industry. These include:

1   *Euphoria.* Tourists are welcomed, with little control or planning.
2   *Apathy.* Tourists are taken for granted, with the relationship between both groups becoming more formal or commercialised. Planning is concerned mostly with the marketing of the tourism product.
3   *Annoyance.* As saturation in the industry is experienced, local people have misgivings about the place of tourism. Planners increase infrastructure rather than limit growth.
4   *Antagonism.* Irritations are openly displayed towards tourists and tourism. Planning is remedial, yet promotion is increased to offset the deteriorating reputation of the destination.

There are myriad examples of regions that have been subject to this cycle in tourism (see also Butler 1980 later in this chapter). As a case in point, Bermuda experienced visitor numbers of some ten times its local population in 1980 (600,000 people) in an area approximately 21 mi^2. This type of tourist-to-local ratio is indicative of the conditions that have led to social conflict. Although visitation has its economic rewards, what the host country gives up to attract tourism dollars cannot be measured in economic terms. It is no accident that the most vital and creative parts of the Caribbean, for example, have been precisely those that have been most touched by tourism (Chodos 1977: 174). The oft-quoted claim of Evan Hyde, a Black Power leader in Belize in the early 1970s, that 'Tourism is whorism' (Erisman 1983: 339) reflects the frequent claims that tourism leads to conflict between locals and hosts. Such has been the case in the Okavango Delta, Botswana, where heightened levels of hotel, lodge and airport development from private external ownership have led to a loss of autonomy, sense of place, and declining access to local resources, as well as a climate of racism in the tourism industry based on the divide between the black population and white tourism operators, with the latter refusing to hire the former into top management positions (Mbaiwa, 2003).

A notable impact of tourism on traditional values is the demonstration effect (Britton 1977; Hope 1980; Mathieson and Wall 1982), where local patterns of consumption change to imitate those of the tourists, even though local people only get to see a side of tourists that is often not representative of their behaviour displayed at home (e.g. spending patterns). Alien commodities are rarely desired prior to their introduction into host communities and, for most residents of destination areas in the developing world, such commodities remain tantalisingly beyond reach (Rivers 1973). The process of commercialisation and commodification may ultimately erode local goodwill and authenticity of products, as identified by Britton (1977):

> Cultural expressions are bastardized in order to be more comprehensible and therefore saleable to mass tourism. As folk art becomes dilute, local interest in it declines. Tourists' preconceptions are satisfied when steel bands obligingly perform Tony Orlando tunes (and every other day the folklore show is narrated in German).
>
> (Britton 1977: 272)

This is the case in Zanzibar, according to Gössling (2002), who writes that young Zanzibaris clamour to identify with Western lifestyles by drinking beer, wearing sunglasses or adopting similar styles of clothing. The tourist lifestyle has thus gained superiority in Zanzibar, leading to a situation where tourists are allowed – even expected – to act inappropriately (e.g. topless bathing), and where local people are influenced to change their traditional ways in mirroring tourist behaviours. In many cases there is scepticism over the benefits of ecotourism. In Shenzha County, Tibet, residents are concerned that ecotourism development will destroy the natural environment of the region as well as disrupt folk customs and culture (Tang *et al.* 2012). Ryan *et al.* (2000) have explored this terrain in concluding that there is a culture of consumerism driving ecotourism culture, and that hedonism is seen to be more important than learning in the ecotourist experience.

The fragmentation of culture occurs on many levels within destinations, most notably from the standpoint of prostitution; crime; the erosion of language in favour of more international dialects; the erosion of traditions, either forgotten or modified for tourists; changes to local music and other art forms; food, in the form of a more international cuisine; architecture; dress; family relationships (e.g. young children earning more than their parents from toting bags at airports); and, in some cases, religion. In recognising the potential for social impact in a tourist region, Ryan (1991: 164) has identified a number of key points, all of which may be used as indicators or determinants of impact. These are as follows:

1   the number of tourists;
2   the type of tourists;
3   the stage of tourist development;

4   the differential in economic development between tourist-generating and tourist-receiving zones;
5   the difference in cultural norms between tourist-generating and tourist-receiving zones;
6   the physical size of the area, which affects the densities of the tourist population;
7   the extent to which tourism is serviced by an immigrant worker population;
8   the degree to which incoming tourists purchase properties;
9   the degree to which local people retain ownership of properties and tourist facilities;
10 the attitudes of governmental bodies;
11 the beliefs of host communities, and the strengths of those beliefs;
12 the degree of exposure to other forces of technological, social and economic change;
13 the policies adopted with respect to tourist dispersal;
14 the marketing of the tourist destination and the images that are created of that destination;
15 the homogeneity of the host society;
16 the accessibility to the tourist destination; and
17 the original strength of artistic and folkloric practices, and the nature of those traditions.

Central in the attempts to secure cultural resiliency is the recognition that both hosts and guests need to be sensitive to one another's way of life. Stronza (2001) observes that we need to understand the dynamics behind both tourists and residents, and their cross-cultural interactions, in appreciating the nature of tourism. While most research on the topic of socio-cultural impacts usually examines the impacts that tourists have on local residents, Stronza examined a number of studies where residents were found to take advantage of tourists. In one case, locals took pleasure in toying with tourists who were characterised as, 'relatively ignorant of local conditions, and thus often appear incompetent, ridiculous, gullible, and eminently exploitable' (Howell 1994: 152, cited in Stronza 2001: 273). This work resonates with research by Fennell (2006) on the theory of reciprocal altruism, and the cost–benefit relationships that fail to take place between residents and guests (refer back to Chapter 7). This dynamic has been further touched upon by Carrier and MacLeod (2005), who discuss the socio-cultural context of ecotourism from the perspective of an 'ecotourist bubble', not unlike Cohen's (1972) concept of the environmental bubble. The former is distinguishable based on the belief that the ecotourist's interaction with the destination – the purchase of ecotourism as a commodity – induces a type of ignorance that clouds the social relations that bring it into existence. The authors recount a discussion with an environmentalist who said that she was careful not to disturb any of the fragile plant life in her wanderings about Antarctica. The tourist was blind, however, to the consequences of the infrastructure and operations that got her to Antarctica and back to the northern hemisphere (discussed further below).

It is indeed an unfortunate reality that the least emphasised pillar in sustainable tourism research is the socio-cultural component (Robinson, 1999), which could be so important in framing our perspective on economics and environment. The socio-cultural component has fallen through the cracks because of a propensity to study the motives and behaviours of tourists only and to neglect the more passive recipients of the tourism industry (Chambers 1999, cited in Stronza, 2001). Sustainability in peripheral locations is also undermined because it is often wielded as an ideological tool which empowers those who support it, usually external agents, to take control over resources based on their own criteria of sustainability (Cohen 2002). Sustainability, argues Thaman (2002), must be rooted within the socio-cultural value set of a distinct group. Inserting external groups with foreign models and paradigms into a destination only serves to reinforce the differing goals and values that both parties live by.

It is also an unfortunate reality that sustainability comes only as a result of physical conditions that limit exploitation. This seems to be the case with the island of Niue (population: approximately 1900 in 1998) because of its poor quality beaches, remote location and high cost to get there. (See Orams 2002 for a discussion of the constraints to the whale watching industry in Tonga, tied strongly to isolation.) With only 1,729 visitors to Niue in 1998, competition from other islands in the region has prevented this nation from reaping larger-scale benefits from tourism (de Haas 2002). This prompted de Haas to conclude that if the tourism carrying capacity of the island was to be met or exceeded, a loss of authenticity, language, customs and dress would soon follow because of the emerging reliance on tourism and the inability to be resilient under conditions of change.

As ecotourism continues to diversify and exploit relatively untouched regions and cultures, there is the danger that a cycle of events similar to that identified by Doxey will occur. The lessons from the Caribbean model of tourism development, for example, are that the industry must tread lightly in securing an equitable relationship between how the industry is planned and developed and the needs of local people. Britton (1977) recognised the importance of small-scale, local architecture, tourism zoning, gradual growth, reliance on locally produced goods, joint ventures and a diversification in the market, in releasing the Caribbean from metropolitan domination. Armed with this knowledge, and experience, it is indeed encouraging to see that some have made a concerted effort to reclaim their cultural past in recognising how transformative the tourism industry can be. Hospers (2003), for example, writes that the intentional move away from industrialisation and mass tourism in Sardinia in favour of small-scale bottom-up ecotourism and cultural tourism has allowed Sardinians to reclaim their agricultural and cultural past. Having its identity removed in the post-World War II era through top-down industrialisation, tourism has infused a new energy and regional flair to Sardinia which has stimulated many innovative new products to support tourism.

Ecological impacts

Concern over the ecological effects of tourism mounted during the 1960s and 1970s (Pearce 1985), through the realisation that the industry had the capability of either moderately altering or completely transforming destination regions in adverse ways. The *National Geographic Magazine* as far back as the early 1960s (Cerruti 1964) was asking whether Acapulco had been spoiled by overdevelopment, while Naylon (1967) discussed the need to alleviate some of the stress caused by a high concentration of tourism in the Balearic Islands and the Costa Brava in Spain. Pollock wrote that although tourism had begun to play an important role in the economy of Tanzania, 'the vital necessity for game conservation in the interests of ecology, tourism, game farming and ranching, and for moral, aesthetic, philosophical and other reasons has been recognized increasingly both at national and international levels' (Pollock 1971: 147). Others have commented on the physical impacts of tourism in city and regional environments, including Harrington (1971), who observed that the unregulated development of hotels in London threatened the quality of life in the city, and Jones (1972), who makes reference to tourism development as a classic case of the battle that exists between conservation and preservation on the island of Gozo. Crittendon (1975) illustrates that while tourism has transformed much of the world's natural beauty into gold, the industry may have planted the seeds of its own destruction.

Sensitivity to environmental issues in tourism studies gained a tremendous boost in the mid-1970s from the efforts of Budowski (1976), Krippendorf (1977) and Cohen (1978). Budowski identified three different 'states' in tourism's relationship with environmental conservation: conflict, coexistence and symbiosis. He felt that tourism's expansion resulted in an unavoidable effect on the resources upon which it relied, and therefore felt that the relationship at the time was one of coexistence moving towards conflict. Krippendorf was one of the first to write on the importance of planning, and the dispersion of tourists and tourism developments, as a means by which to minimise impacts; while Cohen reviewed the work to date (academic and non-academic) on tourism and the environment. He speculated on the apparent 'mood of the day' by insisting that there was indeed a distinct difference between development for purposes of improvement and aesthetic appeal versus the vulgar, undesirable and irreparable damage created by modern tourism.

More research on the ecological impacts of tourism emerged in the early 1980s from Krippendorf (1982), who, like Budowski, recognised that the resource base acted as the raw material of tourism, which through improper use and overuse loses its value. Krippendorf cited ski-slopes, holiday villages, camping and caravan sites, and airfields as examples of developments that when fully functional seem to subsume the environment forever for their own uses. Travis (1982) suggested in his review of literature that while most studies on tourism concentrated on the economic benefits of tourism, there was also a tremendous range of topics related to its negative impact, including pollution, crowding and congestion, damage/destruction of heritage resources, land use loss, ecosystem effects, loss of flora and fauna and increased urbanisation. Concurrently, Coppock (1982) identified similar areas where tourism had an adverse impact on nature conservation in the UK. These were identified as loss of habitat, damage to soil and vegetation, fire, pollution, and disturbance of flora and fauna. In the 1980s, books started to emerge that dealt with the impacts of tourism, including Mathieson and Wall's (1982) work on economic, social and ecological impacts.

Tourism research on ecological impacts further intensified throughout the 1980s on the basis of a wealth of information surfacing on the relationship between tourism and conservation, and the need to address how best to overcome tourism's negative impacts. In a special edition on tourism in the *International Journal of Environmental Studies*, Romeril (1985) wrote that concern for the environmental impacts of tourism has come on the wings of a broader global concern over the conservation of natural resources generated by the United Nations Human Environment Conference of 1972, the World Conservation Strategy of 1980, the Report of the Brandt Commission (1980) and the Manila Declaration on World Tourism in 1980, which stated that:

> The use of tourism resources could not be left uncontrolled without running the risk of their deterioration, or even destruction. The satisfaction of tourism requirements must not be prejudicial to the social and economic interests of the population in tourist areas, to the environment and above all to natural resources which are the fundamental attractions of tourism and historical and cultural sites. All tourism resources are part of the heritage of mankind.
>
> (cited in Romeril 1985: 216)

In the same edition, Pearce (1985) reproduced a framework for the study of environmental stress that was established by the Organisation for Economic Co-operation and Development (OECD) in 1981, and included stressor activities, the pressure resulting from the activity, the primary environmental response and the secondary human response to stress. Four main examples were identified in this framework related to permanent environmental restructuring, generation of waste, tourist activities and effects on population dynamics, as shown in Table 8.1. The importance of understanding the constituents driving excessive levels of pressure in destinations is illustrated more recently in the case of the *touristification* (the number of tourists New Zealand can accommodate over a measurable time frame) of Queenstown, New Zealand (Page and Thorn 2002). Overcrowding and overdevelopment in Queenstown are placing a great deal of pressure on the local authority, which has prompted McLaughlin (1995: 90, cited in Page and Thorn 2002: 235) to suggest that:

> Queenstown is in danger of becoming so successful as a tourist resort that it risks losing itself as a town and irreparably damaging the landscape which not only draws its international clients . . . it's not change *per se* which frightens some residents, but the pace and magnitude of change and the location of development.

(See also Puppim de Oliveira 2003 for a table illustrating many of tourism's shocks on the environment; and OK 2006 for a description of 28 impact variables used to assess the pressure that ecotourism activities have on forests.)

One of the most complete overviews of the history of ecological concern in tourism was written by Shackleford (1985). His review of tourism and the environment suggests that the International Union of Official Travel Organisations, or IUOTO (the precursor to the WTO, now the UNWTO), had been working with the environment in mind since the early 1950s, through the efforts of the Commission for Travel Development. From 1954 onwards the protection of heritage was an agenda item for this organisation. Subsequent work by the IUOTO led to the recommendation by its Fifteenth General Assembly that world governments implement the following 1960 resolution:

> The General Assembly, considering that nature in its most noble and unchanging aspects constitutes and will continue increasingly in the future to constitute one of the essential elements of the national or world tourist heritage. Believes that the time has come for it to deal with the problems raised by the dangers threatening certain aspects of nature . . . Decides consequently to recommend to all IUOTO Member Countries to exercise increased vigilance regarding the attacks made on their natural tourist resources.
>
> (Shackleford 1985: 260)

Other examples of environmental impact research in tourism in the 1980s include work by Farrell and McLellan (1987) and Inskeep (1987), in a special edition of the *Annals of Tourism Research*. Their research suggests that planning and policy are critical components of a more ecologically based tourism development strategy for the future (more on policy in Chapter 10). For example, Inskeep (1987) writes that determining the carrying capacity of tourist sites is an important factor in the planning and design of appropriate tourist facilities, a concept around which Mlinarić (1985) built a discussion on tourism and the Mediterranean (more on carrying capacity below).

Up to and including the 1980s, few models had attempted to study tourism impacts from an ecological standpoint. This notion is reinforced by Getz (1986) who identified three ecologically based frameworks in an analysis of over 40 tourism models. These included a comprehensive model by Wall and Wright (1977), the OECD model mentioned above, and a unique model by Murphy (1983), who made an analogy between the tourism industry (locals, the industry and tourists) and predators and prey interacting within an ecosystem. Although Getz's work was completed some years ago, Dowling (1993) reports that little had changed up to the 1990s with respect to the creation or implementation of tourism development models from the environmental disciplines. Fennell and Butler (2003) point to the fact that because it is largely social scientists making inferences on ecological matters, there is much uncertainty with respect to the ecological impacts of tourism. They also point to the fact that there is little natural science research emerging

Table 8.1 _A framework for the study of tourism and environmental stress_

Stressor activities	Stress	Primary response: environmental	Secondary response: (reaction) human
1 _Permanent environmental restructuring_			_Individual_ – impact on aesthetic values
(a) Major construction activity • urban expansion • transport network • tourist facilities • marinas, ski-lifts, sea walls	Restructuring of local environments • expansion of built environments • land taken out of primary production	Change in habitat Change in population of biological species Change in health and welfare of man Change in visual quality	_Collective measures_ • expenditure on environmental improvements • expenditure on management of conservation • designation of wildlife conservation and national parks • controls on access to recreational lands
(b) Change in land use • Expansion of recreational lands			
2 _Generation of waste residuals_ • urbanisation • transportation	Pollution loadings • emissions • effluent discharges • solid waste disposal • noise (traffic, aircraft)	Change in quality of environmental media • air • water • soil Health of biological organisms Health of humans	_Individual defensive measures_ local residents • air conditioning • recycling of waste materials • protests and attitude change towards tourists • change of attitude towards the environment • decline in tourist revenues _Collective defensive measures_ • expenditure on pollution abandonment by tourist related industries • clean-up of rivers, beaches
3 _Tourist activities_ • skiing • walking • hunting • trail bike riding • collecting	Trampling of vegetation and soils Disturbance and destruction of species	Change in habitat Change in population of biological species	_Collective defensive measures_ • expenditure on management of conservation • designation of wildlife conservation and national parks • controls on access to recreational lands
4 _Effect on population dynamics_ Population growth	Population density (seasonal)	Congestion Demand for natural resources • land and water • energy	_Individual_ • Attitudes to overcrowding and the environment _Collective_ • Growth in support services, e.g. water supply, electricity

Source: Pearce (1985)

from the tourism journals to aid in the continuing struggle to come to grips with the tourism impact dilemma, with the result being that impacts are often anticipated but not controlled (see also McKercher 1993b).

An excellent addition to the literature on the environmental impacts from tourism comes from Newsome *et al.* (2002), who identify a whole range of different types of impacts, their sources and regions in which these take place. The authors note that sources of impacts cited in the literature include trampling (vegetation, microbes, soils) access roads and trails, built facilities and camp grounds (camp sites and firewood) and water edges (river banks, lakes and reservoirs, coastal areas and coral reefs). The book recognises the importance of looking at impacts from a bio-geographical perspective, by identifying specific ecoregions, including mountains, caves, arctic-alpine environments, tropical realms and arid environments. Other works have identified a range of recreational activities and their associated positive and negative impacts along the following lines: habitat change/loss, species change/loss, aesthetics, physical pollution, soil change/ damage, noise pollution, conflicts, energy/water usage, local community and revenue versus costs (see Tribe *et al.* 2000). Weaver and Lawton (2007) argue that a large percentage of studies on the ecological impacts of ecotourism focus on the effects that the ecotourism industry has on wildlife. And in these studies, distance between ecotourists and wildlife is the critical variable affecting increased levels of stress on fauna. The following few examples serve to illustrate the nature of these impacts.

A persistent problem in ecotourism is justifying it as a more ecologically sound practice despite the fact that it, like mass tourism, often involves long-haul travel and associated high usage of fossil fuels to satisfy hedonistic ends. In this regard, Gössling (2000) argues that far from being the low-impact and non-consumptive development option that it is often advertised as, tourism-related use of fossil fuels has an overall significant and detrimental impact. This is especially true of lesser developed countries (LDCs) which rely on long-haul travel from the industrialised nations. Gössling says for a two-week package tour in a LDC, the country itself is responsible for 24 per cent of fossil fuel use (ground transportation, cooking, cleaning, cooling, heating, and so on), with the rest (76 per cent) coming from air travel – which contributes almost 90 per cent of the trip's overall contribution to global warming (in consideration of nitrogen oxides). In related research, Lynes and Dredge (2006) have identified four key environmental issues, including air emissions, noise emissions, congestion and waste, that stigmatise the airline sector. This stigmatism has motivated airline companies to institute tougher environmental management systems in generating more public confidence (and more market share!) in the sector. Ecotourism, therefore, may become more acceptable if airline company 'A' can demonstrate higher environmental protocols than 'B', 'C' or 'D', in the same way that a hybrid car sends the same 'responsible' message to automobile consumers.

Newer approaches are being investigated to assess the carrying capacity and impacts of ecotourism. Alam (2012) looked at the impacts of ecotourism through an assessment of the CO_2 emissions of visitors to three forests offering opportunities in the UK. The estimated CO_2 sequestration and emissions were plotted against visitors in estimating the maximum allowable visitor number (MAVN). Numbers of ecotourists beyond the MAVN indicated unsustainable and carbon-intensive ecotourism. Alam found that the New Forest was unsustainable and carbon-intensive, Cwmcarn was vulnerable, and Coed Y Brenin was sustainable based on the results of the model. As CO_2 emissions is one of the sticky points of ecotourism; that is, opponents argue that ecotourism as a sustainable enterprise will always be constrained by the fact that ecotourists fly long distances to get to their destinations of choice, this method has strong potential for safeguarding sustainability based on numbers of visitors.

Although often touted as a non-consumptive and non-invasive form of tourism that generates significant community economic benefits, SCUBA diving has been described as ecologically destructive (Hawkins *et al.* 1999) or relatively benign. Badalamenti *et al.* (2000) observe that the creation of marine protected areas in the Mediterranean has increased the number of divers using the area and contributed to a series of benthic impacts from the activities of divers and boat moorings. In order to measure the effects of diving, Walters and Samways (2001) during almost 15 hours of observation witnessed 129 accidental, 38 deliberate (but non-anchoring) and 55 anchoring contacts (i.e. holding on to something to remain steady in the water). Most of the accidental contacts were from fin kicks (73.6 per cent), with much less from instruments, knees, hands or elbows, and with only 1.6 per cent of these contacts resulting in any discernable damage. Different amounts of contact with the substratum were detected for inexperienced, experienced and very experienced divers, along with photographers. Research has shown that briefing divers on environmental awareness and appropriate behaviour can reduce the level of contact with coral reefs. With reference to the study site, the authors determined that the reef could sustain increased numbers of divers based on the levels of contact, and damage, detected.

In a study of the participants of a range of nature-based tourism activities in Australia (swimming, boating, fishing, diving, windsurfing, sandboarding, four-wheel driving, camping, bushwalking, horseback riding and sightseeing), Priskin (2003) found that although tourists were aware of environmental impacts from such activities, they perceived the impacts of these activities to be *less* than the perception of the researcher (Priskin herself). Furthermore, those who participated in some activities deemed more harmful, like fishing and four-wheel driving, perceived the activity to be less harmful than the perceptions of non-participants. Based on these results, Priskin argues that much more education on the potential impacts of nature-based tourism activities is needed to help minimise environmental impacts. (See also Daigle *et al.* 2002, for an overview of how wildlife viewers and hunters differed in their beliefs about the benefits derived from their respective activities.) In a related study, Nyaupane and Thapa (2004) found discrepancies between the perceptions of negative and positive impacts from ecotourism as compared to traditional tourism in the Annapurna Conservation Area Project of Nepal. While ecotourism may be theorised as minimising negative impacts and maximising benefits for local people, the authors discovered that residents of an ecotourism area perceived fewer negative and positive impacts from tourism than the traditional area for tourism. More specifically, while the ecotourism area was in fact perceived to minimise negative impacts, there is the belief that economic benefits are not maximised. In contrast, other forms of traditional NBT were thought to incur more negative environmental impacts, but at the same time generate more money for the region. These studies point to the importance of better understanding the psychological and behavioural impacts of ecotourism (Powell and Ham 2008).

Barter *et al.* (2008) examined the differences in the behaviour of Penguin Island pelicans (*Pelecanus conspicillatus*) at two different stages of incubation when approached by a researcher according to three measures: (1) behaviour during the approach; (2) pre-flight initiation distances; and (3) behaviour after the approach. The authors found there were major behavioural changes elicited by the penguins, and also that habituation was observed over the short term. The authors recommend that people stay at least as far back as the longest pre-flight distance recorded during the most sensitive phase of the breeding season.

Similar research has been conducted in New Jersey on the impacts that ecotourists have on a variety of bird species in this region. Burger *et al.* (1995) report that birds are not consistent in their responses to human intrusions, and identify ecotourists as having the potential to disturb birds at all times of the year. According to the authors, this is a result of the fact that ecotourists are interested in the breeding, wintering and migration patterns of birds. For this reason, they have the potential to interrupt incubation, scare parents and chicks from nests, disturb foraging, disrupt the prey-base, force birds away from traditional habitats such as beaches, forests and open fields, trample vegetation and overuse trails. Burger *et al*. felt that ecotourists and birds can coexist but only as the result of careful management of the resource, where each setting and species demands careful study and monitoring. They suggest the use of the following measurements (Burger *et al*. 1995: 64): (1) *response distance*, the distance between the bird and the intruder at which the bird makes some visible or measurable response; (2) *flushing distance,* the distance at which the bird actually leaves the site where it is nesting or feeding; (3) *approach distance*, the distance to which one can approach a bird, head-on, without disturbing it; and (4) *tolerance distance*, the distance to which one can approach a bird without disturbing it, but in reference to passing by the bird tangentially.

In the Great Barrier Reef World Heritage Area (GBRWHA) there is concern that the dwarf minke whale is too curious. Mangott *et al*. (2011) have investigated the relationship between dwarf minke whales and the tourism industry and found that the whales' tendency to approach swimmers and boats, although attractive to tourists, may be a problem in the making. There is concern that the absence of caution on the part of whales will lead to boat strikes and entanglement in nets. And for tourists, there is concern that sooner or later there will be an injury or even a death because of this close interaction. Management options include: (1) bans; (2) space-time closures to tourism operators; and (3) regulation and education of activities. The second option is highly recommended by the authors.

There is also concern over the protection of burrunan dolphins at Port Phillips Bay, Australia. Howes *et al*. (2012) assessed the effectiveness of the Ticonderoga Bay Sanctuary Zone, designed to give resident dolphins an area of refuge from various anthropogenic stresses, including tourism. The authors observed 104 swim-with-dolphin tours and found that tour operators contravened minimal approach distance regulations in all observed encounters. Furthermore, operators did not in fact exercise caution when encountering dolphins in the sanctuary zone itself. Recommendations were made for the move away from passive management strategies to harder core management approaches that include enforcement of transgressions in minimising impact to the dolphins.

Higham and Bejder (2008) write that evidence points to the fact that dolphins at Shark Bay, Western Australia, are not as successful in mating as in other adjacent areas because of the abundance of research and tour boat vessels, even though the industry is well managed. In response to these issues and related concerns, the Minister of the Environment, based on detailed consultations with stakeholders, decided to reduce the number of commercial dolphin-watch licences, and introduce a moratorium on any increase in research activity in the region. Higham and Bejder liken this to a paradigm shift in the commercial tourism industry, in light of efforts to secure the long-term and sustainable future of the industry.

Although the vast majority of work examines impacts from the negative context, and justifiably so, policy-makers and academics need to be aware of the fact that impacts occur along a continuum, and that these are not discrete occurrences but rather determinable along social and ecological lines. The following section on carrying capacity serves to illustrate this point.

Carrying capacity

Increasingly, researchers and practitioners have begun to recognise the dangers inherent in accommodating an increasing number and diversity of experiences for a growing consumer-based society. It is in an agency's best interests to be aware of and sensitive to the broad range of different user groups in a setting and their various needs. Over time, managers have learned that sound planning and development of public and private lands must be viewed as the best way to ensure the safety of the resource base first, even over the needs and expectations of participants. These types of issues have been raised and debated extensively through the literature on carrying capacity.

The concept of carrying capacity is not new. Butler *et al.* (1992) argue that for some time people have worried about their excessive use upon stocks of game and other renewable resources, as suggested by this sixteenth-century poem:

> But now the sport is marred,
> And wot ye why?
> Fishes decrease,
> For fishers multiply.

In the strictest ecological sense, species maintain a balance between birth and death, and predator–prey relationships within an ecosystem. It is the human factor and the manipulation and exploitation of resources that offset this balance. Generally speaking, the concept of carrying capacity can be loosely defined on the basis of the following four interrelated elements: (1) the amount of use of a given kind; (2) a particular environment can endure; (3) over time and; (4) without degradation of its suitability for that use.

In the early 1960s the concept was applied recreationally for the purpose of determining ecological disturbance from use (Lucas 1964; Wagar 1964). However, it was quickly discovered that an understanding of ecological impact might be achieved only through the consideration of human values, as evident in the following passage:

> The study . . . was initiated with the view that the carrying capacity of recreation lands could be determined primarily in terms of ecology and the deterioration of the areas. However, it soon became obvious that the resource-oriented point of view must be augmented by consideration of human values.
>
> (Wagar 1964: 23)

Typically, environmental impacts can be objectively measured through an analysis of ecological conditions, as noted above. In the outdoor recreation literature, a value judgement has been placed on the term 'impact', denoting undesirable change in environmental conditions (Hammitt and Cole 1987). Concern lies in understanding the type, amount and rate of impact on the resource base through recreational use. A campsite, for example, may be severely impacted over time by accommodating high levels of use. Significant changes may occur to the ecology of the site as evident through the compaction of soil (e.g. exposing roots and increasing erosion), vegetation (e.g. using both dead and live tree limbs for the construction of fires, and trampling saplings), wildlife (e.g. habitat modification and animal harassment) and water (e.g. the addition of human waste and chemical toxins to the aquatic environment). The heaviest impact to a campsite, however, occurs during the first couple of years of use, and impact subsides over time as the site becomes hardened (see Figure 8.1). These data provide strong evidence to suggest that new campsites should not be developed, and that the use of existing ones ensures the least amount of disruption to the resource base.

From the sociological perspective, carrying capacity becomes much more dynamic and difficult to measure. The complications arise when considering the level or limit to the amount of use which is appropriate for a specific resource. Owing to the nature of the

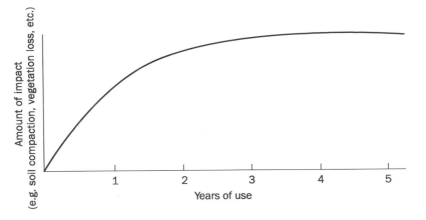

Figure 8.1 Impact on recreation sites

Source: Hammitt and Cole (1987)

resource as a subjective, perceptual entity, different types of users will have different resource needs and expectations. Consequently, the tolerance of these user groups (e.g. jet-boaters and canoeists) to one another will vary. To compound the matter further, the tolerance of individuals within groups (intragroup tolerance) will also vary. To take canoeists as an example, each individual within this recreational group will also have certain expectations. Encounters with other canoe parties (or other user types), the density of use (the number of users per unit area) and the perception of crowding (the behavioural response to such encounters) will differ for these individuals over space and time.

Plate 8.1 Wilderness users are wise to use existing campsites rather than create their own in relatively untouched park regions

Plate 8.2 The impact of park users on the environment takes many forms

Researchers and managers have argued consistently that the goal of recreation management is to maximise user satisfaction (Lucas and Stankey 1974). Despite this agreement, past research has generally failed to document the empirical relationships between use levels and visitor satisfaction deemed necessary for the development of evaluative standards for the management of a resource. Shelby and Heberlein (1986) measured perceived crowding and satisfaction through the importance of use levels and encounters in their analysis of river rafters, canoeists, tubers (people who float down rivers on rubber tubes), fishers, deer hunters and goose hunters in western USA. Use levels provided an objective measurement that evaluated how many people were using the resource. Encounters were determined by having a researcher follow groups and count the number of contacts they had with others, or by simply asking users to report contacts with others. The authors hypothesised that:

1 As use levels and encounters increased, perceived crowding would increase.
2 As use levels and encounters increased, satisfaction would decrease.

They found that higher use levels (the number of people using a resource) do not always make people feel more crowded. There was a stronger relationship between contacts and perceived crowding. Generally, people felt more crowded as contacts increased for all activities except rafting when compared with use levels. This is expected because the number of people one actually sees should have a greater impact than the overall number using the area. Crowding means too many people, but use levels and contacts do not entirely explain feelings of crowdedness. (See Musa *et al.* 2004 for a discussion of the problems that crowding has created in achieving sustainable tourism in Nepal's Sagarmatha National Park. See also Singh and Mishra (2004) for a description of ecotourism in Manali

Plates 8.3 and 8.4 Those who maintain a strict definition of ecotourism would suggest leaving plants and animals alone altogether, which means a refusal to pick them up

of the Himachal Himalaya). Tourism in the region has increased by 270 per cent per decade over the last 30 years contributing to an abundance of undesirable environmental impacts in the absence of good planning.

Shelby and Heberlein also operationalised use level and encounter variables to test whether or not satisfaction decreased with increasing levels of use. Results suggest that recreationists were just as satisfied at high use levels as they were at low use levels. In fact, in all cases low-use-level visitors were not significantly more satisfied than high-use-level visitors. A number of authors, including Shelby and Heberlein (1986), Pitt and Zube (1987), Stankey and McCool (1984) and Graefe *et al.* (1984a), indicate that the weak relationship between satisfaction and perceived crowding occurs for a number of differing normative/perceptual reasons. They offer the following as explanations for this poor relationship:

- *Self-selection.* People choose recreational activities they enjoy and avoid those they do not. There is an expected high level of satisfaction, regardless of the use level, because people will select experiences they will enjoy.
- *Product shift.* Users may change their definitions of recreation experiences to cope with excessive encounter levels. As a result, they may remain satisfied as contacts increase. In addition, the contacts themselves may play a role in changing the definition of the situation (hikers seeing more people in the wilderness may change the definition of the experience).
- *Displacement.* Individuals who are truly sensitive to high-density relationships may have already moved out of the environment being studied to a less intensively used area, being replaced by those less sensitive to high density.
- *Multiple sources of satisfaction.* Satisfaction is a broad psychological construct. The number of other people is only one of many things that might affect satisfaction or dissatisfaction.
- *Rationalising.* Recreationists may make the best of even a bad situation, focusing on positive aspects and minimising those that are less pleasant. People who complain about the number of others on a river are still likely to have a good time by learning to ignore the negative aspects of seeing others.
- *Activity-specific influences.* Response patterns to contacts with others may vary according to the types of activities and behaviour encountered. An individual may be quite tolerant of contacts with hikers and extremely intolerant of contacts with off-road vehicles. The extent to which one type of use impacts another depends upon the social and personal norms visitors use to evaluate the appropriateness of specific behaviours.
- *Conceptualisation and measurement of satisfaction may be inadequate.* The multidimensional character of experience, by definition, makes the likelihood of high correlations between a unidimensional overall satisfaction scale highly unlikely. Research is beginning to show that people can be satisfied and dissatisfied with their experience at the same time. Graefe *et al.* (1984b) found that 71 per cent of visitors to a Recreation Wilderness Area considered their trip excellent or perfect. However, 41 per cent also included the comment that they experienced at least one dissatisfying incident during their visit.

The above seven variables illustrate that the measurement of an individual's level of per-ceived crowding/satisfaction is difficult to attain. Recreationists may adjust to a dissatisfying situation through a product shift, adapt to the situation, rationalise the experience, or displace entirely from the site. The social and personal normative values that an individual might use to evaluate a site are unique and specific. This, coupled with inadequate measures of user satisfaction, may create a tremendous void between what managers feel they know about human-resource relationships and what they do not.

Although carrying capacity provides a quick, easy and inexpensive means to manage protected areas, or other units, it suffers from a range of problems that render it less useful. Farrell and Marion (2002) detail these shortcomings as the inability to: (1) assess and minimise visitor impacts; (2) consider multiple underlying causes of impacts; (3) facilitate different management decisions; (4) produce defensible positions; (5) separate technical information from value judgements; (6) encourage public involvement; and (7) incorporate local resource uses and management issues (see also Hovinen 1981; Wall 1982; Stankey and McCool 1984; Haywood 1986; O'Reilly 1986). The findings of Butler *et al.* (1992), in an extensive review of literature, concur that the concept of carrying capacity requires adept management. No mythical figure exists for limiting the amount of use in an area; rather, different cultural and natural areas have different capacities. Instead, research has leaned more in the direction of normative values in understanding the needs of different types of users. Normative approaches provide information on specific user groups about appropriate use conditions and levels of impacts related to individual activities. In doing so they provide information (either qualitative or quantitative) which may be used by natural resource managers to establish management standards (Shelby and Vaske 1991). For example, it is not necessarily acceptable to suggest that 413 people be allowed to use a park over the course of a weekend. Although many parks and protected areas still maintain a numerical limit in controlling numbers, it becomes the task of the park manager to know the levels of expectations, satisfaction, dissatisfaction and crowding of different types of users (i.e. motorboaters are likely to be able to withstand more use of the resource than canoeists, while canoeists would probably perceive the appearance of motorised craft as a threat to their experience).

The job of managing services and activities at a site, therefore, becomes a significant task. Park personnel must be receptive to queues – not only from the physical resource base (e.g. plant trampling and garbage) but also from visitors – when establishing regulations of where and what people can do. Pitt and Zube (1987) illustrate that once a resource manager has determined that the implementation of some form of recreational use limitation is necessary any of three overlapping courses of action need to be considered.

- *Site management techniques.* Site management techniques focus on improving the environment's ecological capacity to accommodate use. This involves surface treatments (soil management) designed to harden the site where use occurs, and includes approaches that channel circulation and use into more resilient parts of the environment. Also, capital improvements may be developed in underutilised portions of the environment to draw people out of overused areas.
- *Overt management approaches.* Overt management approaches aim at direct regulation of user behaviour. They take several forms: (1) spatial and/or temporal zoning of use (decreasing conflict of incompatible uses such as cross-country skiing versus snowmobiling); (2) restrictions of use intensity (decreasing the number of users in the environment through the closing of trails); and (3) restrictions on activities/ enforcement of user regulations.
- *Information and education programmes.* An alternative to heavy-handed overt methods: (1) Informing users about the recreational resource, and current levels of use; (2) making the users more sensitive to the potential impacts their behaviours might have on the environment; and (3) giving the manager and the users a chance to exchange information concerning user needs and management activities (e.g. brochures

to describe entry points to users and usual intensity of use of different trails in order to distribute users more widely).

The regulation of visitor behaviour is a common approach to addressing management problems at recreation sites (Frost and McCool 1988). Such regulations often go beyond prohibitions on litter, alcohol, noise and so on, and directly restrict what tourists can do at a site, where they may go, and how many may be in an area at a certain time (overt management approach). Therefore, a tourist who wants to maintain a high degree of internal control might perceive the level of regimentation as too high within a certain opportunity, thus eliminating that alternative from consideration (see Chapter 10 for more information on regulation).

In a study of Glacier National Park, rangers were given the task of managing visitors to this region as well as protecting the eagles as an endangered species (each autumn recreationists come to view feeding bald eagles). Restrictions on use included prohibitions against entry where the eagles congregate, restrictions on automobile movement and parking, and close-up viewing available only at a bridge and a blind, but only with the accompaniment of a naturalist (acting as an interpreter and distributing brochures to visitors). With this in mind, the goals of the research were to understand how visitors responded to the current level of restrictions on behaviour, and how such factors as knowledge of the rationale for restrictions influenced these responses. The authors found that 88 per cent of the visitors said they were aware of the park's restrictions, and almost 90 per cent of these visitors felt that such restrictions were necessary, with only about 3 per cent feeling that they were not. Of the visitors who were aware of restrictions, 56 per cent felt that these had no significant influence on their experience. Almost 32 per cent felt that restrictions facilitated their experience, and 12 per cent felt that restrictions detracted from the experience. When such restrictions were correlated with the concept of protecting the eagles, results indicated that visitors overwhelmingly support closures that minimise negative impacts on eagles. Only 4 per cent of visitors perceived the opportunity to view eagles as a higher priority than eagle protection.

This study illustrates that visitors may have prior expectations for a certain degree of social control. The authors felt that visitors were likely to view management actions as acceptable and the regulations as enhancing attainment of certain outcomes, such as learning about nature (Frost and McCool 1988). Visitors who viewed the restrictions as unacceptable may ultimately be displaced. In addition, visitors were further impressed because they knew where and why closures and restrictions applied. This fact verifies the importance of the interpretive programme as a complement to management actions that regulate visitor behaviour (see Chapter 6 for more information on interpretation).

The Butler Sequence

One of the most notable uses of carrying capacity in the tourism literature was developed by Butler (1980), who modified the product life-cycle concept to apply to the life cycle of tourist destinations (Figure 8.2).

Butler's basic premise was that increases in visitation to an area can be followed by a decrease in visitation as the carrying capacity of the destination is reached. Destination areas are said to undergo a fairly uniform transformation over time, from early exploration and involvement through to consolidation and stagnation, as the structure of the industry changes to accommodate more visitation and competing resorts. The implications of this research are such that planners and managers need to be concerned with any sustained decline in the ecological quality of the destination, as this will ultimately spell the demise of the development due to waning attractiveness. This is a good example of a

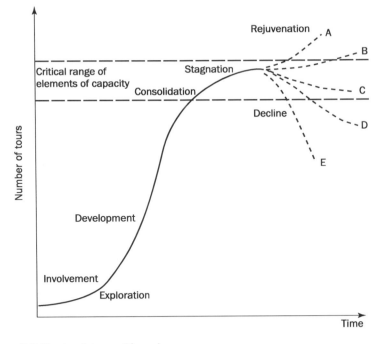

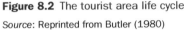

Figure 8.2 The tourist area life cycle

Source: Reprinted from Butler (1980)

conceptualisation that applies to the social, ecological and economic implications of tourism in a particular setting.

Researchers have focused on deriving empirical measurements of the evolution of a destination, especially island environments (Meyer-Arendt 1985; Cooper and Jackson 1989; Debbage 1990; Weaver 1990). The utility of the life-cycle concept has implications in delineating carrying capacity limits, and the social and environmental complications of 'overusage' in tourism destinations. Clearly defining the nature and characteristics of use of these areas must be a priority.

The Galapagos Islands of Ecuador is a case where carrying capacities have been considered as a means by which to control impact through the limitation of numbers of tourists on a yearly basis. The problem identified in the Galapagos is that despite the limitations on numbers of tourists visiting the islands, visitation annually increases beyond these limits because tourism is seen as the solution to the economic despair in this developing country. De Groot (1983) and Kenchington (1989) call attention to the fact that: (1) patrol boats do not always control tourism numbers on the islands effectively; (2) the official limit of 90 tourists on an island at a time is often overlooked; and (3) the number of tourists is still increasing. These researchers suggest that tourism numbers have been controlled ineffectively and inappropriately through airport capacity limits rather than by limits set in accordance with ecosystem sensitivity defined by park planning and management. Thus, even in well-known and highly significant areas, problems of overuse and visitor management still arise. Wallace (1993) feels that it is the growth of the private sector which has been instrumental in dictating the course of action in the Galapagos. Park officials have found it difficult to enforce levels of acceptable use, zoning and the distribution of permits owing to understaffing and other, broader, political issues. The result is that park managers do not feel as though they are in charge of the operations of the park (see Case Study 8.1).

CASE STUDY 8.1

Ecotourism in the Galapagos Islands

Located 1,000 km off the west coast of South America, and straddling the equator, the Galapagos Islands, comprised of some 120 islands, is one of the world's most iconic eco-tourism destinations. This stems from the rich history of the islands which played host to the English naturalist Charles Darwin on his epic journey aboard HMS *Beagle* from 1831–36. Darwin and Alfred Russel Wallace (who was on the other side of the world in the Malay Archipelago) both almost simultaneously developed the theory of natural selection. The observations which Darwin made in the Galapagos Islands, particularly on the different species of finches, were instrumental in his formulation of the theory. Due to the international significance of Galapagos, the archipelago was designated as a wildlife sanctuary in 1935, but this was not enforced until much later in 1959, a hundred years after the publication of *Origin of Species*. Twenty years on, the islands were classified as a UNESCO World Heritage Site and a marine park was established in 1986. Throughout the 1960s, 1970s and 1980s, tourism steadily grew in Galapagos, consistent with the international growth of ecotourism, to the point where researchers began to call attention to the seriousness of environmental impacts from tourism on the region.

De Groot (1983) was one of the first to describe the negative impacts of tourism on the Galapagos in the academic literature. Among his many observations as a guide over a two-year period, he felt that the system of two guides which had developed in the Galapagos was ineffective. The educated, bilingual naturalist guides he felt were valuable, while the auxiliary guides who are not required to have extensive training and speak multiple languages often do not follow regulations. In the late 1980s, Kenchington (1989) noted that tourism to the islands was doubling every five years. This level of growth was encouraged by the Ecuadorian government, which refused to set limits on growth because of the importance of foreign exchange. This led to the development of two types of tourism in Galapagos: one which places heavy emphasis on nature and natural resources (ecotourism); and the other which is poorly organised, basic, unregulated and run by individuals who have little respect for biodiversity (mass tourism). This latter form of tourism in particular is placing a great deal of strain on the natural history of the islands.

The main reason for the escalating numbers of tourists is money. With a 40 per cent poverty rate in Ecuador, the Galapagos Islands have emerged as a gold mine for Ecuadorians. Ecotourism is thus seen as one of the strongest growth industries in the country. In 1997, the Galapagos provided income for an estimated 80 per cent of people living on the islands. Although dated, 1991 figures illustrate the value of tourism to the Galapagos to be about US$32 million, which had jumped to US$60 million by 1996. However, the open-entry philosophy of Galapagos tourism has already proved to be problematic. Price competition has led to a transfer of economic returns to foreign companies rather than to local people. Cruise ships, which are largely owned by foreign interests, are making the most money with very little of this going back to the community. Furthermore, the jobs required to support ecotourism call for a high level of skill. Naturalist guides, as noted above, must be able to speak two or three languages and have specialisation in natural history. This has proved difficult to attain for Ecuadorians and thus created the demand for guides from other countries. These issues, coupled with the development of two airports in the region, have paved the way for unprecedented numbers of tourists (over 100,000 as of 2006, http://www.livetravelguides.com/south-america/ecuador/the-galapagos-islands/galapagos-ecotourism-and-manag/; accessed 8 December 2006) to

continued

help make these new developments cost effective, when many experts in the past suggested that half that number was an appropriate carrying capacity for the islands.

More specifically, a number of environmental problems from tourism have been identified by commentators. These include albatross, sea lions and turtles swallowing plastic bags (thinking they are jellyfish), the harvest of black coral for souvenirs, garbage found throughout many islands and coves, the introduction of sport fishing as an economic alternative under the guise of catch-and-release conservation, the introduction of non-native species, distress to nesting bird colonies, erosion and overuse of some trails. The human impact is also a concern, as suggested by Honey (1999), who observes that the annual growth rate of humans on the island is sometimes as high as 10 per cent, all of whom occupy 3 per cent of the archipelago. Coupled with a culture of diminishing respect for the natural history of the islands, the increased demand for new infrastructure is indeed significant. Most pressing, however, according to Honey (1999) and Atwood (1984), is the introduction of non-native species which have direct and indirect impacts on plants and animals that have no natural defences.

Some authors have observed that there is cause for optimism in this region. The development of a special law (in 1998) for the conservation of the Galapagos, with representatives from government, non-governmental organisations (NGOs), industry and tourism, was designed for the long-term protection and profitability of the archipelago. Honey (1999) reports that this legislation is designed to support residents, stabilise populations on the islands, set aside an additional 2 per cent for human settlement, set aside more land for conservation, extend the zone of protection in the ocean, and ban industrial fishing for specific species. Although the law does not regulate tourism, it does limit tourism infrastructure (but not tourism numbers); it gives rights to local people to become involved in the tourism industry and stresses the importance of environmental education in schools. The park entrance fee of US$100 is one of the highest anywhere in the world and this money, along with funds donated to the conservation fund, has helped to stabilise many park programmes. In addition, the recent certification of many large tour vessels that carry passengers among the Galapagos Islands, under the SmartVoyager programme, has provided more optimism for the region. This voluntary programme, which is a joint venture between the Rainforest Alliance and the Corporation for Conservation and Development (CCD), an Ecuadorian group, is designed to support tour groups that tread lightly on this ecosystem. Boats are evaluated on a number of bases, including wastewater, fuel, docking, and minimising the introduction of foreign species.

Hoyman and McCall (2013) have recently documented the successes and challenges of the special law in reference to ecotourism. Based on interviews with community leaders, they report that there is widespread optimism and support of the laws, but there are many impediments to moving forward. Implementation has been hampered by weak institutions in the Galapagos, a lack of leadership on many different levels, the political nature of decision-making itself, and by the failure to enforce provisions of the Law. Ecotourism in the Galapagos, they argue, is the stock case of how delicate and intricate the balancing act can be between conservation and economic development.

Websites on Galapagos ecotourism:
http://www.columbia.edu/cu/sipa/PUBS/SLANT/FALL98/p26.html
http://www.ecotourism.org/observer/062001/certification.asp
http://www.american.edu/TED/GALAPAG.HTM
http://www.rainforest-alliance.org/news/archives/releases/sv-tours.html

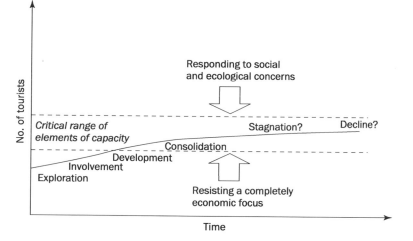

Figure 8.3 A sustainable ecotourism cycle of evolution?

Such has been the case in the Maldives where tourism pressure has been measured according to solid waste disposal (physical capacity), water quality (environmental capacity) and tourist perceptions (psychological carrying capacity), according to Brown *et al.* (1997). Through a survey of tourism resorts, tourists and interviews with officials, the authors found that degradation was a function of a rapid rise in visitor numbers. The response by government was to expand facilities and disperse tourists through more development in different areas, but this has not led to any decline in environmental degradation suggesting that the carrying capacity of the region has been exceeded.

The general nature of Butler's model has rendered it applicable to any tourism setting. The model's significance lies in the fact that in many cases tourism properties suffer from this sequence of rapid development and later decline, just like commercial products in general. Given the preceding discussion on sustainable tourism and alternative tourism (AT) (Chapter 1), it is worth while reconceptualising Butler's model by taking into consideration how such a cycle would, *or rather should*, proceed under ideal hypothetical sustainable tourism conditions. Figure 8.3 attempts to do this, emphasising the relative importance of economic, social and ecological variables in establishing reasonable and long-term levels of carrying capacity within ecotourism destinations. The model illustrates that destination areas will respond to the competing economic and social and ecological demands in ways that respect the integrity of the resource base and local inhabitants. The overall level of visitation is intentionally kept below the identified level of acceptable use, over the long term, with potentially minor increases in use consistent with the ability of the environment to absorb such increases. Price mechanisms would therefore be implemented to ensure that acceptable financial gains are realised from the enterprise.

Preformed planning and management frameworks

Given the constraints of the carrying capacity concept discussed above, theorists have been active in developing a series of preformed planning and management frameworks designed with the purpose of matching visitor preferences with specific settings in parks and protected areas. The ultimate aim of these frameworks is the protection of the resource base, but also to ensure that people are able to enjoy their recreational experiences in managed settings. Examples of these models include the Recreation Opportunity Spectrum (ROS), Limits of Acceptable Change (LAC), the Visitor Impact Management (VIM) process and the Visitor Activity Management Process (VAMP) (see Payne and Graham 1993 for a good description of these frameworks).

Despite the relative success of these models in the realm of outdoor recreation management, there has been only a gradual use and acceptance of these frameworks by tourism researchers. This has generally been the result of the fact that these frameworks have not been developed specifically for tourism. In response to this deficiency, Butler and Waldbrook (1991) adapted the ROS into a Tourism Opportunity Spectrum framework designed to incorporate accessibility, tourism infrastructure, social interaction and other factors into the planning and development of tourism. Subsequently, this framework has evolved into ECOS, or the Ecotourism Opportunity Spectrum (Boyd and Butler 1996), incorporating access, other resource-related activities, attractions offered, existing infrastructure, social interaction, levels of skill and knowledge, and acceptance of visitor impacts, as the means by which to plan and manage ecotourism *in situ*. Other theorists (Harroun 1994) have utilised VIM (Loomis and Graefe 1992) and LAC to analyse the ecological impacts of tourism in developing countries for the purpose of inducing decision-makers to ensure that an acceptable management framework is instituted prior to the tourism development process. Farrell and Marion (2002) developed the Protected Area Visitor Impact Management (PAVIM) framework based on an application to protected areas in Chile, Costa Rica, Belize and Mexico. Although it is beyond the scope of the book to discuss the specifics of this model, in general it:

> provides a professional impact identification and evaluation process, represents cost effective and timely means of managing visitor impacts, and may also better integrate local resource needs and management capabilities and constraints into decision-making. [It] permits rapid implementation and management of visitor impact problems, as a form of triage, if necessary, but also may be used to identify management opportunities and prevent visitor impacts, and can be used in combination with pre-existing frameworks like carrying capacity.
>
> (Farrell and Marion 2002: 46)

The environmentally based tourism (EBT) planning framework by Dowling (1993) is another such model developed specifically for tourism. This model is grounded in the environmental disciplines and recognises that sustainable tourism planning can be accomplished only through a strong linkage between tourism development and environmental conservation. The EBT determines environmentally compatible tourism through the identification of: (1) significant features, including valued environmental attributes and tourism features; (2) critical areas, those in which environmental and tourism features are in competition and possible conflict; and (3) compatible activities, which include outdoor recreation activities considered to be environmentally and socially compatible. The EBT is based on five main stages and ten processes (Figure 8.4).

In general, the objectives stage of the model is important in that it involves the setting of the parameters of the study through discussions with government, local people and tourists. It also involves consideration of existing policies affecting the study region, and the relationship between use and supply as they relate to tourism. In the second stage of the model both environmental attributes (abiotic, biotic and cultural features) and tourism resources (attractions, accessibility and services) are assessed and integrated into a categorisation of sites. In the third stage, an evaluation of the significant features, critical areas and compatible activities, and the relationship of these to each other, is made and involves

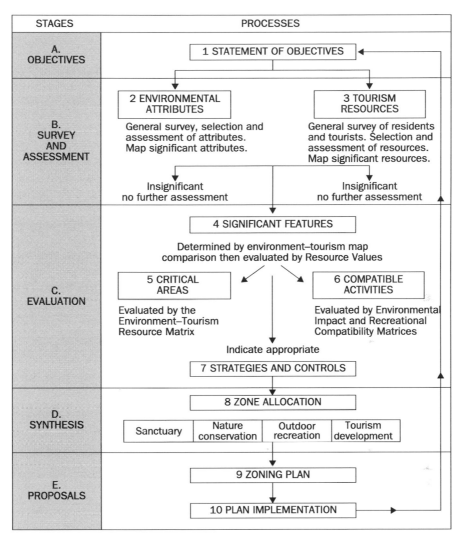

Figure 8.4 The environmentally based tourism planning framework

Source: Dowling (1993)

an overlay of both tourism and environmental attribute data. In stage 4, the identified significant features, critical areas and compatible activities are matched with zones (i.e. sanctuary, nature conservation, outdoor recreation and tourism development) and nodes, hinterlands and corridors identified at earlier stages of the project. The end product of this stage is a map identifying the region's environmental units within the various zones. In the final stage, the process is presented as part of an overall regional management plan. Discussions with resource managers are further required, and associated amendments to the plan in accordance with other land uses, in order for the tourism-environment plan to be implemented. The uniqueness of such a framework is its environmental foundation, the incorporation of tourist and local opinions, the process of achieving tourism environment compatibility, and the fact that it presents itself as one of the only sustainable tourism planning models in existence.

Assessment of ecological impacts

As identified in an earlier section of this chapter, a number of positive changes occurred in tourism-environment research in the 1990s that provide a better understanding of how the tourism industry has potential to compromise the integrity of the natural world. In particular, research has focused on specific impacts in a number of different settings (see, for example, Barnwell and Thomas 1995; Farr and Rogers 1994; León and González 1995; Price and Firaq 1996), but also on the design of better ways to quantify and assess these impacts.

Goodwin (1995) writes that because tourists are bigger consumers of resources than local people, careful environmental assessments of all new tourist developments are essential in documenting unsustainable usage. Getting tourists and other resource users to acknowledge their imprint on the destination is another point. For example, in a study on the perceptions of various stakeholder groups (tourists, entrepreneurs, and locals) in relation to their level and responsibility of impact, Kavallinis and Pizam (1994) discovered that tourists perceived entrepreneurs and locals as being more responsible for such impacts (over themselves), and that local people considered themselves to be more responsible for ecological impacts, relative to the other two groups.

In general, researchers have concluded that parks and protected areas need better indicators of tourist impacts. Buckley (2003) suggested the use of measures sensitive to seasonal cycles, longer-term trends, extreme events and internal patterns; the prioritisation of conservation values according to type of use; and the ability to differentiate between tourism impacts from other sources of impact. Such measures have also included the use of environmental impact assessments (EIAs) (Green and Hunter 1992), both inside and outside of parks. Although EIAs have been the topic of discussion for many years in land use and planning, it is only more recently that they have been incorporated into the tourism development process. In general, the function of an EIA is to identify impacts that are of a non-monetary form, and thus enable developers to use resources efficiently in achieving a reasonably sustainable product over the long term. In addition, the importance of community well-being is an important consideration in the process of conducting EIAs in that there is a move to reconsider them not as a function of the built and natural environments but rather as a process that encompasses the needs of people in these settings.

Mitchell (1989) acknowledges that operationally, EIAs sometimes fail owing to limitations in researchers' basic knowledge; that is, cause and effect relationships of the social and ecological conditions of the system under study (they also fail to effect change because they are ignored!). He stresses the value of pure research in addressing this lack of information. However, before such 'hard' data are collected a substantial amount of 'soft' data must be amassed in order to direct the more qualitative aspects of EIA. Green and Hunter (1992) emphasise the importance of methodological approaches such as the Delphi technique in allowing a subjective assessment of tourism developments by a series of stakeholders likely to be affected by the development. (The Delphi technique incorporates the use of successive rounds of a survey in gaining consensus on a particular issue. Surveys are repeatedly sent to experts in this process of reaching a consensus.) Typically, people involved in the Delphi include experts in various fields, such as planners, tourism officials, academics, engineers, environmental health officers and so on, but also local residents and other affected stakeholders. Only after such information is collected, according to Green and Hunter, should more formal aspects of the EIA occur. This process gives people the opportunity to aid in the course of identifying potential impacts that might not be recognisable to the planning team. The incorporation of stakeholder sentiment is thus widely acknowledged to be an essential aspect of planning and development. Resident attitudes and geographical information system (GIS) (soft and hard aspects of planning, respectively) have been used by Inbakaran *et al.* (2004) for the purpose of understanding the regional distribution of resident attitudes towards future tourism development in Victoria, Australia. The authors find significance in their work on the basis of the fact that planners can identify those community residents who are more positive or negative to the potential for tourism impacts.

CASE STUDY 8.2

The fate of Mexico's Mayan heartland

Ceballos-Lascuráin (1996b) contrasts the development of tourism in the north and south of Quintana Roo, a state in Mexico's southeastern Yucátan. In the north, Cancún predominates as a mega-project of the 1970s, attracting more than 2,000,000 tourists per year. The social and ecological impacts of this development have been well cited, with beaches and lagoons being heavily polluted owing to a lack of appropriate sewage management and the creation of a marginalised economy between the few who are able to capture economic rewards from tourism and the many who have quite literally been displaced from traditional industries. Ceballos-Lascuráin writes that by contrast, the less populated region of southern Quintana Roo, without the beaches to support sun, sea and sand tourism, has recently embarked upon a plan to develop a sustainable ecotourism industry (including archaeological tourism). Daltabuit and Pi-Sunyer (1990), however, report on the state of tourism development in neighbouring Chiapas (also part of the traditional Mayan heartland) and paint a very different picture of 'local' tourism initiatives. Using the example of the archaeological centre of Cobá, the authors illustrate that local people have had their land appropriated for tourism development without their approval, and been told that their opinions are meaningless in the tourism development process. Daltabuit and Pi-Sunyer caution that the international community will have to recognise that tourism (in relation to proposals to further develop the Mundo Maya region for tourism purposes) is not a benign and positive force for conservation and cultural heritage. Rather, it occurs as a function of political and economic factors, in a top-down fashion, that is often at odds with the needs of local people.

Conclusion

Tourism research has been successful in identifying a tremendous range of social and ecological problems brought about by the tourism industry. This dialogue has become so intense that organisations are willing to go to great lengths to protect the cultural and ecological integrity of some of the world's busiest ecotourism sites. The British think-tank The Centre for Future Studies (CFS) argues that far from being a right, ecotourists should enter a lottery for the privilege of visiting sites such as the Great Barrier Reef, Athens, the Florida Everglades and Croatia's Dalmation coast (Sulaiman 2006). Since these destinations are adversely affected by commercial interests, the only realistic way to preserve their integrity is to limit numbers or declare off-limits some destinations altogether. CFS points out that the Great Barrier Reef in Australia is suffering from too many tourists in association with rising sea levels from global warming. In response, tourism industry officials along with scientists involved with the reef argue that the reef is adequately protected with little effect from tourism. In fact, the 800 tourism operators who make a living off the reef are said to be the biggest defenders of the reef and regularly cooperate with marine park officials in monitoring impacts. The Great Barrier Reef generates AUS$5 billion in income per year from almost 2 million visitors. The rationale for maintaining tourism in this region, from the commercial standpoint, is clear. Much is a stake, there-fore, when we attempt to establish limits on the consumption of natural or cultural assets. In theory such demands are easy to make, but when we involve the livelihoods of so many people opposition is bound to be intense by those most affected – and regional and national economic development units will be leading the charge when it comes to the protection of commercial interests.

Summary questions

1   Discuss ways in which the tourism industry can have a transforming and dislocating impact on the social fabric of host communities.
2   What is carrying capacity, and how are social and ecological carrying capacities sepa-rate but related?
3   List some examples of different site, overt and educational management styles.
4   What are the implications of Butler's sequence for tourism development? Why is carrying capacity such an important aspect of this model?
5   What is a preformed planning and management framework, and why are these reported to be better at managing human impacts in natural areas than traditional carrying capacity techniques?

9 Economic impacts and marketing of ecotourism

In this broadly based chapter, the literature on ecotourism and economics is examined with emphasis on predicting the global economic impact of ecotourism. The chapter also includes a discussion of leakages and the multiplier effect, revenues in parks and the economic value of land. The section on marketing investigates the role that marketing plays in the development of ecotourism experiences, as well as how the literature addresses the need for such studies to be accurate in their projections of the market now and in the future. The final section on new technologies illustrates how such knowledge has allowed operators in remote locations to capture part of the market, and demarketing is described as a technique that actively dissuades people from purchasing a product. Parks and protected areas use this technique when visitation levels exceed the capacity of the park to absorb tourist industry pressure.

The economics of ecotourism

Based on a review of the literature on the economic impact of the global ecotourism industry, there is general consensus that ecotourism is expanding even faster than the tourism industry as a whole (see Lindberg 1991; McIntosh 1992; Hawkins in Giannecchini 1993; Luhrman 1997), with upwards of 20 per cent of the world travel market as ecotourism (Frangialli 1997). Other commentators, however, are more sceptical and feel that it is more logical to view ecotourism's growth in a site-specific manner than as a general overview (Horneman and Beeston in Tisdell 1995). Jenner and Smith (cited in Goodwin 1996) estimate that ecotourism had a global value of $4 billion in 1980, $5 billion in 1985 and $10 billion in 1989. They forecasted ecotourism's value to be $25 billion in 1995, and $50 billion by the year 2000; whereas Tibbetts (1995–6) illustrates that of the $2 trillion that tourism generates annually, $17.5 billion is from ecotourism. In 2003, ecotourism is said to have earned US$25 billion globally (Iturregui and Dutschke 2005). Perhaps more realistic is the United Nations World Tourism Organization's (UNWTO's claim that ecotourism constitutes 2–4 per cent of global tourism [WTO (2002) International Year of Ecotourism launched in New York (online): http://www.world-tourism.org/newsroom/Releases/more_releases/january2002/launch; see also Page and Dowling 2002; Cater 2002]. These numbers indicate the disparate views on the economic impact of ecotourism. One of the main reasons for such a discrepancy in quantifying ecotourism is the lack of a clear-cut definition of the term (Hawkins in Giannecchini 1993).

The flow of local money

Foremost in the minds of many local, regional and national bodies charged with the responsibility of tourism development is the importance of earning money. However, all

Figure 9.1 Imports leading to leakages

Source: Goodwin (1995)

such regions are not created equal in terms of their ability to generate and keep money within the economy. Simply stated, while tourism injects money into local economies, the amount of money that stays in that economy is subject to a number of factors. Two concepts, the multiplier effect and leakages, are useful in understanding the impact that money has on an economy.

In general, as new money from tourism enters a local economy it changes hands many times, resulting in a cumulative economic impact that is greater than the initial amount of tourist expenditure. More specifically, direct income or first-round income is the amount of that spent money that is left over after taxes, profits and wages are paid outside the area and after imports are purchased (Getz 1990). The money that remains after these leakages is referred to as secondary income, which circulates successively through the economy creating indirect income and induced income, again with various amounts of leakage occurring (leakage determined by assessing the percentage of money that flows out of the economy (see Figure 9.1)). National multipliers tend to be highest and, according to Bull (1991), have ranged from 2.5 in Canada to 0.8 in Bermuda and the Bahamas. Most of the less developed countries (LDCs), especially island economies, tend to be lower because of the high level of leakage. The effects of foreign investment in a smaller developing economy can be demonstrated by the case of Dominica (Patterson *et al.* 2004). In 1990, the multiplier was found to be 2.1, which was much higher than most other Caribbean nations. However, ten years later in 2000 the multiplier had fallen to 1.45. The main reason for this, according to the authors, was the increase in foreign ownership of tourism facilities in Dominica, or the investment of profits made in Dominica abroad. Comparatively, the income leakage for nature-based tourism (NBT) in the Kuhmo municipality of eastern Finland was 48 per cent, with high leakage attributed to retail trade (e.g. supermarkets and gas stations), and lower levels in accommodation and various forms of nature and culture-based recreation (Rinne and Saastamoinen 2005). It follows that even in the developed world, the multipliers of small towns and counties are often low (in the neighbourhood of 0.25). In the Caribbean, leakages on average are between 60 and 70 per cent and may be even higher in foreign-owned establishments if owners take their profits out of the country, as has often been the case with non-resident and corporate investors (Moreno 2005). In the cruise line industry of Alaska, the import of construction materials for tourism developments and foreign ownership have both contributed to a higher level of leakage (Seidl *et al.* 2006; see also Johnson 2006).

This form of leakage is referred to as import substitution. It is an important concept in the context of ecotourism and sustainable tourism because there is much evidence pointing to the fact that tourism, for example in the LDCs, has been hampered by the fact that management control of the industry lies in the hands of external, multinational interests. Hotels, car rental agencies, restaurants and airlines, all the big money-makers in the tourism industry, are quite often owned by companies that reside outside the destination region, and the destination region relies upon these to export its product through tourism.

It is also the case, however, that leakage can be better managed through proper planning and management. Those regions that can sustain a tourism industry based on the resources that they have at hand will potentially be able to prosper under these conditions. In a study of ecotourism in coastal Belize and Honduras, Moreno (2005) found that leakages can be reduced if foreign business owners live permanently or semi-permanently in their place of business, and choose to spend part of their earnings in the destination region.

Revenue and parks

Tourism is inherently a private-sector activity that capitalises on a market for the purpose of making a profit. A conflict emerges when a profit-motivated enterprise relies on the provision of supply that does not necessarily advocate the same market philosophies. Parks and protected areas, as public entities, provide the cornerstone for the ecotourism industry. Saayman and Saayman (2006) estimated the spending pattern of a typical tourist visiting Kruger National Park, South Africa to be 4,400 Rand (approximately US$598). One-half of this spending was on accommodation, with other categories of spending, in descending order of amount spent, including food; transport; restaurants; beverages; recreation; admission; 'other'; clothes and footwear; souvenirs and jewellery; toiletries; medicine; telephone, fax and internet; and tobacco products. Yet there continues to be debate over whether parks should be operated more as a business in response to shrinking public budgets (private versus public management philosophies). The management of parks has not been subject to the same market principles and philosophies as the private sector. While it is generally accepted that ecotourism in protected areas has positive economic spin-offs (e.g. direct employment, both on- and off-site, the diversification of the local economy, the earning of foreign exchange, and the improvement to transportation and communication systems), there are also associated negatives, including the lack of sufficient demand for ecotourism, which could result in the draining of badly needed funds; the fact that ecotourism may not generate local employment opportunities; leakages may be quite high, as they are in many small and developing regions; and it may not be socially and economically acceptable to charge fees in parks.

The preceding discussion alludes to the fact that one of the most significant issues facing parks and protected areas today is the means of attaining funds for their operation and survival. Many parks agencies that have historically relied on certain support mechanisms have had to consider diversifying in order to maintain good-quality programmes and infrastructure. Sherman and Dixon (1991) illustrate five main ways in which to gain revenue from nature tourism, including:

1 *User fees.* These are usually a reflection of the public's willingness to pay, and in recent years have altered into more of a two-tiered or multi-tiered system with a differential scale of fees, the fee varying according to whether the visitor is a resident or a foreigner.

2   *Concession fees.* In the case of government, fees are charged to private firms who provide tourists with goods and services (guiding, food, etc.).

3   *Royalties.* Souvenir and T-shirt sales provide a good basis of this type of revenue, which is given to the agency as a percentage of the revenue made on the items.

4   *Taxation.* Sales tax, hotel tax and airport tax are examples.

5   *Donations.* Tourists are encouraged to contribute money in an attempt to address a local problem (lack of resources or money for endangered species) and, in the process, aid in the management of a protected area. For example, the popularity of bird-watching as a form of ecotourism has prompted some theorists to support the donation of income from birding festivals and events to conservation programmes in LDCs. Sekercioglu (2002), for example, illustrates that the British Birdwatching Fair raised over US$190,000 in 2000 for habitat protection in Cuba.

Naidoo and Adamowicz (2005) found that a greater than tenfold increase in entrance fees in Ugandan parks is consistent with the general literature on ecotourism (see Maille and Mendelsohn 1993; Moran 1994; Menkhaus and Lober 1996), where willingness to pay study estimates recommend the increase of park entrance fees between 8–30 times the going rate. In regard to the Komodo National Park in Indonesia, Walpole *et al.* (2001) found that ecotourists were willing to pay ten times the current entrance fee to the park. In the case of the Monteverde Cloud Forest Reserve of Costa Rica, the issue of park entrance fees has been widely discussed in the context of the viability of this reserve. According to Aylward *et al.* (1996), in the early 1970s visitors, regardless of their origins, were asked to pay a fee of approximately US$2.30 to gain entrance to the reserve. However, owing to the increasing levels of visitation, and the demands placed on the reserve, a new fee structure had to be developed. In 1995 fees were restructured as follows: less than $1 for Costa Rican students; $1.50 for Costa Rican residents; $4 for foreign students; $8 for non-package tour foreigners; and $16 for foreigners on tours (1996). Such fees have been instrumental in enabling this private reserve to become more economically self-sufficient. But even tiered pricing strategies can be prohibitive. Cohen (2002) recounts the story shared at an international conference whereby children in West African countries will never have the opportunity to view some charismatic animals for which the continent has become famous, because they are found only in game reserves. Local people may not have the money needed to gain entrance to these reserves. Tiered pricing is not the solution, Cohen argues, because it would violate the sustainability imperative that is designed to restrict numbers of tourists.

Chase *et al.* (1998) sampled ecotourists visiting three of Costa Rica's most popular parks, Manuel Antonio, Volcán Poás, and Volcán Irazú, for the purpose of estimating visitor demand elasticities. The authors found that substitutability in visitation exists between parks with similar attractions (i.e. the two volcano parks), where an increase of a pre-existing fee differential can push tourists from one park to another. This is advantageous if park managers wish to reduce overcrowding or stimulate economic development at one park or the other. Furthermore, it was shown that efforts to solve the overcrowding issues at the beach park (Manuel Antonio) cannot be achieved by fee changes at the volcano parks. As such, park managers would have to increase fees at Manuel Antonio or decrease fees at other beach parks.

Tisdell (1995) too has argued that there are some inherent limitations to the implementation of fees in financing ecotourism developments, including the possibility that few people may visit the site and/or if the park is located in peripheral areas. As alternatives, managers might elect to make visitors purchase permits from park offices, require tourist operators to pay visitor fees (as is the case in many destinations), or erect automatic ticket machines in car parks and trail heads – an approach that has worked well in Pacific Rim

National Park, Canada. Laarman and Gregersen (1996) and Steele (1995) concur that pricing holds tremendous power in providing greater efficiency and sustainability in eco-tourism, but is seriously neglected in public policy. These authors identify pricing objectives, pricing strategies and categories of fees in arguing that the user-pays principle and the removal of free access to public lands are perfectly logical today as a means by which to recover costs and, indeed, make money. While it is not necessarily the goal of publicly run natural areas to 'profit', it is for community-run organisations and of course private enterprises. Laarman and Gregersen (1996) offer some guiding principles for fee policies in NBT, as shown in Table 9.1.

Laarman and Gregersen (1996) argue that pricing objectives can be many-sided, and administrators are constantly challenged to set fees in accordance with the resource conditions of the park, the needs of the park staff and the needs of visitors. In Table 9.1 the guiding principles are varied and range from a relationship of fees to general sources of revenue, fees for certain sites, fees only for certain sites, and the management and accounting of fee systems. In some park systems fees that are generated in each of the individual parks go back into a general operational account. The positive spin-off of this is that the money that is generated for this account aids in the maintenance of parks in the entire system. The negative element is that those parks doing the best job (either because they have better administrators or because the park simply generates more visitation) do not get the opportunity to utilise directly the money that they generate for their own purposes. This type of fee philosophy may further decrease the motivation of those working in the money-generating parks, so that they become less conscientious, and may lead those in the money-losing parks chronically to rely on the money generated in other areas.

Steele (1995) illustrates that a policy allowing for open accessibility to ecotourism sites leads to certain economic and ecological inefficiencies. Economic inefficiencies occur if sites allow free entry; they may lose the rental value of the resource. Agencies must also be wary of excess demand where sites are left vacant during certain seasons of

Table 9.1 *Guiding principles for fee policy in NBT*

Principle	Rationale
Fees supplement but do not replace general sources of revenue	Even for heavily visited sites, fee revenue rarely covers total costs, especially capital costs. Heavy dependence on fee revenue reduces visitor diversity and the scope of attractions that can be offered. Yearly fluctuations in fee revenue make fees an unstable income source
At least a portion of fee revenues should be set aside ('earmarked') for sites that generate them	Earmarking increases management's incentives to set and collect fees efficiently. Visitors may be more willing to pay fees if they know that fees are used on-site
Fees should be set on a site-specific basis	National guidelines specify fee objectives and policies, yet management goals and visitor patterns vary across nature-based tourism sites, requiring local flexibility in assessing the type and amount of fee
Fee collection is not justified at all sites	Fees are not cost-effective at places with low visitation demand and high collection costs
Fee systems work best when supported by reliable accounting and management	Administrative decisions about fees require acceptable data on costs and revenues of providing NBT for different sites and activities

Source: Laarman and Gregersen (1996)

the year, and also of costs related to congestion where tourists impact each other on the basis of overcrowding (congestion costs are found to reduce the profit per tourist by lowering tourist demand and raising marginal costs for the supplier). Ecological ineffi-ciencies include consideration of carrying capacities and an analysis of the total volume of tourists and the damage done per tourist (more on this below). By distributing tourists appropriately in space and time, some natural areas, according to Steele, have been able to increase the numbers to the natural area and reduce the overall impact of these tourists. The choices that land managers have with respect to pricing controls and quantity controls (limits on numbers of users, which are employed more often than pricing controls) are also important according to Steele. The use of variable tariffs or tiered pricing is an effi-cient way to increase revenue (see the example of Monteverde in Steele 1995: 85–6), where foreigners are charged a higher entrance fee than locals.

McFarlane and Boxall (1996) write that, historically, many public wildlife agencies get their funds from hunting and fishing licences and general tax revenues. However, owing to the recent decline in many consumptive forms of outdoor recreation (e.g. hunting), and budgets, financial resources for conservation initiatives are dwindling. Their research of 787 birdwatchers indicated that this group shows great promise in supporting conserva-tion efforts in a number of ways. While committed and experienced birders could help to identify wildlife management issues and participate in fieldwork with various agencies, the less specialised birders could aid by improving bird habitat in their backyards and through the contribution of funds. This research demonstrates that conservation agencies must diversify and include innovative schemes to capture the interest and support of the birding population.

The value of land

Land planners and developers recognise that destinations are increasingly demanding business ventures that will add value to raw materials (McIntosh 1992; Theophile 1995). Decisions must be made whether or not (and how) to develop dams, parks and protected areas, mining, or forestry, or other resources, all of which have certain costs and benefits. While many of the large tourism developments of the 1950s, 1960s and 1970s occurred with only financial motives in mind, the sustainability imperative forces developers to be more inclusive of social and ecological concerns – this is especially important for ecotourism.

Munasinghe (1994) has categorised economic values attributed to ecological resources by examining the use and non-use values of assets. He suggests that the total economic value (TEV) of a resource is based on its use value (UV) and non-use value (NUV), which are further broken down in Figure 9.2. In the figure, Munasinghe shaded the option values (an individual's willingness to pay for the option of preserving the asset for future use), the bequest values (the value that people derive from knowing that other people will benefit from the resource in the future) and the existence values (the perceived value of the asset) as a caution to the fact that they are all quite difficult to define. From here, analysts may use a number of non-market valuation techniques to quantify the above values (Munasinghe 1994; see also De Lacy and Lockwood 1992), including the travel cost method and the contingent valuation method.

The travel cost method has been employed by Menkhaus and Lober (1996) in Costa Rica to estimate the value of ecotourism to tropical rainforests, based on a sample of US travellers. This valuation method, according to the authors, 'estimates ecotourism benefits of a protected area based upon observed travel expenses by visitors to an area' (1996: 2). These estimates are then used by decision-makers to address issues related to entrance fees, competing land uses and so on. The argument for the use of such a method is that

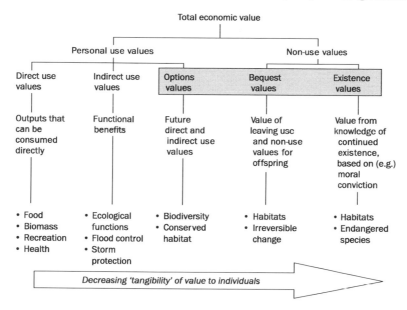

Figure 9.2 Categories of economic values attributed to environmental assets

Source: Munasinghe (1994)

certain goods such as parks are not effectively valued on the basis of entrance fees alone; individuals, it is felt, would be willing to pay much more to experience the good. The measure of the value of a good like ecotourism to a protected area is represented by the consumer surplus. In their study Menkhaus and Lober estimated the consumer surplus to be approximately $1,150, which represents the average annual per person valuation of the ecotourism value of protected areas in Costa Rica for the sample. These authors estimated the annual ecotourism value to Costa Rica's rainforests by this population at $68 million, which is determined by multiplying overall US ecotourist visitation to Costa Rica by the per person consumer surplus.

An example of the contingent valuation method is found in the work of Echeverría *et al.* (1995) in their evaluation of the Monteverde Cloud Forest Reserve of Costa Rica. This technique is an approach whereby 'individuals are asked directly about their willing-ness to pay for some improvement in a resource or for additional use of a resource' (Smith 1990b: 249). This can take the form of an assessment of the willingness of tourists to pay for habitat protection. Echeverría *et al.* (1995) sought to address these non-market values of the visitors to Monteverde in demonstrating that, on average, foreigners were willing to pay $118 of their own income to secure reserves such as Monteverde. This figure is comparable to the willingness of British Columbia residents to pay money to increase the scope of the protected areas in that province.

Tongson and Dygico (2004) help us to further understand how willingness to pay studies contributes to the generation of more park revenue in the face of reduced govern-ment spending and foreign assistance. In addressing the perennial problem of funding at the Tubbataha Reefs Natural Marine Park in the Sulu Sea, Philippines, the authors document the steps taken to arrange for the development of a two-tiered pricing scheme for foreign and local reef divers. Based on a willingness to pay survey of 239 divers, it was found that a mean willingness to pay was estimated at US$41.11 for the entire sample, where local divers are required to pay US$25 and foreign divers US$50 during the dive season. The average annual fee collections cover approximately 28 per cent of annual recurring costs and just over 40 per cent of core costs (e.g. enforcement). The authors concluded that user fees are a direct means of including the public in conservation, and that in addition to the revenue generated from the fees, the process has functioned as a permitting and regulatory instrument to control visitor numbers and activities.

In valuing places like parks and protected areas we should realise that there is a significant degree of grey area between the ideologies of business and those of conservation. Profit and preservation, therefore, are often construed as a partnership, according to Carey (in Giannecchini 1993), that is analogous to an arranged marriage, not one based on love. Giannecchini notes that:

> Their [the tourism industry's] customary goal of quick optimum profits is in direct conflict with long-range goals of protection and conservation. This does not mean that the only, or even primary, relationship between the tourism industry and conservationists must be adversarial. But it does mean that whatever laudable, environmentally sound policies and goals the industry articulates, they will remain subsidiary to the demand for profit. Therefore, if the tourism industry becomes the principal force in the development of ecotourism, it will almost certainly be detrimental to long-range environmental concerns.
>
> (Giannecchini 1993: 430)

A key factor in the need to prove the value of ecotourism to those advocating other land uses has been the economic value of the resource in question. Researchers have looked at ecotourism as a way to demonstrate that, in light of other competing types of land use, ecotourism presents itself as an effective form of land use. Saarinen (2005) corroborates this claim by suggesting that NBT in wilderness areas is fast taking the place of more traditional activities in places like northern Finland, where such activities (e.g. forestry and reindeer husbandry) can no longer support the local economy in the same way that tourism can. The economic advantage of working in ecotourism over other more traditional activities has been documented by Serio-Silva (2006) in his study of the monkey islands of Mexico. Families that rented out parcels of land for the breeding of cattle earned an annual income of US$450 per family. In contrast, boat operators taking tourists on monkey tours were able to earn US$1,482.80, on average, per family per year. It was concluded that not only is the money more substantive in boat tours but those working in this industry were not forced to migrate away from their families for the purpose of generating further income to support the family unit. This newer data substantiates older studies claiming that ecotourism was an important option in Africa because it earned more money per unit area than agriculture and it was more environmentally sustainable (see Western and Thresher 1973). In these older studies it was estimated that each lion in Amboseli National Park was worth about US$27,000 per year, and each elephant herd US$610,000 per year, which far exceeded the amount of money that could be gained through hunting these animals (McNeely 1988).

Other theorists have documented the value of a species to their own survival, and to the survival of a park. Naidoo and Adamowicz (2005) found that the number of bird species within a park was a strong positive predictor of the potential for that park to be visited: the more birds in a protected area, the more attractive the supply to tourists. The implications for ecotourism and conservation are that the more habitats we conserve the more money we can make through ecotourism. This is because species numbers are often a function of habitat size (see MacArthur and Wilson 1963; Diamond 1975), whereby a greater number of bird species, in this case, can be found in larger areas. The benefit of securing larger tracts of wilderness – quantity – are evident, but also remaining tracts of old-growth forest – quality – which also attract more species per unit area than secondary growth (Naidoo and Adamowicz 2005). Carr and Mendelsohn (2003) concur that there are very high economic benefits associated with protecting mega or charismatic attractions like coral reefs.

These figures point to an entrenched narrative surrounding the notion that animals must pay for their own survival; that is, wildlife must have some utilitarian function. There are two main purposes driving this perspective. The first is that the revenue generated from ecotourism in parks should help alleviate poverty and compensate local people for the loss of access to species. The second reason for making wildlife pay is to help generate money for conservation. It is often the case that state-funded conservation programmes (parks and protected areas) do not provide adequate funds to manage parks effectively (Infield 2002). Ecotourism should help in both cases, as noted above.

This form of analysis brings to mind the ways in which people value objects such as wildlife. Clearly there is the economic focus, yet examples of this to some degree are required as a means by which to rationalise the existence of something, like ecotourism, as a new land use. Kellert (1987) suggests that attitudes towards wildlife as a commodity or utilitarian object became less widespread in American society between 1900 and 1975. He believes, however, that this attitude change has occurred in only some parts of society. Those who are elderly, rural, in a lower socio-economic category, and who live and work in resource-dependent occupations still harbour negative perceptions of wildlife. It is often difficult for those who are not in one of the aforementioned categories to conceive of such perceptions. Although the point is not made in Kellert's research, one wonders about the impact that urbanisation, education and wealth have in shaping the more positive attitudes towards wildlife. Ecotourism is, or has been, conventionally considered an activity pursued by those who are better educated and wealthier than the 'average tourist' (as identified in Chapter 2), which would, according to the work of Kellert, lead one to believe that these individuals have a better chance of appreciating wildlife. Those who are providers of the ecotourism experience in the LDCs – assuming that they are not part of the privileged elite of that country – must acquire a 'personal conviction that land managed for wildlife is land ultimately more satisfying, attractive and enjoyable for people' (Kellert 1987: 228).

Theoretical and practical realties

Economists argue that ecotourism cannot accomplish its lofty tenets; for example, biodiversity conservation, lower impacts, provide incentives to protect land, and stimulate community economic development, because we do not understand or control all of the costs and benefits that are needed in considering the efficiency of market exchanges. Isaacs (2000) writes that the ecotourism industry is constrained by the fact that it produces a number of negative costs imposed on others, which are often ignored in the operation of the market. These costs, termed externalities, include, for example, damage to the ecology of an area (e.g. animal harassment and habituation, habitat loss) despite efforts to safeguard these species and their habitats in the first place. Isaacs writes that it is difficult to internalise or take into account these external costs because economic theory tells us that the incentives to ignore externalities are quite strong. Accordingly, it is rational, in an economic sense, to impose costs on third parties if this contributes to the satisfaction of our own self-interest. Isaacs suggests that there are three broad ways to reduce the negative costs of ecotourism in the long run, each with associated limitations based on economic theory. These three include government enforcement and administration, private ecotourism entrepreneur and monopolistic competition, and moral suasion.

Government enforcement and administration

As noted above, governments are often viewed as the most effective body to maintain environmental quality with respect to public goods (see Geist 1994). But given the multiplicity of conflicting goals within government, attempts to uphold the goals of ecotourism may be hampered by a number of problems. Unproductive political behaviour is a concern, where natural resources may be used for other purposes (e.g. roads) instead of conservation. This includes corruption of duty based on fraud and abuse from the money generated through ecotourism. It may also be that regulations to protect wildlife are seen as costly and unenforceable, which may induce members of the tourism industry to compromise these resources on the basis of their own self-interest. Furthermore, government officers may find difficulty in understanding the principles of ecotourism and be less likely to adhere to them. Finally, the avoidance of property rights and the proper distribution of resources by government create an unfavourable view of ecotourism by the citizenry, whereby some are advantaged with perhaps many more disadvantaged.

Private ecotourism entrepreneurs: monopolistic competition

Given the regulatory and bureaucratic challenges in government, above, Isaacs (2000) writes that efforts have been made to transfer environmental assets to the market. There is uncertainty however in regard to how successful service providers will be in balancing profit with the goals of ecotourism – with the former usually taking precedence over the latter. Isaacs notes that: (1) perfect competition (i.e. no economic profit in a perfectly competitive market based on identical goods); and (2) maximum profit (i.e. the market as a monopoly with a single supplier selling completely unique products) models are unsuitable as tools to examine the performance of ecotourism. More suitable is the monopolistic competition model, 'where slightly differentiated products are sold in an oligopolistic market, one with only a few suppliers and no significant barriers to market entry' (Isaacs 2000: 65). Initially, product distinctiveness and marketing attract customers and allow for high returns. However, successful ventures will attract other providers who will offer a similar product which will in turn erode the profit margins of the original supplier from the competition. Such is the case in Costa Rica, where the originator of the Canopy Tour (zip lines between raised platforms in the upper canopy of the rainforest) found increasing competition from many others who realised that they could capitalise on the same activity. (See also Shepherd 2002, who documents how ecotourism has gone wrong in Phang Nga Bay, Thailand, because of 11 competitors to the original SeaCanoe company, all of whom maintained far lower standards than SeaCanoe.) Isaacs observes that in the case of ecotourism, destinations have their own special biophysical attributes (e.g. the Grand Canyon) that help to maintain distinctiveness and the essence of monopoly, which ultimately restricts the level of market entry that would otherwise reduce market share. The key, according to Isaacs, is whether or not ecotourists recognise the degree of difference between rainforests in two different destinations (e.g. Costa Rica and Venezuela), such that both will become competitors in a single ecotourism market.

In the face of decreasing profits from competition, Isaacs writes that firms may do one of two things. They may increase profits or increase demand, both of which may detract from the firm's faithfulness to the tenets of ecotourism. By reducing costs, efforts to minimise external costs are diminished by deliberately not taking steps to, for example, reduce pollution or remove garbage. While the service provider benefits by avoiding these costs, such are typically borne by a third party (e.g. community adjacent to a protected area). In the case of increasing demand, the service provider appeals to a wider market by offering additional services or increasing carrying capacities in existing programmes. What eventuates is an industry that has parallel characteristics of mass tourism (larger numbers of tourists less inclined to learn about the environment), involved in nature-related activities in ecologically sensitive areas (see Chapter 3).

Moral suasion

Isaacs' (2000) final argument, moral suasion, concerns the willingness of individuals to voluntarily curtail their exploitative actions in safeguarding ecological and sociocultural resources. He observes that there is the possibility that principled service providers in ecotourism may make profit secondary to societal well-being. However, these individuals are more the exception than the rule, and even though they have endeavoured to act in this altruistic manner as the leader of their firm, research supports the notion that successors may in fact revert to profit maximisation as the main organisational goal. Economic theory supports the notion that excessive costs that are not consistent with one's competition will ultimately lead to the provider's financial demise. In a climate of heavy competition, economics is very much about efficiency in maximising gains. As for tourists, Isaacs suggests that individuals are more likely than not to 'violate the rules, stray from the path, or damage sensitive amenities intentionally or unintentionally' (2000: 67) in their self-interested pursuits (refer to Chapter 7).

The theoretical foundation provided by Isaacs, above, helps us to better understand the many challenges in balancing ecotourism's profit motive with other components like conservation and community well-being. This balance, or lack thereof, has been recognised by many other theorists who have explored it in detail in many different settings. For example, the significant local additions to income from wildlife tourism based on sea turtles and whales in Australia were sufficient enough to produce support not only for ecotourism but also for the conservation of these species (Wilson and Tisdell 2003). The nexus of economic well-being and conservation, however, was found to be a bit more complicated by Winkler (2006). Winkler developed a bio-economic model of open access land and wildlife exploitation for the purpose of investigating if Integrated Conservation and Development Projects (ICDPs), with ecotourism as the basis, could amplify conservation (habitat and wildlife) along with the well-being of local communities. Although ecotourism can enhance the motivation to conserve, it often fails because of a range of externalities and information deficits (e.g. a lack of knowledge about animal population dynamics). Winkler found that conservation of wildlife and habitat is elevated in communities if the share of ecotourism benefits distributed to local people is also elevated. But because revenues from ecotourism are often not enough, ecotourism benefits must be augmented with a system of taxes on land use and subsidies on crop production. He found that the subsidy on agricultural output and the tax on land use ensure that conservation goals are met, and that any revenues from ecotourism are essential in funding this tax/subsidy regime, with the ultimate end of raising the overall income of the community. This approach is consistent with work by Barrett and Arcese (1998), who argue that ICDPs based solely on the use of wildlife are unlikely to be workable over time.

CASE STUDY 9.1

Whale sharks at Ningaloo, Australia

Topelko and Dearden (2005) write that over one hundred million sharks are killed annually from fishing, bycatch and other anthropogenic causes. While they are one of those groups of species feared and vilified, they are also objects of fascination. Topelko and Dearden argue there are over 500,000 divers per year who photograph, swim with and feed sharks, and this is due to changing attitudes. Traditional/instrumental values are starting to give way to educational values based on new knowledge on the threatened status of sharks, which has in turn created a new type of appreciation and value of sharks that has stimulated the development of an ecotourism industry (Fennell 2012c). In this vein, Topelko and Dearden (2005) argue that it is essential that new tourism strategies develop that emphasise the protection of sharks. The biggest constraint to this idea is said to be the lack of economic incentives to encourage the reduction of fishing that would slow the decline of these species. Government intervention is urgently needed in those countries where the practice of shark fishing is taking place.

One species of shark that has garnered huge interest as an ecotourism attraction is the whale shark *(Rhincodon typus)*, particularly in Australia. Jones *et al.* (2009) found the core market of whale shark tourists to be international visitors (40.2 per cent of all visitors), followed by Australian visitors aged 35 and older (25.1 per cent). Interstate visitors represented 14.9 per cent of this latter figure. Although whale shark visitors stay half as long as other types of tourists, they spend more. In fact, those visitors highly motivated to visit the region for whale shark tourism purposes, termed 'enthusiasts', spent 21.6 per cent more money than tourists visiting for the same amount of time regardless of the presence of the whale shark tourism. Whale shark tourists spent an estimated AUS$6 million in the Ningaloo region of Western Australia alone in 2006.

Whale shark tourism has been so popular that operators have had to develop a code of ethics (Mau 2008). Standards include:

- a 10:1 ratio of divers to whales;
- divers must stay at least 4 m away from the tail of whale sharks, and 3 m away from the head and body;
- vessels are not allowed closer than 30 m from whale sharks;
- a speed limit of 8 knots applies in a 250 m contact zone.

Other researchers have developed a sustainability assessment framework for the whale shark industry at Ningaloo (Rodger *et al.* 2011). This assessment has three main components: The first is designed to improve the sustainability of the industry itself. Second is the development of an auditing system as part of licensing marine wildlife tourism operations. And third is a mechanism to help determine the sustainability of proposed marine wildlife tourism ventures. A number of key ecological and environmental indicators used include, for example, species, threatened status, group dynamics, ages of species, known behaviour, population, and feeding and habitat information. There are also operational and social assessment criteria including type of activity, accessibility, frequency of interaction, guidelines and regulations for focus species interaction, guide training and interpretation. The framework is deemed helpful for the overall purpose of identifying issues of concern and knowledge gaps in reference to a targeted marine species like the whale shark.

One of the issues that has emerged with ICDPs is that international donors – those who often spearhead ICDPs – view the private sector as lying outside of a 'romanticised' view of local community (Riedmiller 2001). The exclusion of small enterprises however is seen to be detrimental to the identification of genuine stakeholders in the community. In an investigation of private sector investment in the Chumbe Island Coral Park in Zanzibar, Riedmiller found that the involvement of small-scale fishers, shell collectors and seaweed farmers in conservation management affirmed their role as members of the private sector, which further contributed to increased awareness by these groups about the importance of conservation. Improved security of land tenure is not only good for business, but also good for sustainable environmental practices.

Investigating ecotourism stakeholders in Belize, Kroshus Medina (2005) found that local people demonstrated a preference for self-employment in ecotourism rather than wage labour (i.e. they chose to be entrepreneurs rather than employees) because of the benefits derived from the former in comparison to the latter. However, given the many constraints in business knowledge and financial resources local people are certain to be disadvantaged over transnational corporations which have many more tools to out-compete small scale ventures. This element of employment was investigated by Wunder (2000) who examined five case studies in the Cuyabeno Wildlife Reserve in the Ecuadorian Amazon region for the purpose of determining links between tourism participation and income, and income incentives and conservation. Following from the literature, Wunder hypothesised that autonomous ecotourism operations would trigger larger local income than the paternalistic model based on a dependency on external tour operators. Based on his empirical evidence, he found that the autonomously operating communities were in fact *not* better off than those receiving a salary – contrary to the beliefs of non-governmental organisations (NGOs) and other tourism theorists. Reliance on the autonomy model may therefore lead to unrealistic expectations on the part of the community and mislead investments. Wunder also found that conservation and environmental awareness were in fact raised because of involvement in tourism, which acted as an incentive, 'for a new rationality in traditional resource use' (Wunder 2000: 477). This is supported by Leader-Williams (2002: 1801) who writes that:

> Where local people earn significant income through participation in nature-based tourism or sustainable use, livelihood strategies can shift away from unsustainable use. However, where tourism and other forms of incentive-based use cannot give local communities what they need in the way of livelihoods, they will continue to invest in activities that do not support conservation efforts or even threaten them, for example by investing in livestock in areas of wildlife-livestock conflict.

In assessing the use of biodiversity conservation money to help fund community-based ecotourism projects, Kiss (2004), suggests that those ecotourism programmes and sites that are not charismatic like mountain gorilla or canopy tour walks will rarely generate revenue at a scale that will induce incentives for conservation. This is because such areas are under great pressure for the use of land and biological resources. Often there are times when natural habitats are opened up for tourism purposes, they are manipulated in ways that disrupt the integrity and resiliency of such areas, which suggests that ecotourism project sites need to be identified not only on the basis of community need but also on conservation needs (Kiss 2004). (See Stronza 2007 for a discussion of how economic incentives in ecotourism alone may not be enough to induce conservation behaviour, where more money and therefore the capacity to purchase more tools stimulate further exploitation of resources.)

The foregoing corresponds to work by Sinha *et al.* (2012), who have investigated ecotourism's contribution to livelihoods of the communities around the Kanha tiger reserve in central India. The authors found that the overall employment potential of tourism in the region was low with little direct impact on household incomes. Only 19 per cent of all the residents in seven villages surveyed (0.71 per cent of the total population of 150 villages around the reserve) were involved in tourism. Two villages between 6 and 8 km away from the reserve had no residents involved in tourism whatsoever. Those villages closest to the entry gates reaped most of the benefits of tourism, and furthermore, those in higher caste positions within society tended also to reap more benefits from the tourism industry than those of lower socio-economic standing. The reason for the disparity is linked to the lack of capacity of local people to have a voice in the governance structures of society. The fact of the matter is that economic benefit plays very strongly in decisions to support ecotourism projects or not. Communities need to know if they have the potential to benefit from such ventures either directly or indirectly. If community members find that they are being deprived of the benefits of ecotourism, they may react in ways that have resonance beyond the community itself (Taylor *et al.* 2003; Vincent and Thompson 2002).

Market-based solutions to environmental problems

In Australia, a company by the name of Earth Sanctuaries Limited (ESL) is attempting to disrupt the traditional mindset of biodiversity conservation by stressing that conventional public approaches to conservation must give way to private enterprise (Sydee and Beder 2006). Such a move comes on the heels of neoliberalist policy development during the 1980s demanding less government intervention and more privatisation, markets, property rights and individualism. Since 1985, ESL, a publicly listed company, has set about proving to the world that the integration of conservation with the marketplace is the best way to save endangered species. Their 'no nonsense' or 'common sense' approach to conservation includes the purchase of habitat, the eradication of feral animals in these reserves and the reintroduction of native (especially endangered) species. The ultimate goal, according to Sydee and Beder, is for ESL to purchase reserves in all of Australia's physiographic land regions and to protect all 100 of her endangered species. The huge amount of money required to accomplish this end comes from a number of sources. The list includes ecotourism, food and beverage sales, accommodation, gift shop, native plant nursery sales, functions such as weddings, conferences, education programmes, film and photography, consulting services on such things as fence-building and feral eradication, contract services for building, captive animal sales, wildlife sales for reintroduction into the environment and donations (Sydee and Bedel 2006: 88). This form of modern conservation, according to Sydee and Bedel, places managerialism at the forefront of decision-making whereby "'conservation" is synonymous with efficient expert management of resources. It is anthropocentric and instrumental rather than ecocentric and ethical'.

But isn't there something inherently wrong with a conservation model that is driven by consumerism, profit and economic growth? If we are to incorporate this model into our broader efforts to develop sustainably, are there not a series of issues that surface in the main? These are discussed by Sydee and Bedel at length, and have been summarised as follows:

1 If endangered species are placed into private hands, then a monopoly has been created over species and ecosystems.
2 Private conservation could lead to dwindling access to wilderness areas, where prices to private reserves increase dramatically in the face of the possibility of unhealthy global environmental assets – the possibility of the 'enclosure of the commons'.
3 There is the issue of producing a 'nature commodity' out of common heritage, where ESL markets and sells Australia back to itself.

4 Decisions to create, where to locate and what activities to programme of Australia's heritage are done without any community or public participation – unless individuals elect to become shareholders of the company.

5 Related to issue 4, above, is the absence of local knowledge and expertise in making decisions about specific species and habitat.

6 Profit as a prime directive has led to the sale of many of ESL's properties, which compromises initial efforts to conserve species and habitat. In this regard, conservation is supposed to be in perpetuity, not at the whim of the market.

In principle, ESL's initiatives are innovative in the sense that they eliminate many of the bureaucratic issues that often constrain conservation. The fact is, however, conservation can only work if there is the guarantee of stability and planning for the long term (Beder 1996).

Another innovative example of conservation that has stability and long-range planning at its core is the Haliburton Sustainable Forest (HSF) in central Ontario, Canada. As noted on its website (http://www.haliburtonforest.com/cons.htm), this 60,000 acre privately owned forest is an excellent example of the balance between the short-term requirements of a successful operational business and the long-term requirements of sustainable resource use and conservation. In just over four decades the forest has been transformed from a run-down forestry holding to a flourishing, multi-use operation which contributes economically and environmentally to the long-term stability of the surrounding rural community, while providing employment as well as environmental benefits to owners, staff and the public at large. The innovative nature of the enterprise has resulted in it being named as Canada's first 'sustainable forest' under the stringent guidelines of the International Forest Stewardship Council (FSC).

What is unique about the HSF is that it operates under a 100-year plan to return the forest into a healthy ecosystem by removing, through age-old techniques, the weakest stock and allowing the most genetically robust species to thrive. In order to accomplish this main end, the HSF has opened its doors to a number of non-consumptive and consumptive activities that help to keep the forest economically viable. Activities include wildlife viewing; a canopy tour; a submarine tour; a wolf centre, including a pack of wolves in a 15 acre enclosure; a planetarium with astronomy programmes; mountain biking; fishing, hunting; snowmobiling; dog sledding; outdoor education; events; and an ecolog home construction programme using a hemlock stand that had blown over in a tornado.

While ESL unabashedly emphasised profit as its main motivation, the 100-year plan holds precedence for the HSF. Activities take place to sustain the forest economically, but these activities never get in the way of the main overarching goal: to bring the forest back to a healthy state. So, snowmobiling and canopy tour programmes and numbers have been deliberately limited, for example, for the purpose of maximising the experience of participants within and between activities.

Marketing

Mahoney (1988) has suggested that tourism marketing differs fundamentally from the marketing of other types of products in three important ways: (1) tourism is primarily a service industry, where services are intangible, and quality control and evaluation of experiences are more difficult to envision; (2) instead of moving the product to the customer, the customer must travel to the product or resource; and (3) people usually participate in and visit more than one activity and facility while travelling. Therefore, tourism-related businesses and organisations need to cooperate to package and promote the tourism opportunities available in their areas. Certainly, marketing must lead to realistic expectations on the part of ecotourists, which includes the provision of information that is complete and responsible (Cutas *et al.* 2011).

As is the case with the conventional travel markets, marketing research in the area of ecotourism has begun to flourish. The work of Ingram and Durst (1987) stands out as one of the earliest studies in this area, through their development of a bibliography on marketing nature-oriented travel in developing countries and a subsequent publication that looked specifically at promotion of nature tourism in developing countries (Durst and Ingram 1988). From their research they concluded that there was a need to improve the operations of tourism offices in the developing world. They also found that although marketing is common in the developing world, countries that are not promoting their natural attractions are likely to miss the opportunity to capitalise on the growing ecotourism market.

Ryel and Grasse (1991) suggest that the two cornerstones to effective marketing in ecotourism are the attraction for tourists (which includes biodiversity, unique geography and cultural history) and tourism infrastructure to support the industry. Given these criteria, agencies are free to undertake the necessary marketing requirements to attract tourists. These authors imply that it is important to attract or market to the 'right' clientele. This includes 'born' ecotourists (those who have a built-in predisposition towards nature and nature travel), and 'made' ecotourists, who can be identified as representing latent demand or those who are unfamiliar with this form of tourism but who can be attracted through effective marketing. Ryel and Grasse suggest that a basic approach to marketing the ecotourism product is through: (1) the identification of the characteristics of a desired group; (2) appropriate advertising; (3) careful crafting of the advertising message; and (4) the development of a mailing list.

In Australia, the Office of National Tourism undertook a market study of ecotourists in an attempt to target ecotourism products more effectively (Commonwealth of Australia 1997). This research involved a number of focus groups involving ecotourists (actual and potential) from across the country. In general, the study found that ecotourists are seeking the following: (1) areas/attractions of natural beauty; (2) small groups away from crowds; (3) some level of interaction with the environment; (4) interaction with like-minded people (see also Fennell and Eagles 1990); (5) some degree of information and learning; and (6) fun and enjoyment. More specifically, the research uncovered three broad ecotourism market segments. These are as follows:

1   *Impulse.* Characterised by nature-based day trips away from the main tourist destinations and mainly booked locally by both domestic and international tourists. The level of activity on these tours varies widely.
2   *Active.* Characterised by younger and middle-aged professionals who generally book in advance. There is a skew to domestic tourists, although there could well be potential for growth through international marketing, infrastructure and product development.
3   *Personalised.* Essentially older professionals (or retired) who expect to be well looked after by the operator. This segment is skewed to international tourists who book overnight ecotours before arriving in Australia (Commonwealth of Australia 1997: 4–5).

In 1994, the Canadian provinces of Alberta and British Columbia published a comprehensive marketing study (ARA Consultants 1994) which focused on assessing current and future demand for ecotourism in these regions. The authors of the report sought information from a variety of sources in formulating their study, including a literature review of ecotourism, a travel trade survey, a general consumer survey of the residents of seven major metropolitan cities in Canada and the USA, and a survey of experienced North American ecotourists. Ecotourism was defined as 'vacations where the traveller would experience nature, adventure, or cultural experiences in the countryside' (1994: 1–5), and hence treated cultural tourism, adventure tourism and ecotourism as being synonymous. One of the main conclusions from this research was that 'the ecotourism market is definitely expanding to encompass not only those with specialist skills seeking strenuous adventure, but also clients seeking activities supported by amenities, a higher level of services, and requiring lower levels of specialist skills' (1994: 6–1). This finding is no doubt a result of the methodology employed in the study. Other conclusions of the study are as follows: for the general consumer group, 77 per cent said they had an eco-tourism vacation, with the remaining 23 per cent expressing interest in ecotourism travel. This led the researchers to conclude that from the general consumers of the seven cities surveyed, the actual market ranges from 1.6 to 3.2 million visitors, and the potential market is approximately 13.2 million ecotravellers. To two of Canada's most beautiful and 'natural' provinces, it is not inconceivable that travellers would involve themselves in any type of natural, adventure or cultural experience. This probably extends to hard-path ecotourism trips right along the spectrum to bus tours of the city of Victoria.

The point to be made from this example is that researchers and practitioners must proceed cautiously in making inferences on the size and scope of the ecotourism market. Past research has concluded that ecotourism is a very small part of the overall tourism continuum. Although it is certainly growing, so are the ways in which to classify and define ecotourism, many of which are based on convenience and less on a thorough review of ecotourism and ecotourism types. The research makes the assumption that, apart from the involvement of cultural and adventure tourism types, experienced and inexperienced ecotourists are one and the same. In substantiating this point, Wight (1996) reports that, with respect to the experienced North American ecotourist sample,

> Because client lists were volunteered, it is not known to what degree those firms volunteering names are representative of the ecotourism travel trade. Consequently, it is also not known how representative the respondents are of all ecotourists. However, the sample does represent a large number of experienced North American ecotourism travelers.

> (Wight 1996: 3)

In this case the softest of the soft-path ecotourists (or culture and adventure tourists) are grouped with the hardest of the hard-path ecotourists, groups which are perhaps worlds apart in their attitudes, motivations and benefits sought. This ambiguity is indicative of the activity packages offered by the tour operators from which the 'experienced ecotourism' sample was drawn. Ranked in order of number of packages offered, hiking was first, followed by rafting, canoeing, cycling, kayaking, horseback riding and then wildlife viewing – activities which are more skill-oriented and less learning-oriented in nature (see Chapter 2 for a discussion on comparing adventure tourism). Consequently, it was found that water-based activities were important, but particularly important to the experienced ecotourist (not surprising, and due most likely to the types of ecotourism packages offered); that virtually all ages of adults are interested in ecotourism; and that interest in ecotourism is spreading to less educated groups.

The results of this study have an element of inclusiveness, indicative of the methodology employed. Accordingly, the usage of the term 'ecotourism' has spun out of proportion and been misused as an industry label to capture a larger percentage of the travel market (see

Goodwin 1995), with the main implication of misrepresenting or 'watering down' the ecotourism market. For example, Thompson (1995) writes that a cruise aboard luxury liners, scuba diving and helicopter sightseeing trips over Hollywood are all being touted as ecotourism. Clearly, in such cases ecotourism has been modified, repackaged and 'mass' produced to the point where the line between what ecotourism is and is not is blurred. If this is the approach by the industry, and if ecotourism is to become successful by conventional industry standards, it will have undermined exactly what it set out to accomplish in the first place. The misrepresentation of ecotourism was the source of a recent debate over the TRINET (the tourism research international computer link organised out of Hawaii). For example, Iverson (1997) writes that:

> When I sought to 'dive with the seals' in Kaikoura NZ I was attracted to a nice glossy brochure with eco-labelling. My accommodation host saw my interest and recommended another company, a fellow who 'started it here'. When I met the fellow with this beat-up 4 wheeler hauling his van of dive gear I was wondering if I made the right choice. I had a great time and discovered later that the company with the glossy brochure had been in trouble with the Dept of Conservation for exceeding their permits and was told that they would take 24 people out with 12 permits and put half of them in the water at one time.

Ryan (1997) follows up on this discussion by making similar observations on the state of affairs in New Zealand. He suggests that there are basically two types of ecotour operators: one that constructs a glossy pseudo front, and the other that is small in scale and that happily accepts size restrictions. Continuing the discussion, Oppermann (1997) felt that too many tourists are subsumed under the ecotourism label, and that researchers need to be wary of studies that use wide-ranging definitions of the term. In order to avoid the misrepresentation of the ecotourism 'industry', Mason (1997b) suggests that a system based on self-regulation is suspect and fraught with abuse. He would rather see a form of policing and/or monitoring to avoid the unacceptable behaviour of those operating under free-market conditions. The problem to many marketers, operators and agents has become so bad that they are refusing to use the ecotourism label because it has gained a poor image and means very different things to different people (Preece *et al.* 1995).

A recent empirical study adds weight to the belief that marketing does not always support the core criteria of ecotourism. Kur and Hvenegaard (2012) analysed the marketing strategies of 62 whale watching operators on Vancouver Island according to how these operators supported the three main tenets of ecotourism (education, sustainability and nature-based) in their brochures. They found that the brochures did a good job at supporting the nature-based component of the activity along with education. However, the brochures lacked attention to environmental and economic sustainability. More specifically, there was room for improvement in the support of conservation programmes on the part of operators and tourists.

There are also challenges inherent in choosing the right logo or brand for ecotourism, especially in consideration of the needs of many stakeholder groups. A case in point is the development of an action plan for the Wet Tropics Management Authority (WTMA) in Australia (Watkinson 2002). Representatives from the Aboriginal community, the tourism industry and the WTMA convened for the purpose of assessing the awareness of the Wet Tropics World Heritage Area and its principles. While the study found that there was indeed high awareness of the region and World Heritage status, the logo was found to have low recognition. Recommendations were made for the development of a new brand, which included a new logo, moving away from the existing logo of a cassowary and a leaf, to a new one of a frog and a leaf and the words, 'Australia's Tropical Rainforests; World

Heritage', which was said to be a more unifying image for the region. Indigenous people requested that an Aboriginal artist for cultural reasons should develop the new logo. Some WTMA staff expressed hesitation over the fact that the cassowary was more of a keystone species needing protection, and there were many strong emotional ties to the original image. This process demonstrates that consensus-building among many different groups is not always easy and straightforward.

A further problem for regions in the development and marketing of an ecotourism product is the fact that there seems to be a feeling of unlimited assurance that ecotourism will ultimately resolve any tourism dysfunctions within a region. Tourism decision-makers and planners, perhaps on the basis of overly optimistic forecasts of management consultants, often inherently believe that their jurisdictions will be competitive in a highly competitive global ecotourism market. This has currently become an issue in Canada, where provinces that cannot hope to compete at the scale of other provinces or other countries – owing to their geological and natural features – have unbridled optimism for the sector. There is no question that ecotourism can and will exist in these regions, but regions must be realistic about the success of ecotourism initiatives. The size and extent of the sector will be a function of the demand, but it does not restrict them from creating innovative schemes to attract ecotourists given the attraction base from which they have to work.

As a case in point, the province of Saskatchewan, in association with Industry Canada, commissioned a study on the market potential of ecotourism in the province (Anderson/Fast 1996a, 1996b). The research was designed to gather valuable information from the travel trade (companies currently operating adventure, culture, nature and ecotour trips) on the potential for Saskatchewan as an ecotourism destination. One of the survey questions sought industry response on nine potential ecotour packages to be offered in the province. In only one case did over 50 per cent of the tour companies sampled feel that they could sell the package to their clients. One of the most significant ecotour attractions in the province is the migratory and resident bird populations. However, in the case of the 'Birding in Saskatchewan' tour, only 43 per cent of the operators said that this tour appealed to them, and only 48 per cent said they could sell the package to their clients. The reasons cited for the lack of interest included 'wrong fit', 'too expensive' and 'not stunning enough'. Other findings of the study indicated that a lack of information on the province and lack of demand for Saskatchewan ecotourism products were two most frequently cited barriers confronting the ecotourism industry in Saskatchewan.

A report on the ecotourism opportunities for Hawaii's tourism industry (Center for Tourism Policy Studies 1994) put ecotourism and Hawaii into perspective by acknowledging the fact that Hawaii faces some stiff competition from already well-established global ecotourism destinations (i.e. Latin America, Africa and Asia). They found that the main considerations constraining ecotourism in Hawaii could be summarised as follows (1994: ii, iii):

1   *Physical characteristics.* Hawaii's isolation, limited size and fragile natural environ-ments create conditions which can lead to irreversible environmental harm from the overuse of its natural resources.
2   *Culture and local lifestyles.* Native Hawaiian issues and the preservation of commu-nity values have become increasingly sensitive to land use decision-making.
3   *Competition abroad.* Hawaii faces competition from a number of well-established ecotourism destinations including countries in Central and Latin America, Africa, Asia and the Pacific. Island destination areas include the Caribbean, Mexico, the Galapagos and the South Pacific.

4   *Dependence on a mass tourism economy.* Ecotourism's contribution to Hawaii's economy will be modest in comparison to Hawaii's mass tourism industry and should be viewed in terms of diversification, not substitution.

5   *Popular image.* As an ecotourist destination, Hawaii will have to overcome its sun, sand and surf reputation and the impression that it is too overdeveloped to appeal to outdoor enthusiasts seeking back-to-nature experiences.

6   *Private-sector investment.* Owing to the relatively low-return/high-risk nature of the ecotourism industry, venture capital and investment support for small business enterprise development are limited while liability costs (i.e. insurance) for landowners remain high.

7   *Public policy considerations.* There is presently a lack of policy which formally addresses ecotourism issues, resulting in inadequate support for the development and maintenance of ecotourism resources.

These barriers nicely encapsulate the constraints that Hawaii will encounter in its planning and delivery of ecotourism in the state. Other regions should be equally wary in the development of their ecotourism product. Substantiating ecotourism through inaccurate research methods is as disruptive as the development of an ecotourism industry on the basis of other related, and less ecologically sensitive, forms of tourism.

Marketers typically follow a systematic planning approach that enables them to focus on their organisational goals, and the specific needs of their clientele in association with various tourism products. A typical plan, based on the work of Lovelock and Weinberg (1984), is outlined in Figure 9.3. The first stage of the marketing plan involves the identification of the direction of the agency or organisation and the associated

Plate 9.1 In Cape Breton Island, Nova Scotia, Canada, Captain Marks is one of the most visible whale watch operators – and the most successful

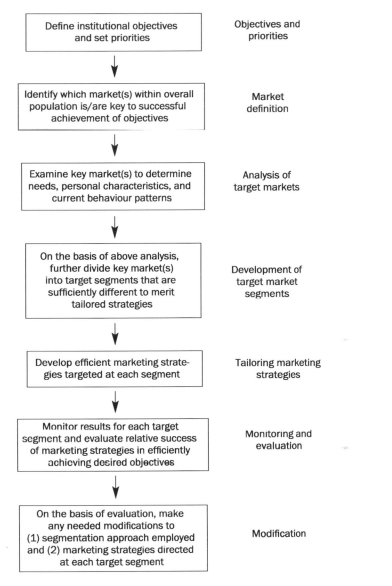

Figure 9.3 The process of market targeting to achieve institutional objectives

Source: Lovelock and Weinberg (1984)

priorities that must be followed. This is followed by the definition of markets that will allow the organisation to achieve its chosen goals. Upon the completion of this stage, the organisation endeavours to examine fully the behaviour, needs and characteristics of the market. Once the markets have been segmented, the organisation can develop specific strategies, which must be tailored to each segment. This is followed by the implementation of a monitoring and evaluation component, with subsequent modification in the future.

The marketing of tourism products is strongly based on market segmentation, as illustrated in Figure 9.3. This has led Weaver and Lawton (2007) to argue that there has been little attention paid to other central components of marketing like promotion and advertising. As such, travel firms do not have the resources to tap the overall travel market, nor the inclination, owing to the magnitude of domestic and international travel. Instead, businesses target certain segments based on the product that they are selling and the needs and expectations of the group to which they wish to sell. Target marketing can occur in four main ways, which often overlap based on the nature of the study undertaken. (See Dolnicar 2002 for a review of the pros and cons of market segmentation in tourism; see also the discussion on ecotourists in Chapter 2.)

1   *Geographically, on the basis of space.* Ryan *et al.* (2000) argue that there may be deep-seated differences between people occupying what we view as ecotourism 'space'. Some may enjoy and admire the space in a general context without any motivation beyond the immediate satisfaction of 'being there', whereas others have specific intellectual reasons for being there in reference to flora, fauna or cultural attractions.

2   *Demographically, based on age, gender, religion, race and so on.* Tao *et al.* (2004) found that North American and Taiwanese ecotourists differed in many respects. One of the main differences between the two groups was the lower age of Taiwanese ecotourists (and lower income) attributed to the fact that the typical Taiwanese ecotourist is male and a recent graduate of university. The authors felt that this was a result of the recent emergence of ecotourism in Taiwan which has spawned a love of nature and love of physical activity in the younger crowd. This aspect of physical activity corroborates the findings of Kerstetter *et al.* (2004), as noted below.

3   *Benefits, which include an analysis of the benefits sought by tourists and the costs they avoid.* Palacio and McCool (1997) used benefit segmentation to classify tourists visiting Belize. Benefit segmentation is defined as, 'the benefits which people are seeking in consuming a given product [as] the basic reasons for the existence of true market segments' (1987: 236). From the completion of 206 surveys focused on 18 specific expected benefits from a visit to Belize, the authors uncovered four discrete segments: 'nature escapists', 'ecotourists', 'comfortable naturalists' and 'passive players'. Comprising 18 per cent of the sample, ecotourists were 59 per cent female and 41 per cent male and stayed on average 6.5 days. This group scored highest on expected benefits related to escape, learning about nature, health through participation in activities and cohesiveness of group, but had similar levels of activity participation when compared to the other three groups. Kibicho (2006) also used benefit segmentation in an effort to investigate how tourists differ in what they want from a visit to Amboseli National Park, Kenya. Of a sample of 131 tourists (58 per cent male, 91 per cent with a university education, with most in the 36–45 age cohort) respondents fell into one of three clusters, including 'Environmentalists', 'Want-it-all tourists' and 'Independent tourists', which in turn were further subdivided into 'ecocentric' and 'anthropocentric' groups based on their environmental attitudes. Kibicho found that tourists with ecocentric views favour more protectionist approaches to park management in Amboseli, while those with anthropocentric views favour the transformation of the park through more liberal views on development.

4   *Psychographically, based on individuals' lifestyles, attitudes, values and personalities.* Silverberg *et al.* (1996) identified six factor dimensions on the basis of their survey of 334 nature-based tourists in southeastern USA (education/history, camping/tenting, socialising, relaxation, viewing nature, information). The authors concluded that psychographic research is an effective way for marketers to better understand

the nature-based traveller. Blamey and Braithwaite (1997) used social values segmentation of the potential Australian ecotourism market as an alternative to benefits or motives, which have been deemed problematic. (See also Mayo 1975, who has suggested that demographics alone are not an effective means by which to segment the market.) By contrast, values are said to be relatively enduring, multidimensional and predictive with respect to the behaviour of individuals. The authors used a Social Values Inventory comprised of societal goals in judging standards used in making judgements about world and community events. Based on a sample of 1,680 Australians registered on the electoral role, four groups of potential ecotourists were empirically identified. 'Committed Greens' as the smallest group, placed faith in the government for wilderness management, believed that all citizens should own and pay for wilderness maintenance, and believed in environmental protection, social cooperation and equity. 'Dualists' were open to new ways to balance economic development and environmental protection, and more trusting of industry rather than government to do the right thing. 'Moral Relativists' were least likely to take an absolutist stance on the protection of nature, and were less supportive of government regulation than Greens, but more so than the last group deemed 'Libertarians'. This final group was the least likely to recycle, and were the least supportive of social welfare. These results led the authors to conclude that the potential ecotourism market includes individuals who are not necessarily sensitive or committed to environmental issues – those perhaps more oriented to softer path forms of ecotourism. The authors also found that even though the potential market was unsophisticated, there was still a willingness to learn about the natural world through the adoption of new rules of ecological etiquette. Blamey and Braithwaite also discovered that there was little opposition to the protection of the environment through institutional means. Finally, they found that how the environment is managed (e.g. entrance fees) is subject to debate, based on the philosophical/political orientations of these groups.

In related work, Kerstetter et al. (2004) developed a motivational and behavioural profile of ecotourists visiting coastal wetlands in Taiwan. One of the main findings of their study was that ecotourists in Taiwan were motivated to visit natural environments for physical health reasons (see item 2, above, on demographics). The authors note that this is not a motivation that is typically found in non-Asian motivational studies of ecotourists, leading them to conclude that we must be more vigilant in taking into account cross-cultural differences that may be important in generating a more universal view of ecotourists. On the basis of environmentally responsible behaviour, the authors derived three types of ecotourists (experience-tourists, learning-tourists and ecotourists) based on factor and cluster analysis. The ecotourist cluster included individuals who responded to a mix of educational and holistic benefits associated with the natural area experience. By contrast, the experience-tourists exhibited less environmentally responsible behaviour, being less likely to accept policies, purchase environmentally friendly products, help maintain local environmental quality, or help others to learn more about the wetland, than their eco-seeking counterparts.

Swarbrooke and Horner (1999) developed a 'shades of green' tourism classification to provide inroads on different environmental tendencies that may be utilised in consumer behaviour research. Tourist environmental commitment was said to be a function of three main factors: (1) awareness/knowledge of environmental issues in tourism; (2) general attitudes towards the environment; and (3) the fulfilment of other commitments in one's life, outside of tourism, such as employment. A large proportion of the population were characterised as 'Not at all green' because of their shallow interest in green issues, with few sacrifices made in the name of the environment. On the other end of the spectrum, the 'Dark greens' represent a small proportion of travellers. This group has a deep interest in specific issues, and make major sacrifices in their lives in relation to specific issues that they (along with society) deem important. This type includes tourists who have been known to boycott hotels that have poor reputations regarding environmental management, or individuals who pay to go on vacation – to volunteer – to work on a community or conservation project.

Demand is very much an integral part of the relationship between the tourism product and the market, and hence Krippendorf (1987) feels that in the future tourism marketers must be more sensitive to the changing composition of the tourist population. He suggests that owing to the changing lifestyles, economic conditions and demographic structure of travellers, 'the market is shifting from manipulated, uncritical "old tourists" to mature, critical and emancipated "new tourists"' (1987: 175). Several years on, this trend has certainly taken shape.

Marketing challenges

Barriers to the development of successful ecotourism businesses have been summarised by Silva and McDill (2004), who feel that ecotourism service providers are amateurs in the tourism sector who are interested in turning their hobby into a business. While it is impressive that these individuals are following their dreams, the problems that they encounter in fulfilling this dream include a lack of awareness of how to source financing, a lack managerial skills and difficulty knowing how best to market their programmes. Silva and McDill (2004) suggest that because supply for ecotourism businesses is often located in publicly owned spaces, it requires a level of cooperation with other businesses (tourism and non-tourism) and the local community (broadly defined). This may be problematic because local people often create barriers to ecotourism development on the basis of local traditions, availability and quality of amenities, and attitudes towards then new business and the arrival of tourists. Government can help the ecotourism sector, but the absence of good programmes will most certainly act as a barrier to success. Silva and McDill (2004) found that there are four categories of barriers to successful ecotourism development: (1) enterprise, which includes the management, marketing, legal and financial elements of running a business; (2) agency, which includes the programmes, marketing, outreach and standards established by various organisations that affect ecotourism; (3) host community, including the attraction base (e.g. forests), infrastructure (e.g. accommodation and food and beverage outlets), and attitudes of local people; and (4) networks, which includes the cooperative relationships, or lack thereof, between those involved in ecotourism (e.g. government and tourism organisations).

Like Silva and McDill, above, Forstner (2004) writes that community-based tourism enterprises in rural areas are often challenged by the inability to effectively market their products because of a lack of knowledge, business skills and resources. Coupled with this is inadequate infrastructure and political power to lobby for policy changes that would improve market conditions. Efforts to offset these constraints have included the use of intermediaries – external support – to market their products more efficiently. Four separate intermediaries are identified by Forstner, including private, associations, public and not for profit, and the strengths and weaknesses of each are described at length in Table 9.2. Forstner concludes by suggesting that the success of community-based tourism is likely to be contingent on the complementary attributes of more than one intermediary in maximising benefits for the rural initiative.

Table 9.2 *Strengths and weaknesses of marketing intermediaries*

Intermediary	Strengths	Weaknesses
Private sector companies	Possess market information and expertise in tourism marketing. Promote product quality. Able to invest in local tourism infrastructure. Provide access to higher-value segments of the market. Have established distribution channels. Provide access to international markets. May offer financial support for other community projects.	Lack expertise and interest in promoting alternatives to tourism. Primarily interested in CBT with higher chances of market success. Discouraged by slow process of capacity-building at community level. Discouraged by risks associated using local service providers. Lack experience in working with local communities. Might take advantage of power imbalance.
CBT associations	Facilitate networking between individual CBT ventures. May integrate individual products into programmes with higher chances of market success. Promote joint marketing. Are more likely to attract funding for marketing tasks. Provide advice on product design. Offer training programmes. Establish linkages with the private sector. Advocate interests of CBT at national level	Lack sufficient business/marketing skills Lack financial sustainability. Possess limited capacity for directly facilitating access to international markets. Face difficulties in practice in treating members equally. May suffer from conflicts between members.
Public sector	Improves local infrastructure. Creates favourable policy framework. Encourages private sector involvement. Employs marketing instruments of national tourist boards. May integrate CBT into national tourism product. Has capacity to integrate CBT into broader programmes for poverty alleviation.	Lacks interest in small-scale tourism operations. Has problems in implementing CBT policies. Lacks capacity for promoting CBT development. Suffers from a lack of coordination within and between public sector institutions. Political interests may subvert CBT initiatives.
Not-for-profit institutions	Have expertise in working with local communities. Can develop capacity at community level. Have information about and access to specific market segments. Advise communities during negotiations with private sector. Lobby for more responsible tourism and interests of local communities. Are able to identify and develop alternative livelihood options. May have capacity to strengthen the role of disadvantaged groups.	Lack sufficient business and marketing skills. Lack professionalism in developing tourism products. May offer only limited support due to dependence on external funding. Values may conflict with commercial viability of CBT ventures. Interests of NGOs and local communities may conflict.

Source: Forstner (2004: 510–511)

New marketing technologies

The Internet is having a direct effect on the tourism distribution system where there is an increasing trend towards direct contact between consumer and supplier. For suppliers; for example, airlines, the motivation involves saving money by not having to pay travel agents or credit card companies a commission for reservations. If an airline ticket costs US$300, the airline would end up making less money going through the intermediaries, whereas it could make the full US$300 by selling direct to the consumer (Gratzer *et al.* 2006). Gratzer *et al.* illustrate, based on a World Tourism Organization Business Council report, that the web may account for one in every four travel purchases by the year 2010. Consumers are now taking a much more active role in their travel choices and options. In a study of online travel purchasers in Hong Kong, Law and Wong (2003) found that of 22.5 per cent of participants who visited one or more travel sites in their study, 6.54 per cent were also purchasers.

In Texas, economic stress and a changing economy have forced farmers to look to alternative forms of income to supplement declining farm incomes. NBT is seen to provide certain advantages in the pursuit of supplementary income given the availability of information technology (IT) to tap into potential markets (Skadberg *et al.* 2005). This is especially true for near urban farm units, which can capitalise on a large proximate market as well as IT infrastructure and services. Like Texas, small, medium and micro-enterprises are making good use of IT for the purpose of overcoming spatial obstacles in rural areas in South Africa (Bourgouin 2002). IT is said to be so important to rural tourism development that it will be essential as a key marketing tool in the face of mounting competition. But despite the importance of the Internet to rural and urban service providers alike, there still appears to be a lack of knowledge of how sectors like ecotourism can effectively use IT for marketing.

Using a database of ecolodge operators, Lai and Shafer (2005) undertook content analysis to investigate the online marketing practices of these organisations (35 operators representing 14 countries). Results point to the fact that most online marketing messages only partially aligned with the principles of ecotourism discussed in the literature. For example, less than one-half of ecolodges provided opportunities to participate in the conservation of biodiversity. This led the authors to conclude that, 'Overall, most of the sample failed to communicate all the necessary elements online to shape the expectations, and influence the attitudes and behaviour of its future clientele' (2005: 156). This finding has been documented by Lück (2002b), who observes that many in the field simply adopt the term ecotourism without an associated change in the way they go about doing business. While the concept of ecotourism became effective in marketing as a tool to increase volume, operational procedures simply stayed the same. Similar findings have been observed by Dorsey *et al.* (2004) who found that: (1) advertisements for ecotourism were not always consistent with the discourse on sustainability; and (2) the pattern of commodification in tourism for the LDCs via print advertising extended also to internet advertising. If ecotour operators are to capitalise on the ecotourism market, they will need to learn how to market their products more effectively, but also implement better practices (practising what they are preaching) if they are to progress beyond the repositioning of products that appear to be eco-friendly (Diamantis and Ladkin 1999; see also Donohoe and Needham 2006).

Demarketing

A recent addition to the ecotourism literature is a focus on demarketing, or the practice of actively discouraging demand for a product (Table 9.3). This is not a technique designed to destroy or eliminate demand for products, but rather to decrease interest because of

Table 9.3 *Demarketing measures identified in tourism and protected area management research*

Marketing mix component	Demarketing measures identified
Product	Discouraging certain facilities that attract 'undesirable' markets Reducing the maintenance of certain walking tracks to encourage use by experienced walkers only Providing safe wildlife observation areas to channel visitor movements Limiting activities either seasonally or entirely Limiting activities by restricting the areas in which they can be conducted
Distribution (place)	Issuing timed entry to visitors based on a specific carrying capacity Introducing booking or reservation systems Introducing permit or licence systems Using an allocation system for visits on a 'first come first served' basis based on social and biological carrying capacities. Limiting the overall capacity of camping and accommodation facilities Limiting total visitor numbers Limiting group sizes Developing a 'Park Full' strategy to encourage use of other destinations Permitting certain activities only under supervision of appropriately educated personnel (accredited commercial operators or park rangers) Making access to fragile areas more difficult while Simultaneously promoting less fragile options
Price	Introducing or increasing prices/user fees (including entrance, parking and camping fees) Discouraging/stopping price discounting practices Introducing differential pricing (where the price increases disproportionately with increasing time spent in the destination) Creating a queuing system to increase the time and opportunity costs of the experience
Promotion	Ceasing/decreasing promotion of product/service (in specific markets) Promoting/stressing restrictions related to product/service Warning visitors of environmental circumstances under which activities may be curtailed Promoting/stressing restrictions or difficulties with travel to the destination Discouraging certain ('undesirable') markets through the style of promotions and the level of information Educating journalists and media regarding appropriate environmental behaviour Promoting/stressing appropriate behaviour/minimal impact behaviour in promotional material Highlighting the environmental degradation that could occur if too many people frequent the area

Source: Armstrong and Kern (2011)

profit margins or quantities. In the case of ecotourism, it is because of too many unacceptable changes to the natural or cultural features of a region. The popularity of some parks means that there is simply too much demand for the resources and attractions found within these areas. One such park is the Blue Mountains National Park in Australia, which is in the position of needing to curtail demand. Armstrong and Kern (2011) identified several demarketing techniques that have been used to militate against ecological and social impacts in parks and protected areas. Table 9.3 is an example of the breadth of these techniques as they apply to the 4Ps (product, place, price, promotion).

Other techniques used by Armstrong and Kern include limiting the duration of activities, closure of park areas, limiting signage, and not promoting areas or certain experiences.

Conclusion

This chapter has taken a broad-based look at economics and marketing in ecotourism. Policy-makers have been mobilised to act in the interests of ecotourism because of the value that ecotourism resources, such as wildlife, have to economic development. Local people have also been provided with incentives to embrace ecotourism on the basis of the benefits they realise in the face of other land uses. There are challenges inherent in this approach, especially if conservation and various forms of wildlife use do not compensate local people who might otherwise use wildlife for their own subsistence purposes. This chapter also discussed the importance of marketing to ecotourism. Truth in marketing is essential in efforts to protect the integrity of ecological and social systems that are impacted by ecotourism activities. As such, marketing must reflect the needs not only of the consumer, but also of the industry, which by many accounts is being diluted by the misrepresentation of ecotourism by operators who have their own best interests in mind over broader concerns.

Summary questions

1   How do alternative forms of tourism, like ecotourism, act as better options in helping to diminish leakages in an economy?
2   Using the example of the Monteverde Cloud Forest Reserve in Costa Rica, how has a differential fee structure been utilised in such an area?
3   What do the travel cost method and contingent valuation method allow economists to do? Why are these important?
4   Some regions boast that 'because wildlife pays, wildlife stays'. What does this mean, and how is it that ecotourism might contribute to the preservation of species?
5   List some ways to segment the ecotourism market.

10 Development, governance and policy

This chapter argues that one cannot fully understand the complexity of issues taking place in ecotourism without a broader discussion on development, governance and policy. After a brief discussion on indicators of development and development in less developed countries (LDCs), the chapter focuses on some of the new research emerging on environmental governance – both from ecotourism and from outside the field. The chapter surveys much of this literature in the context of Glasbergen's (1998) organisational framework of different environmental governance approaches. Effort is made to fit different tourism themes and case studies into this framework, including emphasis on partnerships, community development and social capital. Cooperation emerges as a key theme in this discussion, especially in how it exists as a fundamental component of these various interactive mechanisms. Policy actions are increasingly required to ensure that ecotourism development is consistent with the needs of local people and the environment. The chapter identifies the range of actors who ought to be involved in ecotourism policy development, and it presents a policy framework that encompasses a broad spectrum of issues and elements related to the proper implementation of ecotourism.

Development theory

> I sit on a man's back, choking him and making him carry me, and yet assure myself and others that I am very sorry for him and wish to ease his lot by any means possible, except getting off his back. – Tolstoy

Development theory explores the economic and psychological link that exists between rich and poor countries. The relationship continues to be one that often contrasts meaning and value with respect to the use of natural resources and capital. Tourism is a sector of the world's economy that has tended to underscore and exacerbate the inequalities that exist between core metropolitan nations and peripheral, marginalised economies, and in doing so, perpetuates the division between those who have access to resources and those who do not. To Vogeler and DeSouza (1980), this division has conjured up images of distorted human beings and grotesque folklorism, where the underdeveloped are like children unable to fend for themselves and with little capacity to make decisions. Dependent countries were not considered economic entities until the late 1890s, when a new, 'lofty' responsibility regarding dependent peoples emerged (British Africa encouraged participation in the cash economy). This lofty purpose was coupled with a crude, hypocritical view of natives and administration, with the recognition that external power had a responsibility to develop the resources of dependent countries for the benefit of the whole world.

Some saw the acquisition of overseas dominions as motivated by strategic reasons, while others sought only economic incentives (Brookfield 1975).

Development has been defined according to a number of variables, including number of hospital beds, daily intake of calories, daily intake of animal protein, percentage of population in agriculture, food as a percentage of total imports, export economy, infant mortality, death rate, and aid per capita. Many of these have been criticised owing to the fact that the socio-cultural, environment and economic landscapes of each country are different. For example, food requirements are a cultural phenomenon, where countless generations have formed the habits of food intake seen today. Comparing the daily intake of calories between North Americans and Southeast Asians is pointless. Physiological requirements are different (Mountjoy 1971).

Resources have also been used to gauge the development status of nations, where an abundance of resources is thought to be an indicator of development. However, Mountjoy (1971) points out that even though Brazil has an abundance of resources and Denmark few, Denmark is considerably the more developed. Thus, development should be based not upon potential, but on existing levels of material welfare. Switzerland is a further case in point, where capital is a substitute for natural resource endowment. The progression of development, according to Brookfield (1975), occurred as a result of capitalist countries broadening or diversifying production into new sectors. This was to be achieved internationally through a widening of resource exploitation of LDCs. Brookfield maintains that the most aggressive industrial and trading nations (e.g. Britain) benefited most by putting pressure on other countries to open their trade. Adam Smith's *laissez-faire* philosophy was the true avenue of growth unfettered; it would work towards optimal production.

In recent years a major paradigmatic revision in development thinking has occurred that is presenting a fundamental challenge to the conventional consensus on economic development. This new approach emphasises the basic needs of the poor, and advocates a sensitivity for development at the ground level. Real improvement cannot, therefore, occur in LDCs unless the strategies that are being formulated and implemented are environmentally sustainable (Barbier 1987). Brookfield (1975) emphasises support of this contention: the study of development joins directly with the study of all change in human use of the environment, and provides elements of empirical evidence for infusion with other theories in the task of generating a dynamic human–environment paradigm.

Mountjoy (1971) stipulates that the most important factors in gauging development are people; their number, age, enterprise, initiative, inventiveness, knowledge and willingness to sacrifice. It is not the responses themselves (which are passive), but the way in which they are to be utilised that is deterministic of development. The Western nations' concern for invention is not shared by all societies. Technologically inferior societies exploit single landscapes, within the framework of their social structures, while the Westerner is said to exploit all environments (1971). This is most clearly emphasised by the recent international debates on the implications of global warming for the future viability of the planet. LDCs make a strong case against the more industrialised developed nations for their role in the creation of the pollution (e.g. nitrogen oxide and sulphur dioxide emissions) that has contributed to global warming and ozone depletion.

Tourism in LDCs

Not unlike the tourism literature in general, case studies from the LCDs dominate the literature on ecotourism (Weaver and Lawton 2007). One of the main reasons for this is the potential for ecotourism to act as an agent of economic development for these lower-income communities. It is also the case that the underdevelopment status of countries is what makes these places attractive to tourists from the developed world (Duffy 2006b).

LDCs certainly suffer from a history of colonial domination; what has been referred to as dependent development. A review of tourism literature reinforces the notion that there is a set evolutionary pattern of multinational interests in the tourism sector of the under-developed unit. Dependency, therefore, can be conceptualised as a process of historical conditioning which alters the internal functioning of economic and social subsystems within a LDC (Britton 1982). This dependency occurs not from processes within the LDC's economy, but rather from demand from overseas tourists and new foreign company investment in the LDC. For example, after a potential tourist destination has been identified (on the basis of unique biophysical or cultural conditions), the involvement of the multinational increases. Foreign companies greatly influence the image of a destination country through development and promotion. Such efforts lead tourists to perceive the host country in terms of this image and the nature of hotel accommodation, attractions and other tourist services as publicised. According to Khan (1997), economists argue that mass tourism in LDCs contributes to jobs, quality of life, educational opportunities and better infrastructure. On the other hand, dependency theorists would see this form of development as more for the benefit of the capitalist societies, who have destroyed the local economy through unequal competition.

Domination of the tourism sector in an LDC is most outwardly represented by the foreign ownership of airlines and hotels. However, the grip on the destination region often goes much deeper. Because of the inability of agricultural and manufacturing producers in most underdeveloped economies to guarantee a good-quality supply of goods and services for international luxury-standard tourist facilities, there is a strong reliance on imported supplies for both the construction and the operation of tourist facilities. Also, as Winpenny (1982) writes, middle and senior management levels of tourism developments in underdeveloped countries are often occupied by expatriates. The following scenario of metropolitan dominance serves as an example:

> On arrival in a Caribbean island between plane and hotel, the tourists could well have passed through a terminal building presented by the people of Canada and have been driven along a road either built or improved 'thanks' to Canadian aid, on behalf of the expected Canadian tourists of course. At the hotel you will likely be greeted with rum punches, the first of the few local products to be encompassed by the package. At your first meal, a menu, perhaps designed in Toronto, Chicago or Miami, will provide you with a selection of food imported for the greater part from North America – good familiar, homogenized, taste-free food, dressed up with a touch of local colour.
>
> (Hills and Lundgren 1977: 257)

So, who benefits from tourism in the lesser developed world? It has been found (Economic and Social Commission for Asia and the Pacific (ESCAP) 1978: 40) that where tour packages consist of a foreign air carrier, but include local hotel and other group services, destination countries receive on average only 40–45 per cent of the inclusive tour retail prices paid by the tourists in their home country. If foreign companies own both the airline and hotels, a mere 22–25 per cent of the retail tour price stays in the destination country (refer back to the discussion on the multiplier effect and leakages in Chapter 9). Despite these low numbers, the prospect of generating any level of foreign exchange must be very tempting, as the development cycle is constant. Mass tourism even has the ability to compromise ideology. This appears to be the case in communist Cuba, where there is the necessity of forging joint ventures with international hotel firms to capitalise on what is thought to be the future of tourism in the Caribbean region (Winson 2006).

While Rajotte (1980) feels that tourism is less environmentally destructive than other forms of development that exist in tropical islands, there are still many significant effects that may be attributed to tourism. Many of these impacts are universal, but their intensity and severity are more noticeable in island environments (much of the past research on dependency and development has been linked to island environments). The notion that ecological diversity often comes with size is important. Nation-states of the Caribbean, because of their small size, have ecological systems that are easily disrupted by tourism through yearly visitation levels that exceed the local population of an island (see Hayward *et al.* 1981).

A key area of research for the future lies in the need to document whether or not developing countries like Brazil, Costa Rica, Dominica and the countries of Eastern Africa, undergo the same social, ecological and economic dysfunctions from ecotourism, as outlined above, from the perspective of mass tourism. For example, Kusler (1991) writes that ecotourism in the LDCs is marked by limited dollars, government-directed financing, foreign visitation, foreign ownership of hotels and other facilities, and nonexistent land-use planning. In this sense, ecotourism as an intervention may or may not improve the socio-economic conditions of a country from the example of mass tourism outlined above. It is important to realise, as outlined by Kusler, that the LDCs and their developed-country counterparts are separated by a number of key structural differences. These include finances, the political climate of a country, accessibility to opportunities, discretionary income, and resources for the acquisition and planning of land.

Moreno (2005) feels that the pressure of outside investment translates into a psychology of inevitability about the pace and scale of development, and the associated perception of insignificance regarding the role of local people in ecotourism. Taken out of the hands of the people who live in the destination is the ability to control or mitigate the effects that these developments often have on the resource base, according to Moreno, including higher tourist impacts, the failure of the enterprise to put money into the hands of local people, and marginal efforts to conserve ecological resources. Moreno feels the logical reaction to this scenario is to simply reject outside-funded ecotourism in order to fully control the ecotourism enterprise. This could be done through the identification of strong community leaders, the development of representative institutions, and a common voice to support ecotourism. Unfortunately such efforts are constrained by: (1) the inability to organise to the point of political mobilisation; and (2) whether the mobilised unit can ever be strong enough to influence government policy, in the face of intense development interests from outside. Local development expertise requires time and money in the form of business assistance, training, conflict resolution and affordable loan programmes for small business entrepreneurs. The path of least resistance for cash-strapped governments, unfortunately, is foreign investment. A possible solution out of this loop is the involvement of non-governmental organisations (NGOs), according to Moreno, which can provide the needed support for training and so on, in addition to engendering the capacity for self-determination in moving forward with ecotourism.

The foregoing suggests that we cannot make the mistake of thinking that there is a level playing field when it comes to the development of ecotourism, especially in regard to the LDCs. Van Amerom (2006) argues that the development of ecotourism in South Africa is a function of its relationship with Western countries – where the demand lies. South Africa has a diversity of charismatic fauna in addition to a rich cultural history that is attractive to ecotourists. But because of apartheid up until the mid-1990s, a series of sanctions were put in place to hamper the development of tourism in this region, including condemnation of South Africa on the basis of its apartheid policies; recognition of its role as an instigator of conflict in southern Africa; banning national carriers from flying to South Africa; and halting the flow of South African tourism promotional materials. Once apartheid had been disassembled, the international community embraced South Africa via political and economic support of the black majority government, millions of dollars of support from

NGOs for ecotourism development, positive press coverage, and, perhaps most importantly, favourable travel advisories making it possible for people to feel comfortable in deciding to travel to South Africa. Van Amerom cites the case of the UK's favourable travel advisory to South Africa, based in part on political and economic grounds, which sought to increase travel to South Africa and downplay the fact that many British tourists had been killed or raped while travelling there. After a period of support for tourism in South Africa, Van Amerom argues that the tide may once again turn on this region as a result of conflicts with Western environmental NGOs (e.g. the culling of elephants which has offended animal rights groups), foreign policy decisions (e.g. conflicts with Zimbabwe) and South African domestic policies (e.g. harming the welfare of South Africa's white population mirroring the treatment of whites in Zimbabwe). Drought/deluge cycles in ecotourism are therefore a function of many political, social and economic measures which have sweeping effects internationally.

Pro-poor tourism

Given the many challenges that continue to constrain economic development in LDCs, mechanisms have been proposed to level the economic playing field in tourism and in other sectors of the economy. One of these mechanisms is pro-poor tourism which is an agenda that is geared towards the best interests of local people through the implementation of new and different tourism development strategies that create net benefits to the poor (Roe and Urquhart 2002; PPT 2002; Forstner 2004). Tourism can help to realise this agenda in urban and rural regions through non-farm livelihood opportunities, education and training, healthcare, infrastructure enhancement, sustainable environment management, corporate social responsibility, the creation of employment opportunities, access to markets for the poor and empowerment. Neto (2003) observes that although responsible forms of tourism like ecotourism aim to bring socio-economic benefits to local communities, they are not necessarily designed to alleviate poverty like the pro-poor tourism agenda. Neto argues that although both can be sustainable tourism development strategies, the former focuses primarily on environmental sustainability while the latter has poverty alleviation at its core through the participation of the poorest divisions of society. (See Cleverdon and Kalisch 2000 for a discussion of how the fair trade in tourism agenda may work towards eliminating poverty through control over production and marketing processes.)

Rogerson (2006) observes that South Africa has taken a leading position in the implementation of pro-poor policy during the post-apartheid period, as tourism has been earmarked as a key catalyst for local economic development for the country. Using the example of the Madikwe Game Reserve in the northwest region of South Africa, Rogerson (2006: 54–55) says that this example of rural pro-poor tourism is anchored on three main foundations:

1 To increase the amount of the lodge industry's wage bill captured by local households and to increase the ability of local people to take up employment at all levels of the game lodge industry.
.2 To stimulate a variety of small enterprises capable of taking up contracts inside the game reserve and its lodges.
3 To foster the conditions for the local residents to develop their own lodges in the game reserve in partnership with professional operators and lodge developers from the private sector.

As somewhat of a 'laboratory' for pro-poor policy development, Rogerson feels that emerging developments in South Africa will have relevance for other LDCs in generating momentum on the international scene. However, small-scale nature-based tourism developments in two regions of South Africa (Utrecht and Matatiele), were analysed by Hill *et al.* (2006), who found that the promises of pro-poor tourism have not been realised. This is because despite the belief that the natural resource base acts as a catalyst for development, small towns continue to be marginalised from the major economic centres of South Africa. Additionally, the number of permanent jobs has been limited and there has not been compensation for the loss of jobs following the closure of primary resource extraction industries. The authors observe that although government grants have been important in initiating NBT in these marginalised regions, they fail to become sustainable over the long term because of what might be termed a 'psychology' of grant dependency – too much of a reliance on external funding.

Ecotourism, governance and the politics of development

The previous discussion on development provides ample evidence that it is difficult to separate development from governance and politics. What follows is a discussion of ecotourism from the perspective of governance, institutional arrangements and policy. An attempt is made to match theory with a number of case studies that are found in the ecotourism literature.

Weinberg *et al.* (2002) argue that the conventional view that ecotourism fails because people are not aware of the problem, a lack of technical sophistication, and sustainable community development projects are not market feasible, are all false assumptions. Based on fieldwork at Monteverde in Costa Rica, and Kaikoura, New Zealand, the authors note that the problem really lies in politics. The ecotourism that takes place on the ground in local communities is part of larger political systems that are unable to control economic action, which in turn accelerates the rate of tourism production into something that resembles mass tourism. Because funding comes from outside the community and country, there is pressure to increase tourism for the purpose of demonstrating the profitability of these projects in relatively short periods of time. Ecotourism's fate, according to the authors, is thus tied to the implementation of strong democratic systems of governance in the host country that may not be able to protect the interests of local people first and foremost in the face of exogenous forces.

Duffy (2006b) explains that the politics of ecotourism take place at different scales. At the local scale, ecotourism often leads to costs and benefits in the way that communities are organised. She uses the example of an ecolodge, where some individuals and groups benefit from the development, while many others do not benefit at all. At the national scale, governments have many roles. On one hand, they are responsible for adopting policies that favour the development (infrastructure, air transport, accommodation and so on), while on the other hand at the same time they extend efforts to protect the environment through the development and management of parks and protected areas. At the global level, neoliberal development policies tend to prioritise the needs of the north at the expense of the south.

Duffy's (2006a, 2006b) main thesis is built around the premise that global governance is directly influencing the political landscape of ecotourism. Citing Selby (2003), Duffy contends that global governance can be contextualised in three main ways: (1) it is about dispersing power away from hegemonic centres (like states); (2) about extending and overcoming resistance to liberal democratic values and procedures; and (3) ordering things and people through reason, knowledge and expertise (2006a: 130). Duffy synthesises these ideas on global governance into the following definition as it pertains to her work on ecotourism and Madagascar. It is:

a set of neo-liberal ideas that have been translated into similarly neo-liberal programmes and policies. These policies aim to govern people, resources and activities through complex networks of actors, rather than through a single source and authority, such as a state.

(Duffy 2006a: 130)

More specifically, Duffy (2006a) contends that tourism, and ecotourism more specifically, must be viewed within the wider market oriented, neoliberal, strategy in which they are immersed. Here ecotourism may be used as a tool for economic diversification or outward-oriented growth. And while tourism may be viewed as a vehicle for development, eco-tourism becomes even that much more useful as a vehicle both for development and sustainability. This makes it particularly attractive to organisations like ENGOs, the World Bank and the International Monetary Fund (IMF) that give loans to those jurisdictions that make reforms along the lines of this neoliberal perspective. Once states commit to saving their natural environments, such will pave the way for the stimulation of global business like ecotourism.

In the context of Madagascar, the case study used by Duffy (2006a), the development of a Donor Consortium is an example of the implementation of a global governance strategy. This consortium is comprised of donors, NGOs and the Malagasy state. It includes representation from the USA, France, Switzerland, Germany, the World Wildlife Fund (WWF), Wildlife Conservation Society, Conservation International and the World Bank. Aid is extended to Madagascar in 'exchange' for the creation or expansion of protected areas in the state. But what is confounding, Duffy writes, is that this new preservationist philosophy or ideology exists alongside neoliberalism – where market-oriented strategies based on ecotourism promote the adage that conservation must pay its own way (see the example of Earth Sanctuaries Limited in Chapter 9). Accordingly, a fortress mentality approach is being advo-cated in the conservation efforts in Madagascar, where people are of less importance than natural resources and wildlife (Duffy 2006a). In this regard:

National tourism policies tend to be geared toward the generation of economic growth and the concept of tourism development is almost synonymous with neo-liberal defini-tions of economic growth, westernisation and modernisation for governments, since tourism means employment, balance of payments, regional development and foreign exchange.

(Duffy 2006a: 132)

Duffy is not alone in these claims. Using the example of Celestún, Yucatán and México, Azcárate (2006) provides a rich and detailed account of why development theory has not been successful in the context of ecotourism for this region. Azcárate writes that:

The confidence in a representative way of thinking the *Other*, the romaticisation of non-developed spaces and peoples together with the homogenisation of development, led post-development informed writers to infantilise and reify the very *Other* they were fighting to hear while condemning the development researcher to deconstructivist endeavours and to the theorisation of development in a highly problematic frame of pre-ordained global and local stances.

(Azcárate 2006: 108)

Development as such is not a one-way process, but rather should be viewed as a diverse and complex set of discourses and practices that ought to be explored through multiple local and global ethnographies.

Funnell and Bynoe (2007) observe that institutional arrangements provide the structure that allow various forms of governance to operate, but they also define the power relations that exist in systems like resource management. Defined, institutions are 'regularised patterns of behaviour between individuals and groups in a society, or complexes of norms and rules and behaviours that serve a collective purpose' (Leach *et al.* 1997: 11, as cited in Funnell and Bynoe 2007: 165). Regularised practices performed over time within a society or group become institutions. When actors choose to act in irregular ways, institutions evolve or change (Funnell and Bynoe 2007). As such, institutional arrangements can either encourage or constrain sustainable livelihoods by influencing the manner in which groups use and control natural resources. There are three basic levels that define the institutional framework of ecotourism according to Funnell and Bynoe (2007) as follows: (1) the private sector; (2) state and parastatal organisations that manage and operate environmental assets; and (3) residents who use environmental assets in sustaining themselves. These authors identify four main institutional arrangements for community ecotourism, as identified in Table 10.1.

In ecotourism, much of the literature points in the direction of community-based natural resource management (CBNRM) as the most rational approach to local participatory action and decentralisation, conservation of natural resources, and the ability of local people to earn a living from these resources in the places in which they live. Community-based models and approaches have been discussed at various points throughout the course of the book, owing to the value of these perspectives to the planning, development and management of ecotourism.

Glasbergen's work on environmental governance

Duffy's (2006a, 2006b) work on global environmental governance and ecotourism and Funnell and Bynoe's (2007) investigation of institutional arrangements provides an

Table 10.1 *Institutional arrangements for community involvement in tourism*

Description of institution	Institutional arrangements for ecotourism
Community-based (purest model)	Community-based ecotourism (CBE) that is owned, controlled and managed by members of the community
Joint venture partnership with private sector or NGO	Joint/partnership venture ecotourism with the community or family and an outside business partner or NGO, sometimes based on lease agreement
Private business comprising independent operations; sometimes based on agreement	Privately owned ecotourism enterprise that is owned, controlled and managed by a private tour operator who employs members of communities as cooks, labourers, drivers, gardeners, porters, waiters, maids and tour guides and souvenir vendors
'Non-organisation institution' or user management groups comprising individuals and households who manage their use of natural resources in accordance with community norms and personal understandings of possibilities and limitations of resource exploitation	Family or group ecotourism initiatives within communities

excellent introduction to many of the broad-scale developmental issues that characterise ecotourism. I wish to extend this discussion through a more intensive look at environmental governance based on Glasbergen's work. In doing so, I match Glasbergen's five models to various examples drawn from the ecotourism literature in rounding out the dialogue. Governance in resource management emerges from: (1) the interactions of actors; (2) by encompassing structures and processes through which societal decisions are made and power is exercised; and (3) formally through institutions or informally through norms. Inherent in this concept are aspects of participation, representation, deliberation, accountability, empowerment, and social justice in multilayered organisational hierarchies (Lebel *et al.* 2006, cited in Fennell *et al.* 2008). Important in accomplishing the three aforementioned goals is the precondition of a society that is manageable or open to influence (Glasbergen 1998). This aspect of openness should not be treated lightly. If society is to advance, it must be open to different approaches as well as to the interactive and cooperative relationship between government and other associated entities, especially private parties. For example, Li (2004) documents the barriers and constraints facing the Chinese people as regards community tourism development, where there are no institutional structures in place to counter the power of the state.

In his comprehensive work on environmental governance, Glasbergen (1998) identifies five distinct models in an organisational framework which provide a spectrum of different possibilities in regard to goal-oriented and deliberate intervention within society. These include regulatory control, market regulation, civil society, contextual control and self-regulation, and cooperative management. Although each of these are discussed, it is the last three which will be more fully elaborated upon because of the nature of ecotourism (i.e. the involvement of many stakeholders in the development of policy and practice). This follows from the work of Glasbergen but also other theorists such as Rhodes (1997), who argues that societies can no longer be effectively steered by the state alone, but rather by close interaction between state and society. Civil society, contextual control and self-regulation and cooperative management models are felt to have a significant degree of overlap, as each relies on various levels of cooperation. This overlap is complicated by studies in ecotourism which do not directly adhere to one model over another. In view of this, and the need to strike a balance with respect to the discussion on social capital, community development, partnerships and so on, the flow of the discussion in each section is dictated by theme and convenience.

Regulatory control

In this model, government acts as a regulator of social change (Glasbergen 1998). Legal rules are established with the goal of specifying conduct according to what one must do (prescriptions) and what one must not do (proscriptions). Regulation thus refers to the efforts of an agency (government, international or trade) that has been given the authority to regulate the actions of businesses, like the airline industry, which fall within its jurisdiction (Metelka 1990). In reference to the UK outgoing tourism industry, Forsyth (1995) discusses the link between sustainability and regulation. His findings suggest that tourism businesses are in favour of sustainability; however, he feels that because of the fragmented nature of the industry itself (e.g. hotels, tour operators, travel agents, carriers, associations), different industry sectors have separate aims and varied levels of power, and therefore limited control in efforts to implement sustainable tourism consistently (see also Goodall, 1994 with respect to regulation and environmental auditing).

CASE STUDY 10.1

The regulation of whale watching in Canada

Orams (2002) observes that whale watching globally takes place in almost 100 countries and is estimated to be worth about US$1 billion. The magnitude of the industry is also discussed by Hoyt (2005), who writes that the Atlantic Islands, from north to south, accommodate 1.7 million whale watchers per year spending US$133 million. These figures represent 19.4 per cent of whale watchers worldwide and 12.7 per cent of the total expenditure on whale watching globally. Hoyt notes that whale watching has been on the leading edge of ecotourism as well as the trailing edge. In the latter case, whale watching in regions such as the Canary Islands is characterised as unregulated, unlicensed, 'marine drinking tours', where whales are routinely harassed and misidentified.

In Canada, the impact of ecotourism on the beluga whale population of the St Lawrence River has prompted Blane and Jackson (1994) to call for stricter policies in the control of such vessels. The authors found that the high number of trips/vessels in the area of the whales was affecting their behaviour. Blane and Jackson recommend that the number of operators should be restricted, that low speed limits should be spatially increased and enforced, and that boats should be required to travel around beluga pods in a suitable fashion. Further, the authors suggest that policing should be accomplished by the Canadian government and informally by peers (i.e. operators). Regulations have also been addressed on the west coast of Canada by whale watching operators, government and other interested groups, and a set of guidelines for the industry has been developed. A recent draft of a document by Whale Watching Operators North West advocates the following: (1) stay approximately 100 metres away from the whales; (2) operators should be familiar with the Canadian Federal Fisheries Act; (3) slow down half a mile before arriving at the whales; (4) whales should be approached slowly at any time; (5) when approaching whales from the front, STOP and allow whales to come to you; (6) when approaching from behind, move to the outside of the nearest whale or boat, parallel to the direction of travel at a similar or slower speed; (7) when travelling with the whales, do not alter speed or change course abruptly; (8) never make high-speed runs through the middle of a group of whales or boats; (10) avoid approaching whales that are feeding; (11) try to work with the whales in rotation; and (12) do not reposition your vessel by the leap-frog method.

Market regulation

In the market regulation model, the price mechanism is given a key role in achieving change (Glasbergen 1998). Government's role is as a facilitator of market processes and self-interest is seen not as a problem but rather as a solution: 'Once the financial incentive structure is changed, and environmental goods are rightly priced on the market. . . public and private interests can be reconciled' (Glasbergen, 1998: 6). In environmental terms, the key is to internalise external effects (e.g. environmental degradation from tourism activities). However, it becomes difficult to internalise or take into account external costs because economic theory posits that incentives to ignore externalities are quite strong. Accordingly, it is rational, in an economic sense, to impose costs on third parties if this contributes to the satisfaction of our own self-interest (refer back to Chapter 9 for examples related to this model).

Plate 10.1 Whale watching at St. Andrews, New Brunswick, Canada

Plate 10.2 Minke whale off the coast of St. Andrews

Civil society

Self-confident citizens are given a key role in this model, according to Glasbergen (1998), along with the social ties they spontaneously create through trust and cooperation. The change mechanism is built into the effort to make society more dynamic through communicative rationality – a call for reason and mutual understanding – based on discussion, debate, opposition and social critique. Glasbergen mentions two very different streams of civil society governance. The first of these is the *participatory* stream where agents see government as a positive force that can create the right conditions for the public to address specific issues. An example of this stream is the many 'friends of' groups (e.g. Friends of Point Pelee National Park) that work for the purpose of aiding entities that have specific resource needs. (See Gardner 1993 for a discussion on the voluntary, not-for-profit, autonomous, service driven, and change orientation of environmental not-for-profit groups.) By contrast, the *rejectionist* or *radical* stream is based on the belief that government is generally more supportive of dominant private interests and, because of this, individuals become mobilised as a civil opposition force directed at the state for the purpose of moving an agenda forward. We see this in the form of many fringe social or environmental groups that work against government to affect some level of change which is deemed important for society.

Social capital

The concept of social capital, which has application to this model and the two that follow, may provide the means by which to better understand roles and interactions in communities. It has been defined as the ability of people to work together for common purposes within groups and organisations (Coleman 1988). Inherent in this general description is the notion that for people to come together over a common purpose there must be a great deal of social organisation, shared norms and values, that are activated for mutual benefit. In this regard, social theorists observe that society is founded upon reciprocal relationships between units and that these are solidified through trust and cooperation (see Gouldner 1960; Smelser 1963; Dredge 2006; Wyatt 1977). Ecotourism may be more effectively operationalised through the shared information, knowledge and interconnectedness between different stakeholders responsible for tourism development. While government agencies are often criticised for their 'silo', or independent approach to business, a concerted effort to build bridges between agencies and between sectors within the community creates a stronger foundation for ecotourism development, and the impetus to employ successful models in other community initiatives.

The importance of cooperation as a foundation for trust and mutual benefit in tourism has been discussed by Karlsson (2005; see also Putnam 2000) in a study of 22 small entrepreneurs in Årjäng, Sweden. Karlsson found that the community was characterised by a spirit of cooperation and fellowship among business owners in tourism, where cooperation, even between competitors, is a form of social capital that is important in establishing a positive, healthy business environment. Just as important to Karlsson is the fact that although some communities have this social capital, it can be destructive for those (e.g. tourists) on the other side, where outsiders can be treated to antisocial and immoral behaviour (see also Krishna 2001).

The destructive side of social capital is discussed by Jones (2005) who concludes that although social capital was important in the formation of an ecocamp in Gambia, a series of management issues threaten its longevity. Based on a survey of actors associated with the camp, it was found that although people reported a sense of trust, solidarity, and equality, perception and reality are different. The lack of transparency by those in more powerful positions, coupled with the unwillingness of those in lower positions to challenge these others, has compromised the project. This has been compounded by the passing of power on to family members who do not share the same vision of solidarity for community well-being as their forbearers. Social capital has also been found to be a destructive force by Kontogeorgopoulos (2005), who observes that SeaCanoe has created a great deal of disharmony within the social fabric of the region because of its employee advancement policies. Those who have managed to become guides for the company have been able to enhance their status in the community and greatly improve their lives – in a society well defined by social hierarchies. But this progressive social change, at least for a few, engenders resentment by other local residents, which unhinges the cohesiveness of the community on the basis of competition and conflict.

NGOs

An important aspect of the ecotourism industry is the role that NGOs as not-for-profit agencies play in the development and delivery of ecotourism. Organisations like CI have come to ecotourism quite naturally through their efforts to preserve biodiversity in the LDCs (Case study 10.2). These groups see ecotourism as a vehicle by which to accomplish their ends through community-based initiatives that employ positive commercial alternatives to the extractive industries that have been so devastating to biodiversity. As such, their motives can be thought of as being more park- or education- or ecology-centred than profit-centred when compared to the private sector.

NGO principles are often demonstrated through their involvement in international conservation and development programmes (ICDPs), whereby they leverage large sums of money from donor organisations for the purpose of biodiversity conservation, ecotourism and other community-based projects (as discussed earlier in the chapter). In Barra de Santiago, El Salvador, Ramírez (2005) illustrates that aid for ecotourism development came from a variety of sources. Sources in the country include the America's Initiative Funds-El Salvador, an American-based programme established in 1994 to reduce external debt, as well as from the Salvadorian Environmental Fund, which is a governmental agency that directs international funds for selected environmental projects. Outside sources included: (1) international development aid agencies from Japan, the USA., Spain, Germany, Sweden, the UK and Luxembourg; (2) multilateral assistance from the European Union (EU); and (3) the participation of a number of NGOs who were active in the coordination and facilitation of projects on the ground with community-based organisations. NGO roles included sourcing funds, writing proposals, directing local work, administrating funds, functioning as intermediaries between the community and international donors, and presenting reports.

Despite the many gains made by NGO involvement in LDCs, Brown (2002) suggests that their efforts have been constrained by a failure to understand the complexity of local communities, the level of participation of actors, empowerment and the process of engendering sustainability. (See Butcher 2006 for a good overview of NGOs and sustainability in the context of the non-use of natural capital in ecotourism.) Using the example of empowerment, Brown feels that new institutional frameworks are required which are founded upon new legislation and policies, the reorientation of government and related organisations in society, and the development of new partnerships. Brown uses the example of the reorientation of government departments in environment, economic development and social services which can seriously undermine each other in efforts to combine conservation and development.

Issues surrounding the involvement of external development agencies in ecotourism have surfaced in Kyrgyzstan, where the transition to a market economy has proved challenging (Palmer 2006). As one of the poorest countries of the world, Kyrgyzstan's long-term development plans involve the elimination of poverty and the establishment of a solid reputation in the international community – accomplished in part through tourism. However, given the current state of development, the government of Kyrgyzstan has actively sought the assistance of donor organisations to stimulate the economy and take care of debt. The unfortunate result of this reliance on external development agency funding (i.e. the Swiss Development Co-operation's community-based tourism initiatives) has been a top-down western imperialistic process of ecotourism development that is alien to the local population and tourist operators. This conceptual dissonance between Western consultants and post-Soviet communities has resulted in a series of socio-cultural impacts (e.g. commodification, staged authenticity and institutionalism) that are more typical of later stages of development. In particular, disharmony has occurred where the NGO has implemented local projects without any effort to consult with local businesses, creating a division within society. Critics argue that there appears to be too much of a gap between the daily living conditions of rural people and their values, and these external agencies (Sarrasin 2013). And as Yi-fong (2012) suggests, different groups benefit or suffer in different ways from one another from ecotourism development. Ecotourism often highlights existing inequalities (see Zhuang *et al.* 2011 for a discussion on the potential for an expanded role for NGOs in China; see also Butcher 2005 in reference to how NGOs undermine attempts to secure positive economic development for local people).

Richards and Hall (2000) contend that most attempts at instilling empowerment are distributive in nature. That is, the impetus for power comes from above. Instead they argue that a more fruitful avenue might be generative empowerment, which means that the impetus comes from within. Communities have different social relations and these need to be understood in moving any community development agenda forward. As Richards and Hall suggest, the 'global only becomes manifest where it is rooted in the local'. This means that while global initiatives play an important part in the development of ecotourism (see the section on politics of ecotourism), the local context is where the power relations of the global are seen and felt. Local communities can either succumb to these massive influences, or resist homogenisation effects of globalisation.

CASE STUDY 10.2

Highlighting the NGO: Conservation International

Conservation International is a not-for-profit organisation with two prime mandates: (1) to conserve the earth's biodiversity, and (2) demonstrate that human beings are able to live harmoniously with nature. One of its most significant hallmarks is the creation of the Debt-for-Nature Swap, which gives debt-poor nations a chance to make smart conservation choices. It negotiates to buy their foreign debt in exchange for national conservation initiatives. Native habitats that were in danger of disappearing are now safe havens for thousands of threatened species.

Conservation International's mission in ecotourism is to develop and support ecotourism enterprises that contribute to conservation, and influence the broader tourism industry towards greater ecological sustainability (Conservation International 1997). In order to accomplish these ends, it is involved in a number of national and regional ecotourism development initiatives around the world in countries such as Bolivia, Brazil, Guatemala, Peru, Botswana, Madagascar, Indonesia and Papua New Guinea. CI uses a 'capacity-building' approach in these regions to ensure that ecotourism benefits communities and merges with traditional practices and conservation, through the training of local people via ecotourism workshops within the region. Recently, CI has developed an Ecotravel Center designed to provide information on ecotourism destinations, tour operators and lodgings, and relevant publications and information (http://www.ecotour.org/xp/ecotour/; accessed 12 December 2006).

Contextual control and self-regulation

The contextual control and self-regulation model is based on the critique that society ought to be shaped wholly through government intervention (Glasbergen 1998). Conversely, and based on the theory of self-referentialism – organisations and individuals view the world from their own perspective – stakeholders determine what is most relevant to them based on their own experiences and tactics. As such, external information to the group, such as laws that prescribe certain types of behaviours, do not always have the intended effect because they fail to resonate with the qualities of the subsystem itself. Important here is the change in self-interest. While regulatory and market models are based on individual self-interest, the self-regulatory model applies to the system (group) itself within civil society. The example used in this section is tourism operators.

Tourism operators

One of the significant issues related to the imposition of regulatory measures for ecotourism operators, as a subsystem of society, is the perceived loss of control in the delivery of services and decision-making. In a discussion of the adventure and ecotourism industry in South Carolina, Tibbetts (1995–6) illustrates that many operators in this region favour self-regulation and voluntary guidelines. There is a strong belief by such operators that since they are the ones working in the environment they know how to: (1) manage their affairs without government help; (2) solve their own problems; and (3) take the lead in terms of planning and implementing an appropriate product.

Hiwasaki (2006) investigated three national parks in Japan for the purpose of determining challenges and success factors in generating sustainability in protected areas through community-based tourism. Japan, not unlike many other industrialised countries with large populations, empowers the Ministry of Environment (MoE) to involve stakeholders in park management because of the multiple-use function of parks and because of the complexities of land ownership in and around these areas. Furthermore, the diverse and fluid nature of communities and stakeholders has led to a diversity of conflicting interests contributing to discordant relationships of power and institutional dilemmas. Hiwasaki found that four key success factors are essential in addressing the existing challenges to tourism in protected areas. These include: institutional arrangements, self-regulation through voluntary codes of ethics, high environmental awareness and partnerships. The institutional structure comes in the form of an ecotourism support council established in 2002 that is comprised of the MoE, Forestry Agency, two local authorities, farmers' and fishermen's cooperatives, tourism association, hotel association, guides association, chamber of commerce and industry, and a private enterprise.

The importance of these councils has been reinforced by Yaman and Mohd (2004), who argue that one of the most vital steps in the development of rural sustainable ecotourism in Malaysia has been the establishment of a tourism management committee run by members of the community. Important in this process is the regulation and accountability of these members to prevent the misuse of benefits realised from their participation. Yaman and Mohd observe that this form of cooperation reduces leakages, minimises negative impacts and concentrates benefits locally. Apart from issues of intra-group accountability, there are broader concerns related to group control outside of some external regulatory body, not least of which is larger regional and socio-ecological repercussions. If groups such as tourism operators place personal gain above all else, the effects of their actions can be far reaching. Policy-makers need to decide if tourism operators can mitigate problems of complementarity and conflict, especially when it is the resource base that should be protected first over the needs of tourists and the industry. Today, more than ever, businesses must be accountable to the societies in which they are nested. Given this axiom, it is appropriate that we continue to determine whether or not regulation or self-regulation, or indeed both, might better serve the tourism industry and environment.

Cooperative management

Glasbergen (1998) characterises this last model of governance as the most unique alternative because of its attempt to combine a number of different types of rationality. Complexity and uncertainty are embraced not as challenges but as normal features of problem resolution. In this model, government is merely one of many different stake-holders (heterogeneous sets of actors) responsible for affecting societal change based on mutual dependency, and where symbiosis is achieved through shared definitions, a common desire to tackle the problem and equal participation in dialogue (through relational webs). The cooperative management landscape is complicated by consideration of a variety of different concepts and strategies including cooperation, partnerships, community development and collaboration, each of which are described in detail below. This governance strategy, and associated sub-forms, provides a good example of how this model has evolved over a relatively short period, and how theorists have attempted to compartmentalise different management approaches based on slight variations on select themes. As this governance model has attracted a great deal of interest of late, I include a discussion of more approaches, which are loosely organised into a number of overlapping perspectives.

Cooperation

Cooperation in a tourism context has been defined as 'working together to same end' (see Fennell 2006: 106), or even more broadly as 'working together' (Plummer *et al.* 2006: 501), and can be viewed as foundational to the cooperative management model, but also civil society and contextual control and self-regulation models too; that is, they involve people working in concert. It is not exactly clear how cooperation should be defined in tourism, or other fields, given the multiplicity of definitions used to describe it. Fennell (2006a, 2006b), for example, discusses it in the context of reciprocal altruism whereby agents bear and receive costs and benefits at various times in an ongoing relationship (see Chapter 7); while Moore (1984) suggests that all agents get immediate benefits from the act of cooperation. This argument aside, it is sufficient to suggest that cooperation can involve working together, at a number of different levels, to achieve predetermined ends (see Axelrod's 1984 work on the evolution of cooperation).

An example of how intricately cooperation can be conceived is provided by Tipa and Welch (2006) in their discussion of cooperative management. They argue that cooperative management is based on the interaction between equal partners in decision-making, where each group retains its distinct identity and independence throughout the process. Tipa and Welch argue that this co-management approach, termed 'idealised equality', often fails because it does not take into account that parties rarely see themselves as equal, and because each party has access to differing levels of resources (e.g. funding). One of the earliest examples of cooperation in a co-management structure is Kakadu National Park in Australia, where co-management arrangements were implemented as far back as 1978 (De Lacy 1994). Kakadu is owned by Aboriginal people, it is leased to the national government to be used as a park, the lessors receive in return annual rent, and the Aboriginal owners sit as a majority on the national park board where all policy decisions regarding the park are made (Lane 2001; Cordell 1993). Davey (1993), however, reports that the model breaks down by failing to involve the Aboriginal landowners in policy and planning, by not having majority Aboriginal representation on park boards, and by the lack of training and skills to enable Aboriginal people to better understand their role within the park setting (see also Baker et al. 1992). Whether cooperative management should be characterised as it has been by Tipa and Welch (i.e. involvement of equal partners and the retention of distinct identities), given the more general view of cooperation provided above, is subject to debate, reinforcing the inconsistencies with terminology related to the management of stakeholder interests.

Rozzi et al. (2006) provide an excellent example of the value of cooperation in ecotourism based on their work at the Omora Ethnobotanical Park located near Puerto Williams of the Cape Horn Archipelago region. This park is a private–public arrangement that addresses conservation at three levels, including interdisciplinary scientific research, informal and formal education, and biocultural conservation, with programmes on traditional ecological knowledge, environmental ethics, and ecotourism. To achieve the conservation goals of the project, ten principles have been articulated to make it a success. These include: (1) interinstitutional cooperation; (2) a participatory approach; (3) an interdisciplinary approach; (4) international cooperation; (5) communication through the media; (6) identification of flagship species; (7) education; (8) ecotourism; (9) administrative sustainability; and (10) research. Although the particulars of the cooperative relationship are not discussed in this research, the relevance of cooperation in making these projects a reality cannot be ignored.

CASE STUDY 10.3

The struggle for Kakadu

According to Fox (1996), territories rich in resources provide more options for use. This is the case with Kakadu National Park in the Northern Territory of Australia, which is a World Heritage Site and a place of immense ecological and cultural significance. Kakadu is owned by the Aborigines of the region, who lease it to the National Park Service. Legally, mismanagement of the resource may lead to the tenure being terminated. Ecotourism is thriving in the park, so much so that according to the Aborigines, there is no longer a distinction being made between the effects of mining and tourism; both are contributing to the degradation of the park. Some Aboriginal people recognise that while mining will ultimately end, tourism will continue to grow. The issue of most concern to Aborigines is that their fundamental values are not consistent with current development philosophies in the park. According to Elliot (1991), mining and tourism in Kakadu is a perfect example of the competing and overlapping environmental values that different stakeholders have. For example, would it matter if our actions caused a species to become extinct? Would it matter if our actions caused the death of individual animals? Would it matter if we caused widespread erosion in the park? Is the extinction of a species an acceptable price to pay for increased employment opportunities? One would imagine that miners, the tourism industry, Aborigines, the park service and other interests might differ along these fundamental lines.

Partnerships

A key feature of the cooperative management model is the development of partnerships in pursuing common goals. A partnership can be defined as, 'an on-going arrangement between two or more parties, based upon satisfying specifically identified, mutual needs. Such partnerships are characterized by durability over time, inclusiveness, cooperation, and flexibility' (Uhlik 1995: 14). More specifically, Uhlik (1995) developed a six-stage model of partnership development that concentrates on the conditions that will lead to a successful partnership agreement. These include: (1) education of self and others; (2) needs assessment and resource inventory; (3) identifying prospective partners and investigating their needs and inventories; (4) comparing and contrasting needs and resources; (5) developing a partnership proposal; and (6) proposing a partnership.

The partnerships developed for ecotourism, according to Sproule (1996), must also fit into systems that have been developed at regional and national levels. There are potentially many partnerships that can be struck to facilitate an atmosphere of cooperation and trust, as discussed earlier. Potential partners include: (1) organisations within the established tourism industry, like tour operators; (2) the government tourism bureau and natural resource agencies, especially the park service; (3) NGOs, such as those involved with environmental issues, small business management and traditional community development; (4) universities and other research organisations; (5) other communities, including those with a history of tourism and also those that are just beginning; and (6) other international organisations, public and private funding institutions, national cultural committees and many others. The overall effectiveness of the delivery system is only as strong as its weakest link, and communities intent on the development of a tourism industry will increasingly rely on the positive benefits of partnerships in being accountable to the local and non-local public (Clements *et al.* 1993).

Rather than emphasise what has not worked in the past, the partnership ideal must embrace the present and future needs of the groups involved in any transaction. For those in the realm of parks it may be biodiversity and the establishment of more protected areas. From the perspective of Aboriginal people, decision-making and culture may be issues that top the list. An example of the importance of recognising and placing value in different worldviews of two organisations in a long-standing partnership is in the case of the Wildlife Conservation Society (WCS) and the Capitania de Alto y Bajo (CABI), an indigenous organisation representing over 9,000 people in 25 communities in Bolivia (Arambiza and Painter 2006). For this partnership to stand the test of time, WCS recognises that in order to secure biodiversity conservation they must respect CABI's land-use strategies. On the other hand, CABI views conservation as essential in improving community members' lives through improved health care, education and employment. While one's main mission is not vital to the others', explicit recognition of this at the outset has allowed both parties to deal with disagreements in a straightforward way. In a study of the potential value of ecotourism to the residents of Barra de Santiago, El Salvador by Ramírez (2005), respondents of a survey administered to local people, tourists and summer house owners found that what was most important of a range of factors for a successful ecotourism industry was an organised community, including sound functioning local groups and wide participation by members of the community.

Collaborative management

Tourism researchers have recently explored the area of collaborative planning and management in a variety of contexts (Jamal and Getz 1995), which includes the need for as many stakeholders as possible to participate in tourism decision-making (Gunn 1994); who share resources and cooperate in efforts to problem solve (Bramwell and Lane 2000; Plummer et al. 2006); with the potential to improve interorganisational dynamics (Gray 1989). Examples of the collaborative planning process include work by Medeiros de Araujo and Bramwell (1999) in Brazil; Ladkin and Martinez Bertramini's (2002) work in Cusco, Peru; Hall's (1999) critical look at collaborative tourism planning in Western countries; and Timothy's (1998) work in Yogyakarta, Indonesia (see also Dredge 2006 who discusses collaboration in reference to network theory). In the context of resource management, Tipa and Welch (2006) give greater support to the collaborative process over other approaches which they say allows for discussion about the way power is shared, how it accords value to the aspirations of local people, how it protects decision-making control, and how it places value on agreed upon rules, norms and structures. The intricacies of power, local people, control and structures strike at the heart of many of the aforementioned studies in this chapter on ecotourism. The following few examples, particularly in reference to imbalances in power, serve to emphasise both the prospects and challenges of this type of co-management approach.

Puppim de Oliveira (2003) argues that different stakeholders involved in tourism policy development typically act according to their own interests, power and knowledge. In investigating environmental policy development in three case studies in Brazil, he found varying levels of commitment by local and state governments, small and large tourism businesses, community groups and NGOs, developers and other external actors according to: (1) the building of institutional capacity; (2) establishment of protected areas; (3) investment in environmental projects; and (4) controlling development and tourist flows. For example, large tourism businesses were more in favour of environmental projects where they are less likely to be affected financially. The author also found that the way that policy process was designed, and who was involved, heavily influenced policy outcomes. Alliances between different actors completely changed the nature of policy usually in favour of those whose influence and thought patterns coincided with their own social and ecological concerns. Advocacy coalitions (partnerships and allegiances) are often developed by actors and agencies for their own purposes (Sabatier 1988). This means that policy implementation is not necessarily based on the best interests of society, but rather on individual or organisational self-interest (see Fennell et al. 2008).

Buckley (2003) suggests that while conservation groups aim to use tourism as a tool for conservation, tourism groups like NGOs, government and industry associations typically use conservation as a tool for development. These groups have vastly different portfolios, which may lead to conflicting values – a point emphasised in many places throughout this book. This was a general conclusion drawn by Stein et al. (2003), who found that tourism professionals and public authorities have different priorities for ecotourism development in Florida. The primary emphasis of public authorities was on ecological benefits and resource management, while tourism professionals placed importance on ecotourism as a direct and indirect generator of economic benefits.

Power relations have also been discussed in reference to Amboseli, Kenya, where young economically advantaged Maasai are assuming power as a result of their exposure to the outside world, which has contributed to the erosion of traditional values in the community (Ogutu 2002). The case of Maasailand, Kenya, is illustrative of the fact that community participation has led to inequitable patterns of access to land and control over resources (Southgate 2006). Heterogeneous social conditions are said to weaken the capacity of local people to negotiate for their own well-being. Northern operators in Kenya, aware of ecotourism's huge potential, have been able to successfully negotiate inequitable relationships with local people in the absence of strong institutional support, leading to the loss of control of land and resources with few opportunities to access channels of participation. This led Southgate, citing Jamal and Getz (1995: 190–91), to conclude that 'power imbalances and legitimacy issues related to the stakeholder can inhibit both the initiation and the success of collaboration'. A similar scenario has unfolded in Botswana, where large tourism operators have marginalised local operators and investors, leading to rural development failure and poverty (Mbaiwa 2005). This has also been the case in Ostional, Costa Rica, where outside investors, in the absence of institutional policy and support, have prevented local people from receiving benefits from their involvement in ecotourism (Campbell 1999). The same pattern of exploitation has occurred in Thailand, where outside tour companies have minimised economic benefits of local people living inside Doi Inthanon National Park (Kaae 2006). The problems inherent in managing multiple stakeholders, some with much more power than others, has led theorists to contend that management should be decentralised to the lowest possible level appropriate to the situation and to the dynamics of the group; for example, locally based institutions and authorities (see Sanderson and Koester 2000; Nelson and Gami 2003; and Nagendra *et al.* 2005; Ad Hoc Technical Expert Group on Protected Areas 2003).

Adaptive co-management

A recent addition to the lexicon of governance approaches is *adaptive* co-management, which combines multiple stakeholders (e.g. the State, civil society, industry), the cooperative spirit noted in other models above, and the complex systems view of the world (Fennell *et al.* 2008; Berkes 2004). Bridging on the socio-ecological systems theory approach, adaptive co-management considers policies as experiments from which knowledge is incrementally gained through the feedback mechanism of social learning (cycles of action and reflection); and where flexibility to respond to changes is enhanced (Lee 1993; Folke *et al.* 2002). Armitage *et al.* (2007: 2) observe that, 'as an emergent outcome of these conceptual frameworks, *adaptive co-management* represents a potentially important innovation in natural resource governance under conditions of change, uncertainty and complexity'. As such, adaptive co-management is an evolving process that responds to feedback from socio-ecological systems, and it occurs when responsibilities for a resource are shared by actors who participate in a process which is flexible, dynamic, and oriented towards social learning (Ruitenbeek and Cartier 2001; Folke *et al.* 2002; Berkes 2004; Olsson *et al.* 2004).

Arguing that adaptive co-management is based on a continuum of cooperation and self-interest, and that theoretical ethics provided a solid foundation by which to examine this continuum, Fennell *et al.* (2008) concluded that adaptive co-management without an ethical axis may simply be window-dressing for well-established dilemmas of power and livelihoods. This means that despite efforts to employ our best policy and governance structures, these will be meaningless unless they are grounded in an ethical foundation, which allows for critical reflection that might highlight the ambiguities and tensions between policy and autonomous groups. The characteristics of adaptive co-management; that is, multiple stakeholders, cooperation, complexity, socio-ecological systems, feedback loops for system learning, and the ethical imperative, make it especially relevant to ecotourism. A shift in research with a focus on this area may yield many important conclusions and recommendations.

Tourism policy

A considerable amount of time was spent in Chapter 1 discussing different ecotourism definitions. This was intentional because good definitions give way to good policies, which in turn should lead to good practice: definition, policy, practice (this latter component is dealt with in Chapter 11). The following example serves to illustrate the importance of getting it right when it comes to understanding the basic underlying principles behind ecotourism.

In July 2006, a TRINET discussion (TRINET is an international tourism online discussion group dedicated to tourism scholars) closely examined the following definition of ecotourism developed by the government of British Columbia, Canada, for the purpose of generating input on how ecotourism ought to be defined for policy and practical purposes:

> *Adventure Tourism/Commercial Tourism*
> Commercial recreation, often called eco- or adventure tourism, provides residents and visitors with access to British Columbia's spectacular wilderness through a variety of guided outdoor activities. Specifically, commercial recreation is defined as outdoor recreational activities provided on a fee-for-service basis, with a focus on experiences associated with the natural environment.

The sentiment of many tourism theorists was aptly summarised by Jim MacBeth from Murdoch University, Australia, a former resident of British Columbia, who wrote that, 'the nature of ecotourism and adventure tourism in BC is very much more than the commercial tag put on it by this website', and further that:

> this definition of ecotourism is allowed to stand alone in a government website is simply to display ignorance; you are putting on a 'face' to serious ecotourists and researchers, as well as other policy-makers that brings disrepute to the BC government and to the quality of ecotourism products and operators in BC.
>
> (TRINET, 24 July 2006)

What is the lesson here? Weston (1996) defines commercial recreation and tourism as a service-based industry that is business-oriented, profit motivated, and where there is a market for the manufacturing and delivery of such products and services. Choosing to focus on a business orientation, the market and profit motivation in defining ecotourism is to truly 'show one's hand' in the provision of ecotourism opportunities. At the very least, government should be equally concerned with providing a more balanced view of ecotourism beyond the money that it generates. With this example in mind, the following provides some fundamental remarks on tourism policy, which is followed by a more in-depth discussion of ecotourism policy research.

In a simplistic sense, tourism policy is the identification of a series of goals and objectives, which help an agency – usually a governmental one – in the process of planning the tourism industry. According to Akehurst (1992), however, policy development is much more detailed and can be defined as:

> a strategy for the development of the tourism sector. . . that establishes objectives and guidelines as a basis for what needs to be done. This means identifying and agreeing objectives; establishing priorities; placing in a Community context the roles of national governments, national tourist organizations, local governments and private-sector businesses; establishing possible coordination and implementation of agreed

programmes to solve identified problems, with monitoring and evaluation of these programmes.

(Akehurst 1992: 217)

In this sense, policy is the coordination of many organisations and agencies involved in the provision of tourism services, and the planning, development and management of these groups.

All countries should endeavour to create tourism policy both to guide planning, management and development of tourism consistently throughout a region, as well as to use resources in a wise and efficient manner (Jenkins 1991). For example, Parker and Khare (2005) observe that LDCs usually have foreign investment policies designed to assist investors with development projects. The authors examine a series of policies in southern Africa – rules in regard to expatriate employment, immigration, taxes to be paid, currency exchange and import tariffs – in addition to other restrictions on development.

The sector most responsible for tourism policy creation is government, in either a passive (e.g. legislation is introduced but is not intended to discriminate in favour of the tourism industry), or an active manner, where government takes action to discriminate in favour of the tourism sector managerially, through the creation of objectives and legislative support, and/or developmentally through the establishment and operation of tourism facilities. Governments have given low priority to the establishment of policy in tourism (the first policy reports were established by the European Community in 1986), because tourism development has traditionally been the realm of the private sector (Lickorish 1991; Pearce 1989). If centralised action has not been taken within a country, often regions or localities have taken responsibility, but with uneven results. For example, Niezgoda (2004) observes that in Poland the concept of sustainable tourism has been difficult to implement at local levels because of the fact that community authorities fail to understand the concept as it is set forth in national policy documents. This is especially troubling because as of 2001, it is by law the responsibility of communal, county and provincial governments to develop tourism. This has prompted Niezgoda to recommend the development of a holistic approach to tourism planning involving organisational, legal and political changes that can be understood at all levels. Policy development may also be compounded by the fact that initiatives can be quite divergent as regards tourism and environment depending on the portfolio (e.g. agriculture, the environment, tourism and industry, and transport).

Coccossis (1996) feels that there has not been a good understanding between the complex nature of the environment and tourism (e.g. the synergistic effects of tourism), coupled with a high degree of administrative fragmentation. He suggests that environmental conservation was perceived as a threat to social and economic development; a threat which has only recently subsided as a result of efforts towards the clarification of sustainable development. Without due regard to appropriate development guidelines, the integrity of public and private lands has often been at risk because socio-ecological values are sacrificed at the expense of profit. A key point in Coccossis's argument is that environmental policies are now characterised by a more holistic, ecosystem-based approach (see Chapter 5) that relies less on specific issues and more on a regional integrated perspective. This wider approach involves the following (1996: 10–11):

1 linking development policy with environmental management. As a first step, the review of projects, plans and programmes from an environmental point of view should be instituted;
2 regional-level environmental management schemes which would provide a framework for guiding local environmental management programmes;

3   integration of tourism development and environmental management policies at the local, regional, and national level; and

4   increasing local capacity to cope with environmental issues, particularly in rapidly developing tourist destination areas.

Priority areas identified by Coccossis (1996) for sustainable tourism development in Europe include coastal areas, rural areas, built environment and urban areas, and islands. One of the best examples of policy implementation in relation to hotel development was the execution of measures to control the height of hotels in Bali, Indonesia, during the 1980s. Hotels could not be built higher than the tallest palm trees to avoid the unsightly nature of these developments.

In support of what Coccossis proposes, above, Fayos-Solá (1996) writes that tourism policy has started to change in line with the evolving nature of the tourism industry. Consistent with the paradigmatic shift in the 1980s away from mass tourism to alternative forms of tourism, the industry has given way to supersegmentation, new technologies and an increased sense of social and ecological responsibility. Competitiveness is a key function of the tourism industry and it is based on quality and efficiency rather than quantity, which has been the hallmark of mass tourism. Tourism policy objectives have moved through three distinct generations, according to Fayos-Solá (1996). The first generation was based on large numbers of visitors, the maximisation of revenues from tourism, and the creation of numbers of jobs, indicative of the mass tourism paradigm. The second generation evolved from the economic difficulties of the 1970s, including growth-recession fluctuations. In this stage, social, economic and ecological impacts are better understood, while economic and legal objectives are redefined in tourism's contribution to societal well-being. The final generation shows competitiveness as a key function of the vitality of the tourism industry. In addition, both total quality management and the partnership between public, private and not-for-profit agencies is underscored, leading to a more symbiotic relationship between sectors, as discussed earlier in this chapter. As such, the politics of policy-making for the twenty-first century should go beyond the conventional discussions related to marketing, promotion, tax incentives, accommodation and transportation to more holistic-centred issues related to the environment, social impacts, issues of rational and equitable access (who gets what, when and how), and the international regulation of health and safety issues (Richter 1991).

From a slightly different perspective, Hjalager (1996) argues that innovation must be the key to future policies and regulation designed to limit tourism's impact on the resource base. She feels that policy, when it is considered, is hampered first by the fact that it is governed by ideology and personal sympathies rather than the need for balance, and second by the fact that cost is unlikely to be considered in implementation. Her research touches on three policy control mechanisms: the market, bureaucracies and clans (industry in cooperation with other stakeholders). Bureaucracies are said to regulate tourism in two ways: (1) by directing tourists, locals, and the industry through legislation; and (2) through the provision of tourism infrastructure (see Table 10.2). Central to Hjalager's thesis is the notion that regions must be open to alternative modes of regulation in addressing significant and challenging environmental issues.

Ecotourism policy

If tourism policy has been slow to evolve and unevenly implemented, the same can be said of ecotourism, which often relies on the former in one capacity or another. According to Weaver (2001a), this means that national or regional ecotourism plans are more

Table 10.2 *Innovation's place in tourism policy and regulation*

Instruments	How the environmental effect is achieved	Influence on innovation
Emission standards. Inspection in order to control	Industrial 'end of pipe' norms expressed according to carrying capacities. Development permissions with environmental standards	Control may raise standards, severe control may lead to the development of new techniques. No evidence that tourism will take the innovation lead
Compulsory use of specific energy resources or specific technologies (e.g. district heating or waste treatment systems)	Obtaining economies of scale in environmental management systems	Typically, innovation will not take place in tourist industries, but with specialised plants and suppliers
Zoning in order to limit or control opportunities for development Control of land use	Limiting the developer's activities according to carrying capacities or environmental objectives	Probably none
Zoning of tourists' access to vulnerable resources or areas	Visitor control systems and regulation of tourism by volumes Compulsory methods or training, motivation	Opportunities for development of IT instruments combined with communication/interpretation instruments

Source: Hjalager (1996)

meaningful if related to and informed by a broader national tourism policy, which itself is a manifestation of an overall government policy. And as suggested by Sofield and Li (2003), plans will fail to be implemented if they do not fit the existing policies of government. Compounding this issue is the notion that even with policies in place, governmental agencies responsible for upholding policies have been found to work in isolation from one another, creating a great deal of counter-productivity (Jenkins and Wearing 2003).

In Hawaii, ecotourism policy development initiated as a result of a lack of consensus as to what constitutes appropriate ecotourism development. The Center for Tourism Policy Studies (1994) identified a number of policy recommendations in efforts to develop Hawaii's ecotourism industry, including: (1) a state ecotourism policy and ecotourism development plan; (2) interagency coordination and cooperation with the private sector; (3) increased funding support through direct and indirect means; and (4) active community participation in planning and decision-making. These recommendations are further broken down into integrated planning, public- and private-sector roles, research, land use, conservation, preservation, funding, marketing, operator concerns, socio-cultural concerns, regulation, monitoring and education and training. As a prime American destination, Hawaii is realising that apart from its extensive mass tourism industry, there is an emergent market for ecotourists based on interests in whale watching and the rainforest. Although rough estimates indicate that ecotourism represents about 5 per cent of Hawaiian tourism revenue, there is optimism that ecotourism is on the rise in this state (Marinelli 1997).

Much like the Center for Tourism Policy Studies cited above, Liu (1994) suggests that government policy is one of the keys to supporting the development of ecotourism. Her work on behalf of the American Affiliated Pacific Islands discusses the importance of government policy as a means by which to regulate and monitor the industry in a way that balances protection and restriction without hindering the individual operator. This means a balance between development and conservation, supply versus demand, benefits versus costs, and people versus the environment. In addition, Liu feels that government must play a leadership role in providing the necessary financing, management skills and knowledge so the private sector can operate smoothly and efficiently. This means government must (Liu l994: 8, 9):

1 facilitate efficient private-sector activity by minimising market interference and relying on competition as a means of control;
2 ensure a sound macro-economic environment;
3 guarantee law and order, and the just settlement of disputes;
4 ensure the provision of appropriate infrastructures;
5 ensure the development of human resources;
6 protect the public interest without obstructing private-sector activity with too many regulations;
7 promote private-sector activity by not competing in the business arena with private enterprise; and
8 acknowledge the role of small business entrepreneurs and facilitate their activities.

Liu (1994) has developed a public-sector guide for the implementation of ecotourism, as set out in Table 10.3, which includes points that should be considered in implementing appropriate ecotourism policy.

Increasingly, planners and developers are realising that policy must involve the many stakeholders who stand to be impacted by ecotourism development, as suggested at the outset of this chapter. According to Ceballos-Lascuráin (1996a: 85–91), some of these groups include protected-area personnel, local communities, the tourism industry, NGOs, financial institutions, consumers and national ecotourism councils. Although this is an essential aspect of policy development from the perspective of inclusion, Fennell *et al.* (2001) note that there will be problems in achieving compromise between so many stakeholders in the tourism sector within a particular planning jurisdiction. This is particularly true of ecotourism, which must balance use and preservation, as suggested by Liu (1994) above.

One of the finest examples of policy development involving stakeholder involvement in ecotourism can be found in the Australian National Ecotourism Strategy (Commonwealth Department of Tourism 1994). In this document an integrated approach to the development of ecotourism has been adopted under the belief that development and management of ecotourism is fundamental to optimising the benefits it offers. This strategy is truly a national document in that it integrates the collective opinions of Australians through the implementation of a series of public consultation workshops involving government, industry, conservation groups, educational institutions and community groups. (The stakeholder approach used in this strategy has been outlined by Boo 1992 who suggests that there are several groups that must be involved in the development of ecotourism initiatives.) The intent of the strategy is summarised as follows: 'to provide broad direction for the future of ecotourism by identifying priority issues for its sustainable development and recommending approaches for addressing these issues' (Commonwealth Department of Tourism 1994: 6). The bulk of the strategy concentrates on a variety of issues, objectives and actions, which are essentially steps towards putting policy and planning initiatives into practice. The 12 objectives of the strategy are set out in Table 10.4.

It is the ultimate aim of these directives to move Australia to create an ecologically and culturally sustainable ecotourism industry. The development of this document was instrumental in demonstrating to the rest of the world the advanced state of ecotourism in

Table 10.3 *Policy implementation framework*

Development objectives. Establish economic, ecological, and socio-cultural objectives in consultation with local communities; designate specific areas for ecotourism development.

Inventories. Survey and analyse the region's ecology, history, culture, economy, resources, land use, and tenure; inventory and evaluate existing and potential ecotourist attractions, activities, accommodation, facilities, and transportation: construct or consolidate development policies and plans, especially tourism master plans.

Infrastructure and facilities. Provide the appropriate infrastructure and facilities, avoiding a reliance on foreign capital; establish means to assist the private sector in developing ecotourism enterprises in line with ecological and cultural standards.

Market. Analyse present and future domestic and international ecotourism markets and establish marketing goals; know and understand the market in achieving goals; assist the private sector in its development of marketing strategies.

Carrying capacity. Strive to understand the social and ecological limits of use of an area through appropriate management and research; establish social and ecological indicators of use and impact; implement an appropriate pre-formed planning and management framework.

Development. Establish a development policy giving consideration to balanced economic, ecological, and social factors; form a development plan on the basis of attractions, transportation, and ecotourism regions; assist developers to plan and build ecologically.

Economic. Consider ways to enhance economic benefits; conduct present and future economic analyses; ensure that profits are made, locals benefit, and public revenues are self-sustaining.

Environment. Consistently evaluate the impact of ecotourism on the resource base; link ecotourism with other resource conservation measures (e.g. parks and protected areas).

Culture. Evaluate the socio-cultural impact of ecotourism, prevent negative impacts, and reinforce positive outcomes; empower local people to become decision-makers; conduct a social audit of social impacts.

Standards. Apply development and design standards to facilities and accommodations; facilitate the adherence to standards by providing financial or tax incentives and access to specialists.

Human resources. Promote job creation and entrepreneurship; establish community awareness programmes; provide adequate education and training for local people.

Organisation. Establish a working relationship between public, private and not-for-profit organisations.

Regulations and monitoring. Establish legislation/regulations to promote ecotourism development, through support for tourism organisations, tour operators, accommodation; establish facility standards.

Data system and implementation. Establish an integrated ecotourism data system for continuous operation that provides research and marketing information; identify ecotourism implementation techniques; and collaborate with private industry and educational institutions in implementation.

Source: Liu (1994)

Table 10.4 *Australian national ecotourism strategy objectives*

Strategy component	Objective
Ecological sustainability	Facilitate the application of ecologically sustainable principles and practices across the tourism industry
Integrated regional planning	Develop a strategic approach to integrated regional planning based on ecologically sustainable principles and practices and incorporating ecotourism
Natural resource management	Encourage a complementary and compatible approach between ecotourism activities and conservation in natural resource management
Regulation	Encourage industry self-regulation of ecotourism through the development and implementation of appropriate industry standards and accreditation
Infrastructure	Where appropriate, support the design and use of carefully sited and constructed infrastructure to minimise visitor impacts on natural resources and to provide for environmental education consistent with bioregional planning objectives
Impact monitoring	Undertake further study of the impacts of ecotourism to improve the information base for planning and decision-making
Marketing	Encourage and promote the ethical delivery of ecotourism products to meet visitor expectations and match levels of supply and demand
Industry standards/ accreditation	Facilitate the establishment of high-quality industry standards and a national accreditation system for ecotourism
Education	Improve the level and delivery of ecotourism education for all target groups
Involve Indigenous people	Enhance opportunities for self-determination, self-management and economic self-sufficiency in ecotourism for Aboriginals
Viability	Examine the business needs of operators and develop ways in which viability can be improved, either individually or through collective ventures
Equity considerations	Seek to ensure that opportunities for access to ecotourism experiences are equitable and that ecotourism activities benefit host communities and contribute to natural resource management and conservation

Source: Commonwealth Department of Tourism (1994)

Australia, and it is a model which other governments should follow. For example, the Ecotourism Task Force, in association with Tourism Saskatchewan (the government–private industry consortium that leads tourism development in this Canadian province), has developed a similar document which will act as a foundation for the ecotourism industry in the future. Recommendations include the development of an accreditation process, the development of cooperative relationships with other land-users, and the construction of sustainable ecolodge facilities using state-of-the-art technologies and modelling.

As regions make the decision to plan and develop an ecotourism industry, Ceballos-Lascuráin (1996a) advocates following a basic planning process such as the one identified below. Policy may naturally evolve out of first identifying many of the catalysts and constraints to the development of the industry. In the case of ecotourism, this process must proceed out of government's general development policy and strategy (if such a strategy exists), as noted above.

1   *Study preparation.* Includes the assessment of the type of planning required and the preparation of terms of reference.
2   *Determination of objectives.* Objectives must reflect the national or regional government's general ecotourism policy/strategy, and include development priorities, temporal considerations, heritage, marketing and annual growth.
3   *Survey.* A complete evaluation and inventory of existing resources must be made, especially related to the attraction base. The ultimate aim of this inventory is to link attractions to various market segments and forms of development.
4   *Analysis and synthesis.* This step involves studying the historical background of tourism in the region, analysing constraints to development, legal and risk management considerations, financing, tax incentives, protection of cultural and natural features, and other economic-related variables (e.g. contribution to gross national product (GNP), and complementarity with other sectors of the economy).
5   *Policy and plan formulation.* From an analysis of the synthesis, policies must be structured to reflect the economic, social and ecological needs of the region. Alternative policies should be developed to assess how each fits with the country's overall development policy, from which final policies are developed according to infrastructure, human resources, transportation, inter-sectoral coordination, establishment of councils and committees, tax incentives and subsidies, and the creation of tourism programmes.
6   *Recommendations.* The result is a plan that indicates attractions, tourism development areas, transportation linkages, tour routes, and design and facility standards. Also, recommendations are made for implementation, zoning, land-use plans for the future, economic benefits, education and training, ecological and social impacts, private industry incentives and legislation.
7   *Implementation and monitoring.* Prior to implementation, the policies and overall plan should be reviewed and ratified legally. Formal review periods should be established and committees or corporations should be developed to help implement or guide the implementation of tourism developments.

The prime mandate of many state and provincial tourism authorities is to bolster tourism and ecotourism in their respective regions through the development of new product strategies to capture a larger travel market. Given the marketing premise under which many of these regional bodies operate (they are designed primarily to generate profits through marketing and incentives), the social and ecological elements associated with sustainable tourism and ecotourism often take a back seat. However, with increasing frequency subcommittees are being created within these agencies to deal specifically with ecotourism and sustainable development. For example, in the 1990s the California legislature's Senate Select Committee on Tourism and Aviation convened to address the many problems facing the California tourist industry, including energy shortages, the inhospitality of cities and development overkill in resorts. From this committee's study of these and other problems, it was suggested that a strong ecotourism ethic (see Chapter 7) should be created that could be advantageous economically, socially and ecologically, for the purpose of generating more positive tourism experiences in the state. In part, because of some of these early examples of policy development at the regional level, I sent out an email in 1997 to all North American provincial and state governments requesting information on ecotourism policy and accreditation. Although the response was limited, the overwhelming consensus among regional bodies was that policy and accreditation are 'on the table', meaning that while they have been thought about, little has been done to move forward (see also Edwards *et al.* 1998).

In some cases regions are in the process of stating recommendations to government on how to proceed in the ecotourism sector. For example, Florida developed a report on how to protect and plan for its heritage and commercial assets (Ecotourism/Heritage Tourism Advisory Committee 1997), with the mission of developing, 'a blueprint that identifies goals, strategies and recommendations needed to create a statewide, regionally based plan to effectively protect and promote the natural, coastal, historical, and cultural assets of Florida, and to link these to commercial tourism in Florida' (1997: A–1). The plan focuses on forming strategic relationships with agencies in the community, developing inventories of sites, protection of the environment, education and marketing.

In concluding this section, ecotourism policy may be viewed as being whatever governments choose to do or not to do with respect to ecotourism (see Hall and Jenkins 1995). This becomes a relevant statement when one considers that this may involve action, inaction, decision, non-decisions, choice and process. This means that ecotourism policy should be for non-development in the same way it should be for development, as long as the integrity of people and environment is safeguarded in an equitable manner (Fennell and Dowling 2003). But this is not always the case. The 'policy tied to economic growth' perspective is demonstrated through the institutionalisation of environmental policy in Malaysia, where Hezri and Hasan (2006) argue that policy development can be patchy and haphazard. Malaysia's environmental policy has followed two broad waves of institutionalisation. The first came about during the 1960s and 1970s as a reflection of policy initiatives in the more developed world. The second emerged in response to the Bruntland Report of 1987 and the Rio conference of 1992. But despite the early onset of environmental measures in Malaysia, difficulties have emerged because of a poor response to the first wave, which constructed barriers to attempts to concrete the second and more important wave of institutionalisation. Hezri and Hasan (2006) argue that this problem is a function of path-dependent institutionalisation. This occurs because of sustained economic rewards that make any form of change slow or nonexistent. If a country is doing well economically, 'the mainstream path will form a force hostile to change' (Hezri and Hasan 2006: 47). They use ecotourism as a case where sustainability as a new policy goal has been recast as a tool for economic development. The end result of this is that environmental policy for sustainable development has taken on a business-as-usual mentality in accordance with the first wave of institutionalisation.

Conclusion

Policy has been identified as an important vehicle with which to balance the economic, social and economic elements that are deemed so important to a successful ecotourism industry. Unhappily, though, policy tends to emphasise the economic benefits at the exclusion of social and ecological factors. It is essential that power relationships in policy development do not overwhelm the efforts of tourism interest groups to strike an effective balance in ecotourism planning, development and management. A series of environmental governance models developed by Glasbergen were highlighted for the purpose of describing the many different styles used in conceiving and managing ecotourism. Those that emphasised cooperation among equal partners were deemed more important than those that emphasised self-interest. Future research might endeavour to elaborate on the costs and benefits of each model for the successful implementation of ecotourism at many scales.

Summary questions

1 Why is the link between national or regional policy and ecotourism policy so important?
2 What is the relationship between good definition, policy and good practice?
3 What is governance?
4 List five different environmental governance models that may be used to better understand how ecotourism is planned and managed.
5 What is dependent development, and why is it that LDCs seem to be more dependent than more developed nations?

⓫ Programme planning

This chapter focuses on the practical components needed for the development of ecotourism programmes. This is an area that has received scant attention in ecotourism and tourism more broadly, both in theory and practice, largely because of what appears to be a conceptual difference in how and why products are developed for tourists. Literature from the recreation field is used for the purpose of identifying fundamental differences between programmes and products. If ecotourism service providers are to develop meaningful programmes for participants, then the utilisation of a widely accepted recreational programme planning model will be helpful in generating satisfying experiences as well as profitable enterprises. This includes programme philosophy, planning, design, implementation and evaluation. The chapter also discusses aspects of professionalism that service providers will need to employ in the development of excellent ecotourism programmes.

Planning ecotourism programmes

One of the mysteries in tourism research, at least to me, is the absence of a more thorough treatment of recreation and leisure studies in the literature. By this it is suggested that although recreation and tourism can be viewed as separate fields of study, there is a strong conceptual connection between the two areas. This has been emphasised by a number of authors over the years. For example, Jansen-Verbeke and Dietvorst (1987) and Edginton *et al.* (1980) suggest that in analysing the leisure discipline from the perspective of the individual there is little difference between elements of leisure, recreation and tourism. Tourism therefore is very much a recreational beast, leading some to suggest that the basic motivation for tourism is the human need for recreation (Graburn 1989). This link is further emphasised by Metelka (1981: 90), who defines tourism as 'Free spontaneous activity; synonymous with recreation. An activity done for its own sake, rather than for economic gain'. Subsequent to this, Metelka (1990: 154) defined tourism in three ways: (1) the relationship and phenomena associated with the journeys and temporary visits of people travelling primarily for leisure and recreation; (2) a subset of recreation; that form of recreation involving geographic mobility; and (3) the industries and activities that provide and market the services needed for pleasure travel. (See Mieczkowski 1981 for a discussion of the relationship between leisure, recreation and tourism.) The resultant overlap between recreation and tourism is a topic in the work of Hall and Page (1999), who say that there is increasing convergence between the two concepts in terms of theory, activity and impacts. Perhaps one of the key reasons for polarity is the notion that tourism continues to be examined from marketing and business contexts, while the recreation delivery system still maintains a strong focus on satisfaction (i.e. how the recreation delivery system might better satisfy the needs of people).

Satisfaction and profit

A fundamental difference between recreation and tourism lies in what might be referred to as the product mentality. The product mentality carries with it, as either perception or reality, an emphasis on profit and production, and marketing and business planning. Products may be viewed as goods or as services, where goods are produced and services are performed. Tourism providers often view their offerings as goods in the sense that they are able to stockpile them for use again and again, while tourists view the product more as a service in the way in which experiences are immediately consumed on site. The tourism-as-service perspective is one advocated by Holloway (1994), who notes that tourists cannot inspect their package tour before the purchase. It is therefore a speculative investment involving a good deal of trust on the part of the purchaser. Although subject to debate, there appears to be an overriding emphasis on the profit motive in tourism, which is probably a natural by-product of the marketing and business culture that pervades this field. (The literature supports a vast array of tourism development examples that discuss the value and benefit of tourism almost solely on the basis of economic impact. The same holds true for ecotourism, where there is an ever-expanding base of literature which questions the interests and values of those involved in the business side of ecotourism (see Munt and Higinio 1993; Steele 1993).) This is not only a focus that stems from the private/commercial sector, but also one which has infiltrated the public realm. One of the classic examples is the municipal perspective where decision-makers have become quite adept at promoting tourism to bolster their economies. There is certainly nothing wrong with this type of behaviour. Problems arise, however, when the profit motive becomes the sole overriding motivation of service providers, with little regard for participant satisfaction.

Not for profits (e.g. museums) are also involved in the provision of ecotourism services. These organisations are heavily involved in the delivery of ecotourism programmes *for profit*, which, in theory, is inconsistent with their overall mandates (although not-for-profit agencies may be defined as 'not government', so this opens the door to the profit motive). Ziffer (1989) has suggested that not-for-profit sector agencies often sponsor ecotourism trips for a number of reasons, including member service, donor trips, a source of funding, and for education and research – and making a considerable profit. In a survey of American-based nature tour companies, Higgins (1996) found that the not-for-profit sector has a firm stake in the delivery of ecotourism services, with 11 not-for-profit tour operators (17 per cent of the sample totals) serving just over 20,000 clients. Higgins discovered that not-for-profits use, 'sophisticated marketing campaigns, including direct mail, advertising in tourism publications, and the development of glossy nature-tour brochures' (Higgins 1996: 16). The important role that not-for-profits play in ecotourism was also discussed by Weiler (1993), who found that of 402 tour descriptions (55 tour operators), 336 were private-sector tours and 66 were not-for-profit or university-run tours. Universities are given the status of not-for-profit, yet they sponsored and ran programmes designed to make money. (See Backman *et al.* 2001 for a discussion on the blurring of for-profit and not-for-profit roles in the development of ecotourism programmes.)

A basic aspect of recreation is that service providers are in the business of facilitating, as much as possible, participants (as something different than clients) to have satisfying recreational experiences. Indeed, this is the recreational programmer's prime mandate (Figure 11.1). The figure shows that in the case of recreation, the satisfaction of the visitor is central, with more of a focus on profit in tourism.

The accepted methodology for ensuring a common approach to recreational satisfaction is recreation programme planning. Programming is a process which entails following many simple but well-planned principles, which are not only geared towards the

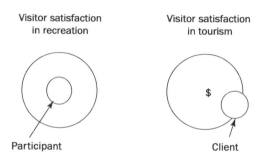

Figure 11.1 Satisfaction in recreation and tourism

satisfaction of participants, but also to better understanding how the agency may be accountable over issues related to capital, equipment, effective use of time and the behaviour of employees. Simply stated, programming is the process of organising resources and opportunities for other people for the purpose of meeting their leisure needs (Searle and Brayley 1993: 214). Recreation programming is also a form of strategic planning. It is dynamic, cyclical and interactive between people and resources, and seeks to implement and evaluate in order to achieve goals and objectives. The importance of programming to recreation is as follows:

> The program is what recreation services are all about. All else – personnel, supplies, areas and facilities, budgets, public relations – exists primarily to see that the program occurs and that people enjoy participating. Planning is the tool that makes programs happen.
>
> (Russell 1982: ix)

In general, the inherent strength of programme planning is that it is one integrated process, broken down into a few broad units, which typically include philosophy, planning, implementation and evaluation. Depending on the model chosen, these units will specify certain aspects of the programming process. However, because there are literally dozens of tasks which need to be accomplished along the way, certain tasks will need to be done in a logical order, concurrent to others, or even in reverse order (due to the nature of the programme). The following model (Table 11.1) is suggested for ecotourism programme planning. Although not exhaustive, it provides an overview of many of the aspects of ecotourism programme planning, as they apply to each unit. In the interests of space, only a few of these will be discussed in brief (see Fennell 2002b for a more in-depth overview of this process), after the following section on programmers and professionalism.

Given that ecotourism is a form of tourism that is supposedly more ethical, responsible, small scale, and community-based than other forms of tourism, it may be useful as a model to underscore the human, service values which are central to the development of sound recreational programming. As this book has attempted to point out, ecotourism is often far from ideal in how it is practised. Programming can help through the development of a standardised process of programme delivery which is drawn from a focus hinged on a participant satisfaction end. If we plan with a profit end in mind only, as outlined in the first case in Figure 11.2, the aspects we need to accomplish in realising that end – those which will need to go into the boxes in both cases of the figure – will be very different.

Hultman and Andersson Cederholm (2006) adopt the foregoing mindset in suggesting that the search for authenticity in ecotourism involves not being treated as customers, but

Table 11.1 *Important aspects of programme planning*

Philosophy	Needs and assets	Planning	Implementation	Evaluation
Philosophy	Understanding	Programme	Green marketing	Why evaluate
Mission	needs	structure	Quality	Who evaluates
Vision	Defining need	Brainstorming	Staff training	Formative
Goals	Needs	Briefing	Public relations	evaluation
Objectives	assessments	Health and safety	Budgeting	Summative
Programme	Assets inventory	Transportation	Itineraries	evaluation
theories	SWOT analysis	Food and water	Implementation	Evaluation models
Programme	PESTE analysis	Clothing/	strategies	Programme
strategies	Understanding the	equipment	Operating in the	decisions
Programme	environment	Permits	environment	Debriefing
approaches		Leadership		Accreditation
		Risk management		Audits
		Interpretation		
		Business planning		

rather as persons in heart-to-heart interactions. 'Cutting edge' programmes, therefore, are not only about the region visited but also about the unique experience at these places. The authors argue that ecotourism can become 'placeless', whereby the value of the experience outweighs the value of the place in which it occurs. The authors argue that this was the basic thesis in Poon's (1993) work on 'new tourism', where contemporary tourism products have evolved into more personalised, tailor-made excursions. This line of thought has been adopted by Gössling (2006), who observes that ecotourists are more amenable to consuming experiences these days than sustainable journeys. Evidence of this rests within a myriad of ecotourism websites where service providers use a variety of powerful messages that focus on the experiential basis of ecotourism. Gössling argues that cornerstone attributes of ecotourism like 'conservation' and 'education' are now subordinate to more hedonistic themes such as 'pleasure' and 'meaning' in advertisements, which give added value and superior experiences. As such, 'Nature no longer exists as a scientific entity [see Chapter 2]; it is now a romanticized playground for experience-interested tourists' (2006: 93). Intriguing in Gössling's discussion is not only that there is a move from meaning (i.e. active education and strong environmental commitment) to pleasure,

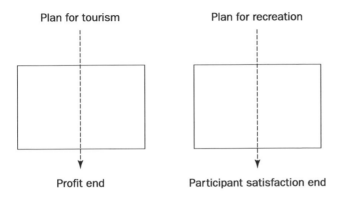

Plan for tourism   Plan for recreation

Profit end   Participant satisfaction end

Figure 11.2 Planning for tourism and recreation

Plates 11.1 and 11.2 No matter where the ecotour occurs, operators and guides must place considerable value on the needs of participants

but also that there is a trend to advertise ecotourism experiences as 'deep' and 'embodied'. This is contrary to the conventional approach to ecotourism, where what was most important in distinguishing ecotourists was the degree to which the individual subscribed to environmental education, environmental commitment, sustainable practices and so on. This may very well be a reflection of the level of knowledge that 'ecotourists' have about this form of tourism, which may in turn be a function of the mixed messages that ecotourists have received over the years regarding what is (almost everything) and is not (almost nothing) ecotourism. What is being suggested here is that ecotourism may thus be seen as an experience that is more exclusive on the basis of novelty and high price, rather than on traditional ecocentric-based offerings. The logical question that must be asked is whether or not these hedonistic offerings are indeed ecotourism.

Programmers and professionalism

Mitchell (1989) provides an excellent overview of a hands-on approach to ecotourism operation and outfitting (see also Chapter 6 in reference to operators, guides and outfitters). Although too detailed to present in its entirety here, Mitchell identifies the following main programme-related aspects in structuring and running ecotours: professional tour guiding, a survey of personal assets, a survey of recreational assets, balancing demand, environmental education, tour planning, pricing, preparing the client for the trip, questions clients ask, packing lists, client medical information, cooking and food, conflict resolution, camp and camp cleanliness, first aid, legal considerations, on-the-trail sitings and evaluation. For those more interested in actually organising and conducting tours, this document is a must. In addition, more specific manuals related to first aid, risk management, leadership, interpretation, recreational programming, marketing and finance, and so on, are recommended to prepare the operator for the task of leading successful ventures in the field.

Tourism service providers coached in proper techniques of programming may be witness to the many tangible benefits, which are generated for the programme and service provider, for other employees and, most of all, for the participant. Also, programmers who may work in companies with as few as two people working in them must often 'wear many hats'. This may mean involvement in administration, public relations, community liaison, marketing, promotion, evaluation, programme design, establishment of goals and objectives, needs assessment, leadership, budgeting, facilitation, teaching and environmental education, scheduling and so on. Because the responsibilities of the programmer run deep, it is essential that they adhere to the following principles (adapted from Edginton *et al.* 1980):

- placing the needs of the participant first;
- commitment to the ideals of ecotourism;
- protection of the participant's rights;
- acquisition of knowledge prior to engaging in professional activities;
- practice of the highest standards of professional service;
- continuous upgrading of professional knowledge, skill and ability;
- operating ethically and equitably;
- maintaining a collaborative relationship with the participant;
- operators must self-regulate their behaviour accordingly;
- contributing to the development of the profession and other professionals.

These guidelines underscore the importance of being professional in the delivery of services, where the customer is valued as more than simply a means to a financial end. Valuing and practising equity, high standards, collaboration with others, commitment, knowledge and protection of rights are critical for success, but also in terms of the communication of such values within the operation.

Professionalism has been viewed in essentially two ways. In one sense, the term implies only that one is paid for a service (e.g. a professional football player). It is, however, the stricter sense of 'profession' regarding conceptions surrounding occupations such as lawyers and doctors, where individuals adhere to professional norms, that is more difficult to define and is the focus of this section (see also Wearing 1995; Sessoms 1991; Wilensky 1964). In recreation, researchers and practitioners have historically found it difficult to agree on a set of professional criteria to guide accepted practice. Professions therefore need to satisfy, and be consistent with, broad criteria in pulling the discipline together, as evident in the following work by Weschler (1962):

1 The profession has a body of specialised knowledge.
2 The profession sets its own standards.
3 The profession requires extensive preparation.
4 Specialised knowledge is communicable.
5 Service is placed over personal gain.
6 The profession has a strong professional organisation.

In recent years university and college programmes in Canada have shown interest in addressing the issue of professionalism in the adventure tourism and ecotourism fields. The adventure Travel Guide Diploma programme of the University College of the Cariboo (UCC) in Kamloops, British Columbia, is one such programme that has been developed to prepare a more highly trained labour force in this area (Olesen and Schettini 1994). These authors suggest that demand for highly trained personnel in the adventure tourism industry will grow at 3.6 per cent (compounded rate), which, comparatively, far exceeds the anticipated 1.6 per cent yearly growth rates of the entire British Columbian labour force. This programme recognises the integral relationship between industry and government (and government-sponsored institutions) in certifying highly trained instructors, guides and operators (see the section on certification in Chapter 7). The UCC (now Thomson Rivers University) programme emphasises the following five main themes as being critical to the success of the programme: (1) technical skills (e.g. wilderness first aid training), (2) business skills, (3) hospitality and service skills, (4) environmental skills, and (5) intimate involvement in industry associations (e.g. student networks with potential employers).

The tourism field will need to be vigilant in considering how best to integrate professionalism in the context of service delivery, especially in regard to the implementation of sustainable tourism measures (as noted in Chapter 1), where there continues to be a void in human resource management and other management practices (Henry and Jackson 1996). A key area in making this a reality is through educational programmes that enable the tourism worker to feel personally competent and confident in his or her work; as Wright (1987: 17) points out, 'it is preferable to establish credibility through capability, competency, efficiency and a professional attitude. One earns respect; one does not establish respect through a glossy diploma on the wall'.

Programme philosophy, mission, goals and objectives

The development of an ecotourism philosophy requires the examination of some basic assumptions regarding the place of ecotourism in the context of the environment, the nature of human activity, the basis of human nature and the nature of human relationships (Malloy and Fennell 1998a). These may further evolve into a series of other questions related to the meaning of leisure for individuals and for society as a whole. Kraus and Curtis (1990) write that the philosophy which leisure service organisations develop is a direct reflection of the acceptance of leisure as an important part of life and the extent to which it meets human needs. These needs may relate to accessibility, happiness from involvement in the pursuit, the consideration of a means–end relationship, rights of the individual, freedom, as well as equity (see Edginton *et al.* 1998). An agency's philosophy, then, is a direct reflection of what the entity values. There are perhaps a series of other core environmental values which would be articulated by ecotourists over other types of tourists as being fundamental to the travel experience. These may include reducing, reusing and recycling; harmony; exploration; multiculturalism; preservation of landscapes; biodiversity conservation; integrity; learning; service and knowledge.

Mission statements are important as vehicles to actuate the philosophical direction of an organisation. These should be bold, inspiring and communicate direction to customers and workers inside the organisation. Campbell *et al.* (1990) suggest that mission statements should be built around a purpose, values, strategy, standards and behaviour. In essence, all of the various things that the organisation will plan on doing will emanate from the mission statement. Kraus (1997) writes that the mission should be reviewed periodically in light of changing conditions, needs and agency capabilities.

Kraus (1997) further writes that goals and objectives, although often used interchangeably, are different in what they aim to accomplish. Goals are designed to reflect the mission of an organisation, and may be viewed as purpose(s) which are often communicated as statements which originate from the mission statement. Objectives, on the other hand, are much more specific, measurable and attainable, depending on the nature of the service provider. They are the steps required to reach specific goals, and are written in a concise way according to the actions to be achieved as well as the outcomes which are expected over a given period of time. Objectives will quite naturally flow from the various goal or purpose statements which are a direct reflection of the agency's overall mission.

Objectives are of two types. The first type includes programme (also called operating, production, input or implementation) objectives, which relate to the means (materials, leaders, facilities, etc.) used to operate the programme, the amount of energy expended in the programme delivery, or matters related to client participation (e.g. to increase ecotourist awareness of endangered plants.) The second type is referred to as performance objectives, which are the measurable end products of the programme that link directly with the mission and goal statements. (e.g. Participants will have identified at least one half of all bird species on the identification list.)

Needs and assets

Understanding participant needs and resource assets is critical in decisions regarding the initiation of new programmes and for the choice to continue with a programme after it has been offered. Needs and assets are not static, but rather change over time. As such, a periodic assessment of these helps to keep the programme timely and efficient. The ecotourism service provider will be met with a variety of needs (see Knowles 1980), which derive from participants, the organisation itself and the community. All of these must be filtered according to the overall requirements of the ecotourism operation, as well as the various sources of need. Based on the philosophical orientation of the service provider, such needs will be operationalised through a series of programme priorities that provide the basis of the ecotourism programme.

Needs assessment

The first step, and one of the most important in the process of developing ecotourism programmes, is the identification of participant needs which the service provider will seek to isolate and later satisfy. Indeed, it should be the prime directive of service providers to strive to meet or exceed the expectations of participants through the programmes that they develop (McCarville 1993). To do this, programmers frequently use needs assessments which incorporate any number of different survey or measurement tools to explore needs. (Tourism planners have often used tourism strategies or tourism marketing plans which focus on aspects of demand from generating regions, as well as on the state of affairs within destination regions, including capacity, attractions and impacts (Veal 1992). The extent to which these effectively address recreational or tourism need, in the manner addressed in this chapter, is open to debate.)

The needs assessment is, 'the process by which the program planner identifies and measures gaps between what is and what ought to be' (Windsor *et al.* 1994: 63; see also Gilmore and Campbell 1996). Such needs may be identified for an individual, but more typically for groups of people (grouped by gender, age, geographic location and so on). Table 11.2 addresses some of the reasons why practitioners conduct needs assessments (adapted from DeGraaf *et al.* 1999). For example, under service orientation, the needs assessment may encourage tourists to provide valuable input into the programme; it may allow for new ideas in creating innovative programming; it may allow the service provider to serve as many people as possible; and it helps the programmer stay in tune with psychological and physiological needs of participants. Depending on the time, situation, resources and experience of the programmer, a number of steps will need to be followed in conducting the needs assessment, including the following general catregories: (1) defining need; (2) identifying study subjects; (3) asking the right questions; (4) collecting the data; (5) analysing and interpreting data; and (6) reporting and using the data for planning.

Table 11.2 *Why conduct needs assessments?*

Reason	Description
A. Service orientation	
Tourist Input	Programming decisions may be influenced positively by allowing tourists the opportunity to share their views
Solicit new ideas	Allows for creativity and innovation in programme planning
Inclusion	Enables the provider to serve as many people as possible, regardless of age, sex, ability, race, religion, education
Meet real needs of people	Helps the programmer stay in tune with the physiological and psychological needs of participants
B. Desire for quality	
Professional commitment	Operators must strive to be ethically based and to exceed standards which may be set for the industry/region
Resource allocation	Helps to inform the operator on where s/he needs to focus (e.g. more staff, better transportation, and so on)
Accountability	Being responsible for our actions and who and what they impact (tourists, parks, local people, other land users)
Increase profits	Important element to ensure that the company makes profits while ensuring tourist satisfaction
C. Programme management	
Manage duplication	Duplicating services within or between operations is a waste of resources and leads to unhealthy competition
Address safety	How to better use facilities, transportation, and better understand hazards in very dynamic environments
Plan a variety of programmes	Help us design and implement the best possible programmes for the specific targets that we cater to
Develop/meet individual goals	To help providers and participants to meet their goals, such as skill development and environmental education

Source: Adapted from De Graaf *et al.* (1999)

Plates 11.3 and 11.4 The needs of ecotourists involve getting close enough for observation and photography

Inventories of attractions and resources

Just as the needs of participants change from year to year and from programme to programme, so too do the base of attractions and physical, natural, recreational and socio-cultural resources which form the basis of ecotours. In view of this, the service provider should regularly inventory attractions and resources in order to document these changes. In general, such an inventory is a systematic compilation of attractions and resources which occur within a tourism destination region, and which demarcate the spatial and temporal boundaries of the ecotour. It is the task of the operator not only to list these potential attractions and resources but also to identify any special significant factors related to these attractions which would allow the operator to generate greater value in programme offerings.

A comprehensive list of general attractions and resources is included in Table 11.3. The table includes the 'presence/availability' of a resource, the 'quantity' of the resource

Table 11.3 *General resource inventory*

Resource	Presence/ availability	Quantity (no.)	Present quality (rank 1–10)	Capacity to absorb tour (low, med., high)	Future potential (low, med., high) and notes
Natural attractions					
Beaches	No	–	–	–	Nothing available, not appropriate
Birding	Yes (etc.)	Many types	8	(medium)	Migratory and resident species (high)
Karst landscapes					
Temperature					
Rainfall					
Average sunshine					
Desert environment					
Forests					
Fungus					
Geological formations					
Hiking trails					
Islands					
Lakes					
Lichen					
Mammals					
Mountains					
Nature trails					
Parks/protected areas					
Plants					
Rivers					
Waterfalls					
Wilderness areas					
Cultural/historical attractions					
Accommodation					
Transportation					
Food and beverage etc.					

Source: Adapted from Fennell (2002b)

(e.g. waterfalls), as well as the 'quality' of the resource for the purposes of the ecotour. It also includes a section to assess the ability of the resource to absorb the demands of a tour, which could include selected composite measures such as temporal and spatial aspects of the tour, levels of impact, levels of use, transportation and so on. The operator may also record a subjective assessment of the future potential of the resource.

Once completed, the information yielded from this process will be an essential component of the programming planning stage. The operator should also search for individuals or groups – human resources – in the community who may be able to help in the development of the programme (e.g. interpretation of aspects of the environment). Every community has people who are knowledgeable about the natural and human history of a region, and who may be willing to share this information with ecotourists. For this reason, the identification of a number of specialists will help put variety in the programmes with the added benefit of building strong support in the community (see Chapter 6).

Programmers can also use SWOT analysis, which is an acronym describing the process of better understanding Strengths, Weaknesses, Opportunities and Threats, in assessing the programme design potential. An example of this process can be found in Kelkit *et al.* (2005), based on their work on Kazdagi National Park, Turkey. Kazdagi (Mt. Ida) is a good example of a deliberate method to increase ecotourism as a sustainable development option in the face of mounting environmental problems. In recognition of the important role that the park can play as an *in situ* gene bank, park managers have eliminated hunting and mining in the region completely in favour of a range of other less intrusive outdoor recreational activities (wildlife viewing, trekking, jeep safaris and picnicking). In an effort to be as proactive as possible in helping to balance conservation with increasing tourism pressure, the management team undertook a SWOT analysis in order to derive strategic proposals for future development. The result of this analysis is located in Table 11.4.

Table 11.4 *SWOT analysis of Kazdagi National Park*

Strengths	Weaknesses
The park has experience with tourism.	Activities not yet available for all people.
Visitors are satisfied.	Local people not supportive of conservation yet.
Attractive natural resource base.	Too much access to the park by car.
Attractive historical resources.	Trails are not well signposted.
An intensive recreation area exists to accommodate tourism.	Car parking facilities are inadequate.
The park is an employer.	Litter and litter collection remain problems.
There are a range of activities for tourists.	
Natural resources not yet threatened.	
Opportunities	**Threats**
Encourage local people to participate in park's protection.	Increasing tourist numbers will place pressure on the park.
The protection of the park's resources will make the area more attractive as a tourist destination.	No strict park regulation makes it difficult to curtail park development.
	The park's budget is inadequate to meet the demands of an increasing flow of visitors.

Kelkit *et al.* (2005) argue that in order for the park to become sustainable, it must not succumb to economic interests alone. While the park is acknowledged as an excellent candidate for ecotourism, the interest in increasing visitation in the face of diminishing local support, an insufficient budget, and few regulations, opens the park up to a scenario that is all too familiar in ecotourism.

Programme design

This section of the programme planning model is the most comprehensive and includes a number of key features including programme structure and logistical considerations. All of these will be discussed in consideration of a variety of sub-themes that fit within each.

Structure

The programme structure involves five main components, which outline the basic shell upon which other programme elements are added (see Table 11.5). These include:

1 *Programme areas.* Ecotourism programme areas are activities which form the basis of the actual programme. From a purely recreational perspective, and at a broader level, these activities include hobbies, music, outdoor recreation, mental and literary recreation, social recreation, arts and crafts, dance, drama and sports and games. Ecotourism programme areas would include, for example, bear viewing, whale watching, stargazing, participation in wolf howls, birding and perhaps other more marginal ecotourism activities like canoeing and kayaking – depending on the motivations of participants.

2 *Programme formats.* Ecotourism experiences can be structured on the basis of one of many different programme formats. A programme format such as 'trips and outings' will be most often used by ecotourism service providers. This is not to say, however, that it is the only one, or that more than one format could not be used for a programme. Ecotourism providers may develop 'instructional' classes for their potential clients, including, for example, slide shows on the birds of Costa Rica. This breakdown of different formats suggests that while certain activities are best presented using one type of structure over another, it is also important to realise that the appropriate format will enable guides and staff to perform at the levels to which they feel most comfortable.

3 *Programme setting.* Even though ecotourism has been a concept intricately tied to wilderness settings, more and more research is emerging to suggest that it is also an urban and rural phenomenon (see Chapter 3). As such, wilderness, urban and rural settings can be matched with at least five sub-settings, including marine, freshwater, terrestrial, aerial, and a combination of these.

4 *Programme mobility.* Forms of transportation may be categorised according to motorisation and non-motorisation. In the case of the former, this may include cruise liners, powerboats and helicopters; in the latter, canoes, rafts, balloons and snowshoes.

5 *Programme lodging.* The success of ecotourism experiences is hinged, in part, on the choice of accommodation used in the programme. These may range according to units that have a non-fixed roof, and those with a fixed roof. Non-fixed roof units include hammocks, tents, caves, tepees, lean-to's, igloos and bivouacs; fixed-roof dwellings include hotels, motels, resorts, bed and breakfasts, recreational vehicles, ecolodges, ranches, huts, houseboats and tree houses (lodging is discussed at more length in Chapter 6).

Once programme areas, formats, setting, mobility and lodging options are selected they may be fixed within a broader programme matrix. This allows for an appreciation of the magnitude of different options available to the service provider, while at the same time providing the basic structure of the programme. Programmers select the most appropriate options for the programme on the basis of availability, what might appeal

Table 11.5 *Programme design matrix*

Programme area	Programme format	Programme setting	Programme mobility	Programme lodging
Nature oriented		*Backcountry*	*Motorised*	*Non-fixed roof*
Bird census	Club	Marine	Cruise liners	Hammocks
Nature games		Freshwater	Powerboats	Tents
Erosion control	Competition	Terrestrial	Aeroplanes	Caves
projects		Aerial	Helicopters	Open shelters
Nature crafts	Trips	Combined	Trains	(lean-to)
Wolf howls		settings	Cars and buses	Tepees
Tree/plant	Special events		Off-road vehicles	Igloos/snow
identification		*Rural*	Snow vehicles	caves
Stargazing	Instructional	Marine		Bivouacs
Nature walks		Freshwater	*Non-motorised*	
Wildlife viewing	Drop-in	Terrestrial	Canoes and	*Fixed-roof*
Hang gliding		Aerial	dugouts	Hotels and
Outdoor	Service	Combined	Kayaks	motels
photography		settings	Rafts	Resorts
Spelunking	Outreach		Yachts	Bed and
Snorkelling		*Urban*	Gliders	breakfasts/Inns
		Marine	Balloons	RV/caravan parks
Adventure		Freshwater	Mountain bikes	Ecolodges
Culture		Terrestrial	Skis and	Ranches
Farm and ranch		Aerial	snowshoes	Huts and cabins
Heritage		Combined	Horse, llama,	Houseboats
(etc.)		settings	camel	
			Hiking/	
			backpacking	
			Wheelchairs	

Source: Adapted from Fennell (2002b)

most to participants, or other issues including the goals and objectives of the programme and the strength and weaknesses of staff and available resources. This may entail a great deal of brainstorming on the part of programme staff, which may also include the development of several programming scenarios. These may be weighted or judged according to any number of criteria, including time, budget, equipment, human resources, risk and so on.

Logistical considerations

Logistical and practical aspects are important to the physical and mental health of participants. This starts with participant briefing either well ahead of the trip date or the day of the trip, depending on the ecotour. This information can be rather specific to the tour, as regards animal checklists and so on, or it may be of a general nature, with regard to passports and visas, vaccinations, health insurance, importing goods, luggage, money and home security. In briefing ecotourists, it is also helpful to rate tours on the basis of terrain type, length and physical exertion required. This provides the needed information enabling ecotourists to find the programme that best suits them physically, emotionally and mentally. Attracting individuals to a tour who are not mentally or physically prepared for the rigours of the trail accomplishes nothing and, in fact, depending on the risk involved, might place individuals at risk.

Reasons to be cautious

As outlined above, there are many reasons why ecotour operators should be cautious in the development of their tours. It is widely recognised among those who programme in the field of recreation that the most hazardous part of a programme is not the activity participation phase but rather the transportation of participants from one site to another (Gilchrist 1998). To ensure that accidents do not occur, service providers must be cautious in who they enlist to drive vehicles, the vehicles themselves which are chosen to transport, and the length of time that drivers are asked to transport participants.

No matter how refined or primitive the programme setting, there must be utmost concern for the manner in which food is stored, served and disposed. It also follows that to keep food safe is to know the environment in which one is travelling. Urban environments will be different than wilderness ones, while hot and humid environments will be different than cold and dry ones. Furthermore, it is not just bacteria and protozoans which need to be considered in keeping food safe, but also ants, mice, raccoons and bears. Operators may need to use airtight containers for food storage in wilderness areas, or they may opt to keep food safe by hanging it from trees out of the reach of nocturnal animals. It should also be noted that guides must be open about cleanliness, by making it a practice to wash hands and dishes often and in full view of participants, to ensure that both parties feel comfortable with the food-handling process.

Stomach ailments are common health problems in travellers. The usual culprits are food and the consumption of contaminated water. In the case of the latter, contamination comes in the form of bacteria (e.g. *E. coli*, *Salmonella typhi*, *Vibrio cholerae*), protozoans (e.g. *Giardia lamblia*, *Cryptosporidium*) or viruses (e.g. hepatitis), which appear to be much more prevalent in tropical destinations (Brock 1979). The largest of these, the protozoans, have been the topic of great interest, especially in relation to the devices designed to eliminate them from drinking-water supplies. Because of their size (10–100 μm, compared to bacteria which are generally 2 μm in size, and viruses which are 0.1 to 0.2 μm), protozoans are more easily captured through various water filtration devices that are on the market. Pathogens in water can be eliminated in one of three ways: boiling, chemicals such as iodine and filtration. The operator should keep in mind that although generally effective, protozoans, like *Cryptosporidium,* are resistant to iodine, and some people react negatively to this substance.

Operators should insist that their guides carry up-to-date back-country first aid guides in the event that first aid is required. There should be the recognition that first aid in wilderness settings is different than first-aid in urban settings. In the latter case, responsibilities surround breathing, bleeding, splinting and safety. In the wilderness, care goes beyond what is presumed to be 'first aid' to other concerns related to exposure, shelter, travel, sanitation, innovation and the administering of proper fluids and food (Merry 1994). Furthermore, guides should be aware of access and egress points while in the field, know where the nearest healthcare facilities are located and how to get there, carry a cellular phone, as well as a list of a number of pertinent contact numbers in the event of an emergency. In addition, guides should be encouraged to seek advanced certification in first aid in order to be as current, and as prepared, as possible.

The range of health risks that travellers are exposed to is extensive considering the number of plants and animals which have the ability to inflict pain or discomfort. Concern exists over the largest and most toxic animals that inhabit the same spaces as tourists (e.g. crocodiles, spiders and snakes in Australia, bears in Canada, snakes in India). In reality, it is the microscopic organisms, as we have seen above, which usually cause most of the problems. The effects of plants and animals may be indirect (their presence may dissuade people from travelling to a region), or direct, such as in the case of a first-hand encounter with a plant or animal. For example, mites and ticks are carriers of pathogens that are the cause of many human diseases such as Lyme disease, Rocky Mountain spotted fever and tularemia. Lyme disease is a case in point, where the secretions from the deer tick, *Ixodes dammini*, infect human hosts with the bacterium *Borrelia burgdorferi* through its feeding behaviour. The deer tick has highly specialised hooked mouth parts that enable the invertebrate to maintain a solid hold on the host. To aid in its feeding, the tick will also flatten itself on the host to avoid being brushed off. Early symptoms of the disease include a skin rash at the point of contact, and later a doughnut-shaped rash. This is accompanied by flu-like symptoms, swollen glands, stiff neck, backache, nausea and vomiting. Later stages of the illness may include problems with joints, inflammation of the heart, neurological problems (temporary paralysis of face muscles), with many of these symptoms highly variable between patients.

Risk management

At its most fundamental level, risk management is a process developed by an agency to avoid costs. Generally, these costs can be as a result of injury, damage to equipment and facilities, as well as wrongful actions. The main sources of risk (i.e. the things that exist in one's programme which contribute to the types of risk outlined previously) include facilities, equipment, the programme itself and people (Corbett and Findlay 1998). More succinctly, risk management may be defined as the, 'formal process of assessing exposure to risk and taking whatever action is necessary to minimize its impact' (National Association of Independent Schools, cited in Ammon 1997: 174). It is thus a sophisticated phrase for a very basic management tool, and the need for an appropriate risk management plan cannot be overstated for all recreation and tourism providers. Naturally, however, in consideration of all these various forms of tourism and recreation, some, like adventure pursuits, require very detailed and specific plans, for, as Liddle (1997) suggests, such activities are based upon the perceived potential for loss, with the occurrence of actual loss which may mean the loss of life for these activities – being unacceptable. In general, the risk management plan will involve a series of steps allowing the operator to identify risks, classify these, develop risk control measures, implement these measures, and monitor and modify (Watson 1996; see Table 11.6).

Potential risks are often subject to an evaluation of either 'low, medium or high', or perhaps some other method such as '1' (low) to '10' (high). For example, under the category of 'location' the site may be evaluated as low, the weather as low, the routes and accommodation as medium, and the facilities as high. Corbett and Findlay (1998) suggest that the next step is to determine the measures which may be employed to control the significant risks which have been identified on the basis of the following four options:

1   *Retain the risk.* Nothing is done because the likelihood of occurrence is low and/or the consequences are slight.
2   *Reduce the risk.* Practical steps are taken to reduce the likelihood of occurrence and/or the consequences, usually by changing human behaviour or actions. This may include designing and following regular maintenance programmes for equipment and facilities, helping staff to pursue appropriate professional development, or posting appropriately worded warnings and rules within a facility.
3   *Transfer of risk.* The level of risk is accepted, but passed on to others through contracts. This may include the insistence that all participants sign a waiver of liability, the purchase of insurance which is appropriate in scope and amount for all activities, or the contracting out to other agencies (e.g. transportation).
4   *Avoid the risk.* Steps are taken to restrict, limit, postpone or cancel certain activities. This may include postponing events in dangerous conditions, not travelling in bad

Table 11.6 *Outline of a risk management plan*

1 General description	6 Participants
(a) Name of program	(a) Number
(b) Type of activity	(b) Skill level
(c) Level	(c) Characteristics
2 Dates and times	7 Leaders
(a) Dates	(a) Number/roles
(b) Times	(b) Qualifications
3 Goals and objectives	8 Equipment
(a) Organisational	(a) Type and amount
(b) Activity	(b) Control
4 Location	9 Conducting the activity
(a) Site/area	(a) Preactivity preparation
(b) Weather	(b) Group control
(c) Routes/accommodation	(c) Teaching strategy
(d) Facilities	(d) Time management
5 Transportation	10 Emergency preparedness
(a) Mode	(a) Policies
(b) Routes/destinations	(b) Health forms
	(c) Telephone numbers

Source: Adapted from Ford and Blanchard (1993)

weather, restriction to participants with certain levels of expertise, insistence on maintaining a 'no alcohol' consumption policy for certain activities, or adhering to an organisation's or region's policies.

Bodansky (1994) observes that risk is a function of the magnitude and probability of harm. It involves estimates of probabilities about whether certain events will occur or not. In adventure recreation, risk is normally understood in two ways. Real risk is the true potential for loss, where no loss means zero risk, and death relates to extreme risk (Priest 1990). Real risk cannot be estimated with certainty at any one time. On the other hand, perceived risk amounts to the best estimation of real risk. Those who have gained extensive experience at high-risk activities such as mountain climbing can better understand the inherent risk of an activity. These options are structured in order to provide enough flexibility for service providers to make decisions which are not only in their own best interests (as regards cost and resource accountability as well as liability), but also in the interests of ecotourists, who will want to know that the operator has done as much as possible to control for the expected and unexpected.

Programme implementation

The effort placed into planning will have merit if it leads to programme implementation. This step of the programming process can include marketing, quality, staff training, public relations, budgeting, implementation strategies, and schedules and itineraries. For the purposes of this section, only implementation strategies will be considered.

McKenzie and Smeltzer (1997) argue that programme implementation can occur successfully if consideration is given to: (1) the resources that are available for the

Table 11.7 *Major steps in Borg and Gall's research and development cycle adapted to programme planning*

1 *Preliminary field testing.* Programme given to a few individuals from the target population, with interview, observation, and survey data collected and analysed.

2 *Main programme revision.* Based on results of preliminary field test.

3 *Main field testing.* Programme given to about twice as many individuals as in step 1. Pre- and post-programme quantitative data collected; results examined with respect to programme objectives.

4 *Operational programme revision.* Based on results of the main field test.

5 *Operational field testing.* Programme given to about twice as many clients as step 3. Interview, observation, and survey data collected and analysed.

6 *Final programme revision.* Based on results of the operational field test.

7 *Dissemination and implementation.* Programme shared with others and implemented.

Original source: Borg and Gall (1989); this table adapted from McKenzie and Smeltzer (1997)

programme; and (2) the setting for which the programme is intended – which is especially relevant to ecotourism. Programme implementation, therefore, may involve a significant outlay of resources, including time, personnel and capital. These resources are best viewed in the context of a specific strategy.

A model discussed by McKenzie and Smeltzer (1997) (Table 11.7), developed by Borg and Gall (1989), is based on a series of field testing and revision stages which allow the programmer to secure a heightened level of programme acceptance and momentum. Its strength lies in the ability to change the programme at various stages as well as to drop the programme altogether without a significant outlay of capital if, for example, participants are greatly dissatisfied with it after step 3. This implementation strategy includes an element of programme building, which is strongly recommended by McKenzie and Smeltzer. While it appears to be about as thorough as one needs to be in the construction of a programme, the authors suggest that the process may be modified or scaled down to suit the needs of the service provider. The importance of the programme relative to other programmes offered, in light of time and other resources, will be an important consideration in assessing the level of detail required for this approach. Having said this, it should be realised that those who follow through with the model in all likelihood stand a greater chance of developing more successful programmes later on.

Programme evaluation

While evaluation is often thought of as a process which is employed at the end of a programme only, it is actually an ongoing process which enables service providers to control the events that go on within the programme setting. Its importance to the overall programme planning process is underscored by Hall and McArthur (1998), who suggest that organisations which fail to evaluate are not committed to: (1) understanding performance; (2) learning from mistakes; and (3) self-improvement. Evaluation can be defined as 'the identification, clarification, and application of defensible criteria to determine an evaluation object's value (worth or merit), quality, utility, effectiveness, or significance in relation to those criteria' (Worthen *et al.* 1997: 5; see also Busser, 1990). Evaluation can take place in the following areas: (1) administration (e.g. planning, organising, staffing, training, directing work, exerting control and inspiring creativity); (2) leadership or personnel (the qualifications, competency

☐ yes ☐ no ☐ n/a Does the operation include accommodation that directly involves and enhances the activities?

☐ yes ☐ no ☐ n/a If portable accommodation facilities (tents/tepee) are used, are they designed for minimum impact?

☐ yes ☐ no ☐ n/a Does the accommodation enhance the 'nature' experience?

Figure 11.3 Accreditation standards for sustainable accommodation

Source: Ecotourism Society of Saskatchewan (2000)

and performance of staff); (3) programme content (meeting needs of participant, scheduling, transitions, goals and objectives, supervision, records and so on); and (4) physical properties (including design, construction and maintenance of facilities and equipment).

A popular evaluative technique which is especially relevant to ecotourism programmes is the evaluation by standards approach. In this form of evaluation, professionally developed standards developed by respected professionals in the field are established to act as levels of achievement or goals, which must be satisfied (Kraus and Allen 1997). For example, rafting operators would have to ensure that participants wear helmets at all times while in the raft. This method focuses more on administrative details than customer satisfaction, and the prescribed standards which are developed often reflect the values of the agency or professionals responsible for their development (e.g. Association for Experiential Education, which has established a whole series of standards for a number of outdoor recreation activities). The same holds true for accreditation. The province of Saskatchewan has recently developed a range of professional standards which prescribe exactly what operators must do to ensure an appropriate standard of care (e.g. considerations of design, ecolodge considerations, enhancing the nature experience) in order to gain accreditation (see Figure 11.3). Farrell and Lundegren (1993) write that this is one of the most common evaluation approaches used by those in the public domain.

Formative and summative evaluation

Based on the previous discussion, evaluation should be employed at stages where it is thought to be most beneficial, including programme development and implementation stages. Evaluation at earlier stages is referred to as formative evaluation, and focuses on how best to initiate the programme. For example, it may be used to fine-tune the informative aspect of a wolf-howling programme at selected points of the survey. Formative evaluation provides immediate feedback on the value of the interpretive tools, which may thus be built into the programme at later dates.

By contrast, summative evaluation provides programme decision-makers and consumers with judgements about the programme's predetermined merit in relation to criteria. Using the example of the wolf-howling programme above, a summative evaluation would seek to determine the degree to which the curriculum improved the quality of the programme and the level of satisfaction of ecotourists, and thus the future value of the programme. The element of continuation is important, because it is a key concept in differentiating between formative and summative evaluations. While formative evaluation leads to decisions about programme development and implementation (modification and revision), summative evaluation leads to decisions about programme continuation, termination, expansion or adoption (Worthen *et al.* 1997).

Evaluation thus equips the service provider with the required information to gauge the significance of the programme. With this information in hand, the service provider has what is needed to decide the fate of the programme on the basis of a number of different options, including: (1) continue the programme with no modification; (2) continue the programme with modifications; (3) end the programme and repeat it at some other time; (4) end the programme and modify it at a later date; or (5) end the programme altogether. Obviously in cases where the programme has been quite favourably evaluated, little or no modification is necessary (although like any true professional the service provider will naturally want to tinker with the programme to make it better). At the other end of the spectrum, negative evaluations would mean that the programme should be shelved or removed completely. However, as Carpenter and Howe (1985) quite rightly suggest, there needs to be a very tight relationship between programme revision and needs assessment (the beginning of the next programme cycle). Without assessing the participants' needs from time to time, programmers will be unable to know if the demand for a terminated programme can be rekindled. The link between the evaluation-needs assessment cycle is well stated by these authors, who write that, 'Program revision allows you to be immediately responsive to your clientele as a result of your evaluation findings, while needs assessment keeps you informed of the changing demands for leisure programs and services, enabling you to remain responsive over the long run' (1985: 172). Programme demise or death for those which are unpopular may simply mean that there may be a time in the future when the programme is once again able to generate enthusiasm.

There are various techniques that theorists have used to assess service quality of ecotourism programmes and attractions. Mohd Shahwahid *et al.* (2013) employed a critical incident technique to assess the satisfaction and dissatisfaction of ecotourists at the Kampung Kuantan Firefly Park, Malaysia in regard to the services surrounding a species of beetle (and its habitat). The respondents identified five service failure domains including dissatisfaction with the quality of attractions and facilities, tourist expectations before the visit, inadequate knowledge of employees, unacceptable behaviour of employees and other tourists, and inadequate responses to tourist needs. The authors also identified the most common recovery strategies used by park employees. In order of importance these included: (1) smiles, apologies and admittance of a mistake; (2) remedial responses like the correction of a mistake; (3) management intervention (warnings to staff); and (4) taking no recovery action at all. On this latter case employees did not admit or just ignored the problem.

Chan and Baum (2007) assessed the perceptions of ecotourists' experiences through in-depth interviews at two ecolodges in Sukau, Malaysia. Both positive and negative experiences were isolated based on the expressive dimensions of their service experience at the ecolodges. The authors used six domains of service experience, as seen in Table 11.8

Positive expressive dimensions that described the ecotourism experience, and the functional or utilitarian aspects of the ecotourism programme itself are as follows: (1) hedonic included wildlife, natural environment, riverboat cruise trips and jungle

Table 11.8 *The construct domains of service experience*

Dimension	Examples
Hedonic	Excitement, enjoyment, memorability
Interactive	Meeting people, being part of the process, having choice
Novelty	Escape, doing something new
Comfort	Physical comfort, relaxation
Safety	Personal safety, security of belongings
Stimulation	Educational and informative, challenging

Source: Chan and Baum (2007)

walks; (2) interactive included wildlife, staff, guides and other members of the group; (3) novelty included wildlife; (4) comfort included wildlife, natural environment and riverboat cruises; (5) personal safety included boatmen and the riverboat cruises; and (6) stimulation included educationally and physically challenging tour components.

Negative expressive dimensions which described the ecotourism experience through the interviews, and the functional or utilitarian aspects of the trip included: (1) don't want/don't like/hate were expressed about noise and the fumes from the boats; (2) uncomfortable/unpleasant were expressed about mosquitos; (3) negative things/frustration/dissatisfaction was expressed about fumes from the boat as well as riverboat operations including congestions; and (4) feeling sad/unhappy/disappointed was expressed over the functional trip-based aspects like not being able to see some wildlife, short duration of cruises, poor sanitation of toilets, poor maintenance of bedrooms, bumpy roads and disappearance of the forest. Chan and Baum (2007) conclude by suggesting that ecotourism services are produced by a number of different aspects, and that management needs to take a holistic and integrated approach to the provision of ecotourism programmes in generating positive perceptions in ecotourists. Added to this is the notion that new, innovative products are needed in nature-based tourism and ecotourism, which ought to be based on flexibility, individuality, hybridity and activity (Saarinen 2005).

Conclusion

At the outset of this chapter it was suggested that there are many commonalities between tourism and recreation, especially from the experiential standpoint. Service is a key aspect that is common to both areas, but there does not appear to be the same level of importance given to service theory and practice in tourism, as there is in the field of recreation, where the service delivery system is designed foremost around participant satisfaction. However, the tourism industry will need to adapt if the services we provide for tourists – as participants, not clients – are to be viewed as something more than a commodity. More and more often, the sophistication and the competitiveness of tourism services demand a different approach to service provision. We can go so far as to suggest that, all things considered equal, the operator who does a better job programme planning will stand a greater chance at achieving financial success. In doing so, however, the by-product is a satisfied participant base. In the twenty-first century, tourists want to live their travel experiences in the same way they live their recreation experiences. This means education, skill development, exercise and self-actualisation. Ecotourism, if planned and managed accordingly, can be an excellent outlet to satisfy these physical and psychological needs.

Summary questions

1   In general, how can needs assessments and assets inventories allow ecotourism operators to help themselves and, in doing so, benefit ecotourists?
2   How is it that programme planning might enable the ecotourism service provider to value other things beyond the profit motive? Will such an emphasis ultimately help the provider remain competitive in the marketplace in the long run?
3   Provide a general overview of the programming planning model. Why is the cyclical nature of this type of planning so important?
4   How do formative and summative evaluation techniques differ?
5   What is SWOT analysis and how is it helpful to programme planners?

⑫ Conclusion

> We come more and more to see through the follies and vanities of the world and to appreciate the real values. We load ourselves up with so many false burdens, our complex civilization breeds in us so many false or artificial wants, that we become separated from the real sources of our strength and health as by a gulf.
>
> (J. Burroughs, *Time and Change*, 1912)

This book has attempted to project a realistic approach to the foundations, concepts and issues of ecotourism as they currently exist. At the outset it was suggested that there is no question that vastly disparate views exist on how ecotourism ought to be constructed in theory and practice, and these positions quite naturally stem from the many different angles that we approach the subject (e.g. economics, ecology, anthropology, park management, marketing, service provision). Different priorities and values do lead to different outcomes.

The approach taken in this book is one that is admittedly more protectionist, which has ultimately led to a more elaborate definition of ecotourism. Such a view is based on the belief that if ecotourism is to survive it will have to do so by adopting a 'greener' agenda. While some argue, as I do, for a stricter definition of ecotourism, some of my colleagues go so far as to suggest that ecotourism cannot survive devoid of a relationship with sustainable mass tourism. We will have to see. If ecotourism does partner mass tourism, has it not undergone a significant transformation? Is it not then mass sustainable tourism with a sprinkling of nature? I still believe there is a distinct market of ecotourists, contrary to what some scholars argue. I just think it is not a very large one (see Chapter 2). I look at the example of Quest Nature Tours, where almost 70 per cent of their clients are repeat visitors. Natural history-based tourism is the primary travel outlet of these tourists, and they seek to enjoy the wonders of the natural world regularly and, importantly, with the same operator who can provide the types ecotourism experiences they seek.

I was told as a graduate student that if ecotourism was to become successful, it would have undermined everything that it had initially set out to accomplish. This we might call the 'Frankenstein Syndrome': We use all sorts of parts to create something new with all good intentions, which later turn out to go all wrong. And there are reasons why the experiment, and outcome, sometimes goes wrong. In ecotourism, like so many other forms of tourism, what we have done is measure success not on the basis of social and ecological harmony, but rather on profit. This is the cold, hard, calculative side of human nature that focuses more on efficiency, productivity, science, technology and the short term. I would argue that we need to focus more on grounding ecotourism in a philosophical rationale, which allows us to consider the bigger picture beyond concerns with our own immediate self-interest. In this latter capacity we open the door to practices that consider families, communities, memories, history, rights and reverence for place (Fennell and Malloy 2007) alongside profit. Furthermore, ecotourism in name does not guarantee sustainability in practice: the two are not synonymous. Ecotourism can be just as unsustainable as other more intrusive forms of tourism if it is not driven by an ethical imperative.

This idea of how success is measured in ecotourism is represented in a series of meetings I attended on the development of the ecotourism product along the Trans-Canada Highway. The idea was that rather than travel through a province, and hence view the highway simply as a medium of transportation, tourists should instead be encouraged to stop and enjoy ecotourism sites along the way. This association grew as a consequence of the Trans-Canada Highway ecotours developed by Environment Canada in the 1970s and 1980s to allow people travelling on the highway to enjoy natural attractions of various regions across Canada. However, whereas the Environment Canada publications were not developed with a profit motive in mind – this was simply a programme designed to strengthen the human–land relationship in Canadians and international travellers – the more recent meetings on the Trans-Canada Highway took on a very different form. There was an overriding emphasis on measuring success only in terms of the economic return from efforts put into marketing the product (Chapters 9 and 11). It made me wonder whether anything would be gained if a person, visiting any given eco-attraction along the stretch of highway, did not spend a dime at a Trans-Canada ecotour site. So, what determines success of ecotourism ventures? People making money, or tourists gaining rewarding experiences, or both? Does the market really reflect the numbers of ecotourists reported these days, or is it simply the industry manufacturing and repackaging the product into ecotourism-like experiences for a public that is not so environmentally friendly? The reader will not find the answer to this question here, but nevertheless it is a question that needs to be addressed. Ecotourism is far too 'sexy' these days to be ignored by an industry bent on profit.

When the chair of the meeting suggested that we, 'take a broader look at ecotourism', including any cultural and natural attractions, I thought about the example of helicopter ecotourism in Hollywood mentioned in Chapter 9. But what about the differences between the helicopter tour in Hollywood and hard-core ecotourist flying to Costa Rica? Which is more intrusive? There are those who argue that the former would in fact be less ecologically disruptive than having people traipsing through sensitive rainforests; after all, those in the helicopter are not touching the ground. But yet the helicopter experience as ecotourism seems fundamentally wrong. The answer could be that a bona fide ecotourism experience includes ecotourists having some tangible, concrete connection to the natural world. This means seeing it, feeling it, touching it – sensing it. This involves a type of learning that cannot occur in the helicopter because there is no opportunity to get 'into' nature. Also, is it the helicopter itself that is the main attraction, or the content of the experience (i.e. the interpretation of Hollywood)? Finally, how does it contribute to conservation? The point is, if a helicopter flight over Hollywood can be ecotourism, then the gates are wide open. The vultures are circling.

The issue of mass tourism as ecotourism identified above pertains to a related problem surrounding resort development. Personally, I worry when I see names like Ramada on ecolodge signs, which is the case in a recently developed ecoresort in the Yucatan Peninsula of Mexico. This concern is not only from the perspective of ecological impact (Ramada has apparently adopted an environmental integrity image in its product development, as have other hotel giants such as Holiday Inn), but rather from a socio-economic perspective. One of the core criteria of ecotourism identifies the importance of local participation and benefits. How will Ramada remain internationally competitive if it is pouring its revenues into the local community and the preservation of biodiversity of the region? There are big questions that need to be asked when the international business community gets involved in such 'local' enterprises. This scenario is more along the lines of what Britton (1977) and others suggest about metropolitan domination of the industry, where it is rational to impose costs on others in the pursuit of individual or organisational self-interest. In my estimation, one of the biggest threats facing the ecotourism industry today, as implied earlier, is the danger that the large hotel and resort corporations will infiltrate the ecotourism industry with less than altruistic ends.

Canadian Pacific Hotels and Resorts, for example, has recently developed more environmentally friendly policies as a result of a formal survey of its thousands of employees (D'Amore 1993). Its 12-point environmental action plan is indeed a laudable example of leadership in this area. Would such policies have been popular 20 years ago? Not likely. Ramada and other hotel chains are responding to what they perceive as a movement in public opinion on the management of hotels. There appears to be a good deal of business savvy in this choice (i.e. they benefit the hotel in terms of lower maintenance and overhead costs, and the client gains peace of mind), but is that all it is? It reminds me of the recent trend for big companies to provide health memberships for workers under the assumption that there is a genuine interest in the health and well-being of employees. The underlying premise is likely that employees get fit and, in the process, are more productive on the job with less absenteeism, with the ultimate end to save the corporation money. In cases such as this where there seems to be a win–win situation it is hard to argue against the fact that the corporation is doing nothing more than treating the worker as a commodity, with the ultimate goal of extracting the maximum yield out of this resource. I do not wish to be too pessimistic, but self-interest, not altruism, is the general rule of the day. One wonders about the lengths to which hotels and other businesses would go to remain successful if resource protection and environmental concerns were not contingent upon financial success and public opinion.

The foregoing examples are important because they expose ecotourism's vulnerable underbelly – and there is evidently much to expose! For example, critics charge that ecotourism: (1) must be ecologically benign and operate at low levels of carrying capacities in order for it to become successful; and (2) may not be sufficient to allow all members of a community to benefit economically. A persistent problem in ecotourism is the resource demand required to move people to (and within) the destination. Can we be considered earth-friendly if we travel half-way around the world while at the same time using vast amounts of fuel to satisfy our hedonistic ends? Perhaps not, but at the very least we can try to support carriers that implement better environmental protocols in minimising their ecological footprint (e.g. more efficient fuels and the implementation of carbon offsetting programmes, the latter of these discussed by Putz and Pinard 1993). As regards point two, above, I argue that community benefit is a function of the mechanisms put into place to either engender trust and cooperation or take it away. Planning and management are critical to the delivery of ecotourism, and must occur through systems of policy and governance that involve informal norms and formal institutions that emphasise participation and representation, among other factors, that are open to influence and change (Chapter 10). In this regard, ecotourism needs to be integrated into a broader sustainable development policy, which constitutes strong planning and management of the resource and the industry. Cooperation also means that there are stable and vibrant linkages made between sectors in the economy for collective benefit. In this, we have a long way to go.

Critics of ecotourism further charge that the definitions and goals of ecotourism are far too lofty; that is, there is very little chance that all aspects of elaborate definitions can be reached. This argument is a valid one that should be further examined. Essentially there are two ways of dealing with it: (1) to agree, and modify the definition to be more forgiving, which is the prevailing norm; or (2) to argue that we must set lofty goals in order to accomplish holistic ends. In support of the latter, the rationale is that the inclusion of certain principles in ecotourism, like learning, sustainability, ethics and local benefits, should induce practitioners to strive to achieve these objectives – in much the same way that sustainable development is very much a process and not necessarily an end. Those who actively pursue these goals should be recognised for their efforts, which is a fundamental premise driving certification systems that have been implemented in various regions – although many of these systems have been criticised for reasons outlined in Chapter 7.

Oscar Wilde said that a critic is, 'a man who knows the price of everything, and the value of nothing'. In my estimation, the real job that lies ahead is not in writing off ecotourism (this is too easy), but rather in understanding that: (1) there are few other options to turn to at present in tourism for leadership; and (2) we need to better understand who we are at the core. As regards the first, I am not sure that some of the newer hybrid terms in tourism like geotourism, wildlife tourism or responsible tourism offer any more hope in fixing the many problems that exist in tourism human–human and human–environment interactions, at least in terms of what currently exists in the literature. These problems are a function of who we are at the deepest level, and until we start to better understand the dynamics of this, no name, label or concept can have any hope of making things better. Ecotourism can still be the most ethical form of tourism, but we do need to temper this hope or belief in full view of some fundamental truisms about human nature from economic, political, social, technological and ecological perspectives. What continues to constrain us from reaching such a point, apart from a serious lack of knowledge on human behaviour itself, is a disconnect between industry, local people, academics and government with respect to a vision for the future – as a collective calling or philosophy (see Weaver and Lawton 2007).

It is indeed perplexing that in an era of mass communication and technology, where information is so readily available, we seem to have vastly different trajectories when it comes to ecotourism in practice and in principle. We have known this for years as a result of publications that have been far-reaching enough to examine the concept without rose-coloured glasses – even during its formative years (see Cater and Lowman 1994). There is as much value in examining the ecotourism 'done poorly' examples as there is in those cases where it is done well. If anything, these case studies give us the opportunity to construct a tension between theory and practice, which might affect the right type of change in this dynamic industry. Some priorities that may help achieve better balance between the competing divisions of ecotourism are as follows:

- Respect should be extended to the interests of individual animals used in ecotourism, beyond efforts to manage resources and populations. New research on animal ethics and the first principle discussed in Chapter 7 should provide guidance on those uses deemed acceptable and unacceptable.
- Respect should be shown for the rights of local and Aboriginal peoples. This may include employing these marginalised populations in middle or upper management roles, providing opportunities to improve their position within the firm, or expressing their unique abilities through ecotourism.
- People need incentives to conserve biodiversity and support forms of development like ecotourism. If models are implemented that take away the ability of local people to benefit, we can expect such models to be less successful than others that are more holistic and integrative.
- Research supports the notion that the development of an ecotourism industry should not be imposed on a community, but rather should start from the ground up. Such an approach emphasises the importance of decision-making from within rather than from outside. External influences (experts) should be encouraged, but not at the expense of the community's integrity and know-how.
- Theorists should further examine how cooperation and trust can be activated to build meaningful relationships between the various actors involved in ecotourism for collective benefit.

- Industry decision-makers should strive to optimise the number of operators and tourists within regions in line with the capacity of the environment to absorb such use. This implies that more research needs to be done in all types of environments to understand limits of use.
- Motivate operators to strive for ongoing levels of competency through environmental management systems, certification schemes and industry-imposed codes of ethics that are normatively conceived and evaluated. Government, academia and industry should work together, where acceptable, to institute fair and equitable guidelines to regulate and/or guide the ecotourism industry.
- Motivate national and regional governments to actively coordinate research with the help of international agencies to identify and rectify situations where the industry has had a negative impact on the resources and people of a region.
- Sponsor research that demonstrates to government and other decision-makers the benefits of ecotourism, especially in the face of other more intrusive forms of land use. Take care in supporting market research that does not over-inflate ecotourism and lead to unbridled optimism.
- Understand that ecotourism is simply another expression of human nature. If we continue to ignore knowledge in areas that would help us better understand the essence of tourist behaviour, we will continue to fail to understand how to navigate the labyrinth of impacts that continue to plague the industry. This suggests that an interdisciplinary approach is essential in filling in many of the gaps that currently exist in ecotourism research.
- Encourage service providers to incorporate innovative ways to educate tourists about the special ecological and cultural aspects of the experience. Guides should be encouraged, and supported, to practise lifelong learning in attempts to be as informative as possible.
- Establish stronger linkages between educational institutions and local people by providing such people with the tools (e.g. diplomas, degrees) to work in the industry. These programmes need to be developed in association with the needs of local people, not necessarily the needs of institutions.
- Educate service providers on the need to follow all guidelines that have been developed to maintain the integrity of socio-cultural and ecological systems.
- The fate of ecotourism may be tied to strong or weak democratic systems of governance, with the former deemed essential – in an integrated capacity – in efforts to protect local people from exogenous forces. Governance models must be inclusive and open to change in supporting people at many levels of the political and economic hierarchy.
- Reinforce to service providers the importance of making ongoing contributions to conservation in the areas that they use and depend on. This could be in the form of financial donations or through activities that rehabilitate or restore protected areas (e.g. planting trees or removing garbage).
- Ensure that ecotourism policy evolves out of representative participation. In Canada, for example, policy is derived from the interests of industry, consultants and government, and generally geared towards growth and self-interest. Part of the problem in this equation lies in the fact that the academic side of ecotourism often represents little more than voices in the wilderness.
- Continue to disseminate and support a comprehensive definition of ecotourism. At present there is little consensus among practitioners and researchers, which is hampering policy and practice (e.g. fishing and hunting as ecotourism).
- Organise international events in ecotourism that serve a purpose beyond the elevation of organisational self-interest. This final issue is explained in more detail below.

The International Year of Ecotourism (IYE) in 2002 provided the first truly global effort to develop guidelines for ecotourism. The Quebec Summit represented the culmination of over 20 preparatory meetings around the world during 2002, and involved over 3,000 participants. Important in the organisation of this event was the involvement of many different stakeholder groups who shared a vision about the future of ecotourism (the Quebec City Declaration on Ecotourism presented in its entirety in the Appendix, lists the involvement of governments, industry operators, trade associations, non-governmental organisations (NGOs), academics, consultants, intergovernmental organisations, communities and indigenous groups, and the responsibilities that each has for the future of ecotourism). Dedication of an international year was a sign that ecotourism: (1) will continue to be a formidable industry power in tourism; and (2) was entering another phase of development beyond basic questions related to ecotourism is a sustainable form of community development, definitions, calls for ecotourism as the 'fastest growing sector of the world's biggest industry', and if it really does work (Fennell 2002c; Weinberg *et al.* 2002). This new phase will need to be different, if we are to expect change, and be led by approaches that emphasise other factors beyond efficiency, productivity, technology and short-term cost–benefit calculations – calculative thinking in the mind of Heidegger (1966). By contrast, Heidegger's concept of meditative thinking supports rootedness and an examination of the broader horizon of possibilities, and arms us with the tools needed to better appreciate the needs of all those who stand to be affected by the ecotourism industry. Denying the importance of these qualities is to take us down the same road of powerlessness, depression, self-interest and instrumental reason (Saul 2001). Despite other intentions, the Ecotourism Summit in Quebec City (2002) reinforced the division between the haves and have-nots because of a tightly controlled programme that served the interests of large governing bodies and NGOs who maintain similar organisational agendas, but not the interests of those who could have benefited most from this meeting. In the absence of meditative thinking we can see why very little has emerged from these meetings to actually *make things better* (see Tepelus 2008, who has written more widely on the main criticisms and achievements of the IYE).

In summary, the quote from Burroughs at the outset of this chapter makes a reference to the fact that we live in a superficial and valueless world. As a major global economic force, tourism is caught in the mainstream of a complex civilisation that tends to emphasise material ends over other more virtuous ends. As part of the tourism industry, ecotourism will have to struggle to identify itself as part of either the conventional front or the alternative front. The coming years will be important in defining the role that ecotourism plays in building a better alternative. If it is more about false burdens, vanities, artificial wants and calculative thinking, then ecotourism, I submit, will not be successful in its true meaning. If, on the other hand, it demands a values-based approach in philosophy and application, then it will have something to contribute to human–human and human–environment relationships and, in doing so, act as a useful model for other forms of development – tourism and non-tourism – in a very complex world.

Summary questions

1 Contrast calculative and meditative thinking?
2 What are some of the main arguments by critics of the ecotourism industry?
3 Why is it dangerous to measure success of an enterprise (like ecotourism) in economic terms only?
4 Why is it important to get 'in' to nature as an ecotourist?
5 Are measures to be more sustainable by hotel chains an example of altruism or self-interest?

**Appendix
Québec City declaration
on ecotourism**

In the framework of the UN International Year of Ecotourism, 2002, under the aegis of the United Nations Environment Programme (UNEP) and the World Tourism Organization (WTO), over one thousand participants coming from 132 countries, from the public, private and non-governmental sectors, met at the World Ecotourism Summit, hosted in Québec City, Canada, by Tourisme Quebec and the Canadian Tourism Commission, between 19 and 22 May 2002.

The Québec Summit represented the culmination of 18 preparatory meetings held in 2001 and 2002, involving over 3,000 representatives from national and local governments including the tourism, environment and other administrations, private ecotourism businesses and their trade associations, nongovernmental organisations, academic institutions and consultants, intergovernmental organizations, and Indigenous and local communities.

This document takes into account the preparatory process, as well as the discussions held during the Summit. Although it is the result of a multistakeholder dialogue, it is not a negotiated document. Its main purpose is the setting of a preliminary agenda and a set of recommendations for the development of ecotourism activities in the context of sustainable development.

The participants at the Summit acknowledge the World Summit on Sustainable Development (WSSD) in Johannesburg, August/September 2002, as the ground-setting event for international policy in the next ten years, and emphasize that, as a leading industry, the sustainability of tourism should be a priority at WSSD due to its potential contribution to poverty alleviation and environmental protection in critically endangered ecosystems. Participants therefore request the UN, its organizations and member governments represented at this Summit to disseminate the following Declaration and other results from the World Ecotourism Summit at the WSSD.

The participants to the World Ecotourism Summit, aware of the limitations of this consultative process to incorporate the input of the large variety of ecotourism stakeholders, particularly non-governmental organizations (NGOs) and local and Indigenous communities:

Acknowledge that tourism has significant and complex social, economic and environmental implications,

Consider the growing interest of people in traveling to natural areas,

Emphasize that ecotourism should contribute to make the overall tourism industry more sustainable, by increasing economic benefits for host communities, actively contributing to the conservation of natural resources and the cultural integrity of host communities, and by increasing awareness of travelers towards the conservation of natural and cultural heritage,

Recognize the cultural diversity associated with natural areas, particularly because of the historical presence of local communities, of which some have maintained their traditional knowledge, uses and practices many of which have proven to be sustainable over the centuries,

Reiterate that funding for the conservation and management of biodiverse and culturally rich protected areas has been documented to be inadequate worldwide,

Recognize that sustainable tourism can be a leading source of revenue for protected areas,

Recognize further that many of these areas are home to rural peoples often living in poverty, who frequently lack adequate healthcare, education facilities, communications systems, and other infrastructure required for genuine development opportunity,

Affirm that different forms of tourism, especially ecotourism, if managed in a sustainable manner can represent a valuable economic opportunity for local populations and their cultures and for the conservation and sustainable use of nature for future generations,

Emphasize that at the same time, wherever and whenever tourism in natural and rural areas is not properly planned, developed and managed, it contributes to the deterioration of natural landscapes, threats to wildlife and biodiversity, poor water quality, poverty, displacement of indigenous and local communities, and the erosion of cultural traditions,

Acknowledge that ecotourism must recognize and respect the land rights of indigenous and local communities, including their protected, sensitive and sacred sites,

Stress that to achieve equitable social, economic and environmental benefits from ecotourism and other forms of tourism in natural areas, and to minimize or avoid potential negative impacts, participative planning mechanisms are needed that allow local and indigenous communities, in a transparent way, to define and regulate the use of their areas at the local level, including the right to opt out of tourism development,

Note that small and micro businesses seeking to meet social and environmental objectives are often operating in a development climate that does not provide suitable financial and marketing support for this specialized new market, and that to achieve this goal further understanding of the ecotourism market will be required through market research at the destination level, specialized credit instruments for tourism businesses, grants for external costs, incentives for the use of sustainable energy and innovative technical solutions, and an emphasis on developing skills not only in business but within government and those seeking to support business solutions,

In light of the above, the participants of the Summit produced a series of recommendations to governments, the private sector, non-governmental organizations, community-based associations, academic and research institutions, inter-governmental organizations, international financial institutions, development assistance agencies, and indigenous and local communities, presented in an annex to this Declaration.

Québec City, Canada, 22 May 2002

Annex I—Recommendations of the World Ecotourism Summit, Québec City, May 19 to 22, 2002.

The participants to the World Ecotourism Summit, having met in Québec City, from 19 to 22 May 2002, propose the following recommendations:

A.   To Governments

1   *formulate* national, regional and local ecotourism policies and development strategies that are consistent with the overall objectives of sustainable development, and to do so through a wide consultation process with those who are likely to become involved

in, affect, or be affected by ecotourism activities. Furthermore, the principles that apply to ecotourism should be broadened out to cover the entire tourism sector;

2 In conjunction with local communities, the private sector, NGOs and all ecotourism stakeholders, *guarantee* the protection of nature, local cultures and specially traditional knowledge and genetic resources;

3 *ensure* the involvement, appropriate participation and necessary coordination of all the relevant public institutions at the national, provincial and local level, (including the establishment of inter-ministerial working groups as appropriate) at different stages in the ecotourism process, while at the same time opening and facilitating the participation of other stakeholders in ecotourism-related decisions. Furthermore, adequate budgetary mechanisms and appropriate legislative frameworks *be set up* to allow implementation of the objectives and goals set up by these multistakeholder bodies;

4 *include* in the above framework the necessary regulatory and monitoring mechanisms at the national, regional and local levels, including objective sustainability indicators jointly agreed with all stakeholders and environmental impact assessment studies, to prevent or minimize the occurrence of negative impacts upon communities or the natural environment. Monitoring results should be made available to the general public, since this information will allow tourists to choose an operator who adopts ecotourism principles over one who does not;

5 *develop* the local and municipal capacity to implement growth management tools such as zoning, and participatory land-use planning not only in protected areas but in buffer zones and other ecotourism development zones;

6 *use* internationally approved and reviewed guidelines to develop certification schemes, ecolabels and other voluntary initiatives geared towards sustainability in ecotourism, encouraging private operators to join such schemes and promoting their recognition by consumers. However, certification systems should reflect regional and sub-regional criteria and build capacity and provide financial support to make these schemes accessible to small and medium enterprises (SMEs). A regulatory framework is needed for such schemes to fulfill their mission;

7 *ensure* the provision of technical, financial and human resources development support to micro, small and medium-sized firms, which are the core of ecotourism, with a view to enable them to start, grow and develop their businesses in a sustainable manner. Similarly, that appropriate infrastructure is established in areas with ecotourism potential to stimulate the emergence of local enterprises.

8 *define* appropriate policies, management plans, and interpretation programs for visitors, and to earmark adequate sources of funding for protected natural areas to manage rapidly growing visitor numbers and protect vulnerable ecosystems, and effectively prevent the use of conservation hotspots. Such plans should include clear norms, direct and indirect management strategies, and regulations with the funds to ensure monitoring of social and environmental impacts for all ecotourism businesses operating in the area, as well as for tourists wishing to visit them;

9 *include* micro, small and medium-sized ecotourism companies, as well as community-based and NGO-based ecotourism operations in the overall promotional strategies and programmes carried out by the National Tourism Administration, both in the international and domestic markets;

10 *develop* regional networks and cooperation for promotion and marketing of ecotourism products at the international and national levels;

11 *provide* incentives to tourism operators (such as marketing and promotion advantages) for them to adopt ecotourism principles and make their operations more environmentally, socially and culturally responsible;

12 *ensure* that basic environmental and health standards are defined for all ecotourism development even in the most rural areas and in national and regional parks, that can play a pilot role. This should include aspects such as site selection, planning, design, the treatment of solid waste, sewage, and the protection of watersheds, etc., and *ensure* also that ecotourism development strategies are not undertaken by governments without investment in sustainable infrastructure and the reinforcement of local/municipal capabilities to regulate and monitor such aspects;

13 *invest*, or support institutions that invest in research programmes on ecotourism and sustainable tourism. To institute baseline studies and surveys that record plant and animal life, with special attention to endangered species, as part of an environmental impact assessment (EIA) for any proposed ecotourism development;

14 *support* the further development of the international principles, guidelines and codes of ethics for sustainable tourism (e.g. such as those proposed by the Convention on Biological Diversity, UNEP, WTO) for the enhancement of international and national legal frameworks, policies and master plans to implement the concept of sustainable development into tourism;

15 *consider* as one option the reallocation of tenure and management of public lands, from extractive or intensive productive sectors to tourism combined with conservation, wherever this is likely to improve the net social, economic and environmental benefit for the community concerned;

16 *promote* and develop educational programmes addressed to children and young people to enhance awareness about nature conservation and sustainable use, local and indigenous cultures and their relationship with ecotourism;

17 *promote* collaboration between outbound tour operators and incoming operators and other service providers and NGOs at the destination to further educate tourists and influence their behaviour at destinations, especially those in developing countries.

B. The private sector

18 *conceive, develop and conduct* their businesses minimizing negative effects on, and positively contributing to, the conservation of sensitive ecosystems and the environment in general, and directly benefiting local communities;

19 *bear* in mind that for ecotourism businesses to be sustainable, they need to be profitable for all stakeholders involved, including the projects' owners, investors, managers and employees, as well as the communities and the conservation organizations of natural areas where it takes place;

20 *adopt* a reliable certification or other systems of voluntary regulation, such as eco-labels, in order to demonstrate to their potential clients their adherence to sustainability principles and the soundness of the products and services they offer;

21 *cooperate* with governmental and non-governmental organizations in charge of protected natural areas and conservation of biodiversity, ensuring that ecotourism operations are practiced according to the management plans and other regulations prevailing in those areas, so as to minimize any negative impacts upon them while enhancing the quality of the tourism experience and contribute financially to the conservation of natural resources;

22 *make* increasing use of local materials and products, as well as local logistical and human resource inputs in their operations, in order to maintain the overall authenticity of the ecotourism product and increase the proportion of financial and other benefits that remain at the destination. To achieve this, private operators should invest in the training of the local workforce;

23 *ensure* that the supply chain used in building up an ecotourism operation is thoroughly sustainable and consistent with the level of sustainability aimed at in the final product or service to be offered to the customer;

24 *work* actively with indigenous leadership to ensure that indigenous cultures and communities are depicted accurately and with respect, and that their staff and guests are well and accurately informed regarding local indigenous sites, customs and history;

25 *promote* among their clients, the tourists, a more ethical behavior vis-à-vis the ecotourism destinations visited, providing environmental education to travelers, professionals and fostering inter-cultural understanding, as well as encouraging voluntary contributions to support local community or conservation initiatives;

26 *diversify* their offer by developing a wide range of tourist activities at a given destination and extending their operation to different destinations in order to spread the potential benefits of ecotourism and to avoid overcrowding some selected ecotourism sites, thus threatening their long-term sustainability. In this regard, private operators are urged to respect, and contribute to, established visitor impact management systems of ecotourism destinations;

27 *create and develop* funding mechanisms for the operation of business associations or cooperatives that can assist with ecotourism training, marketing, product development, research and financing;

28 In relation to the above points, *formulate* and *implement* company policies for sustainable tourism with a view to applying them in each part of the ecotourism operation.

C. Non-governmental organizations, community-based associations, academic and research institutions

29 *provide* technical, financial, educational, capacity building and other support to ecotourism destinations, host community organizations, small businesses and the corresponding local authorities in order to ensure that appropriate policies, development and management guidelines, and monitoring mechanisms are being applied towards sustainability;

30 *monitor* and conduct research on the actual impacts of ecotourism activities upon ecosystems, biodiversity, local indigenous cultures and the socio-economic fabric of the ecotourism destinations;

31 *cooperate* with public and private organizations ensuring that the data and information generated through research is channeled to support decision-making processes in ecotourism development and management;

32 *cooperate* with research institutions to develop the most adequate and practical solutions to ecotourism development issues.

D. Inter-governmental organizations, international financial institutions and development assistance agencies

33 *develop* and *assist* in the implementation of national and local policy and planning guidelines and evaluation frameworks for ecotourism and its relationships with biodiversity conservation, socioeconomic development, respect of human rights, poverty alleviation, nature conservation and other objectives of sustainable development, and to intensify the transfer of such know-how to all countries. Special attention

should be paid to countries in a developing stage or least developed status, to small island developing states and to countries with mountain areas, regarding that 2002 is also designated as the International Year of Mountains by the UN;

34 *build capacity* for regional, national and local organizations for the formulation and application of ecotourism policies and plans, based on international guidelines;

35 *develop* international standards and financial mechanisms for ecotourism certification systems that takes into account needs of small and medium enterprises and facilitates their access to those procedures;

36 *incorporate* multistakeholder dialogue processes into policies, guidelines and projects at the global, regional and national levels for the exchange of experiences between countries and sectors involved in ecotourism;

37 *strengthen* their efforts in identifying the factors that determine the success or failure of ecotourism ventures throughout the world, in order to transfer such experiences and best practices to other nations, by means of publications, field missions, training seminars and technical assistance projects; UNEP and WTO should continue this international dialogue after the Summit on sustainable ecotourism issues, for example by conducting periodical evaluations of ecotourism development through international and regional forums.

38 *adapt* as necessary their financial facilities and lending conditions and procedures to suit the needs of micro-, small- and medium-sized ecotourism firms that are the core of this industry, as a condition to ensure its long term economic sustainability;

39 *develop* the internal human resource capacity to support sustainable tourism and ecotourism as a development sub-sector in itself and to ensure that internal expertise, research, and documentation are in place to oversee the use of ecotourism as a sustainable development tool.

E. Local communities and municipal organizations

40 As part of a community vision for development, that may include ecotourism, *define and implement* a strategy for improving collective benefits for the community through ecotourism development including human, physical, financial, and social capital development, and improved access to technical information;

41 strengthen, nurture and encourage the community's ability to maintain and use traditional skills that are relevant to ecotourism, particularly home-based arts and crafts, agricultural produce, traditional housing and landscaping that use local natural resources in a sustainable manner.

Accessed 18 October 2002
http://www.travelmole.com/cgi-bin/item.cgi?id=81853

Bibliography

Acott, T.G., LaTrobe, H.L. and Howard, S.H. (1998) 'An evaluation of deep ecotourism and shallow ecotourism', *Journal of Sustainable Tourism* 6(3): 238–253.

Ad Hoc Technical Expert Group on Protected Areas (2003) 'Protected areas: their role in the maintenance of biological and cultural diversity', United Nations Environment Programme, Convention on Biological Diversity, Tjärnö, Sweden, 10–14 June.

Adams, J.S., Tashchian, A. and Shore, T.H. (2001) 'Codes of ethics as signals for ethical behaviour', *Journal of Business Ethics* 29: 199–211.

Adams, S.M. (1983) 'Public/private sector relations', *The Bureaucrat*, Spring: 7–10.

Adams, W. and Hulme, D. (2001) 'Conservation and community: changing narratives, policies and practices in African conservation', in D. Hulme and M. Murphree (eds), *African Wildlife and Livelihoods: The Promise and Performance of Community Conservation*, Oxford: James Currey.

Adams, W.M. and Infield, M. (2003) 'Who is on the gorilla's payroll? Claims on tourist revenue from a Ugandan National Park', *World Development* 31(1): 177–190.

Aguirre G.J.A. (2006) 'Linking national parks with its gateway communities for tourism development in Central America: Nindiri, Nicaragua, Bagazit, Costa Rica and Portobelo, Panama', *PASOS. Revista de Turismo y Patrimonio Cultural* 4(3): 351–371.

Akehurst, G. (1992) 'European Community tourism policy', in P. Johnson and B. Thomas (eds), *Perspectives on Tourism Policy*, London: Mansell.

Alam, H.M.A. (2012) 'Maximum allowable visitor numbers: a new approach to low carbon ecotourism development', *Asian Journal of Applied Sciences* 5(7): 497–505.

Alderman, C. (1992) 'The economics and the role of privately owned lands used for nature tourism, education and conservation', paper presented at the 4th World Congress on Parks and Protected Areas, Caracas, February.

Almeyda Zambrano, A.M., Broadbent, E.N. and Durham, W.H. (2010) 'Social and environmental effects of ecotourism in the Osa Peninsula of Costa Rica: the Lapa Rios case', *Journal of Ecotourism* 9(1): 62–83.

Ammon, R. (1997) 'Risk management process', in D.J. Cotten and T.J. Wilde (eds), *Sport Law for Sport Managers*, Dubuque, IA: Kendall/Hunt.

Andersen, D.L. (1993) 'A window to the natural world: the design of ecotourism facilities', in K. Lindberg and D.E. Hawkins (eds), *Ecotourism: A Guide for Planners and Managers*, North Bennington, VT: The Ecotourism Society.

—— (1994) 'Ecotourism destinations: conservation from the beginning', *Trends* 31(2): 31–38.

Anderson/Fast (1996a) *Ecotourism in Saskatchewan: Primary Research*, Saskatoon, Saskatchewan: Anderson/Fast.

—— (1996b) *Ecotourism in Saskatchewan: State of the Resource (Report 1)*, Saskatoon, Saskatchewan: Anderson/Fast.

Anderson, W. (2008) 'Promoting ecotourism through networks: case studies in the Balearic Islands', *Journal of Ecotourism* 8(1): 51–69.

Aperghis, G.G. and Gaethlich, M. (2006) 'The natural environment of Greece: an invaluable asset being destroyed', *Southeast European and Black Sea Studies* 6(3): 377–390.

Applegate, J.E. and Clark, K.E. (1987) 'Satisfaction levels of birdwatchers: an observation on the consumptive–nonconsumptive continuum', *Leisure Sciences* 9: 129–134.

ARA Consultants (1994) *Ecotourism–Nature, Adventure/Culture: Alberta and British Columbia Market Demand Assessment*, Vancouver, BC: ARA Consultants.

Arambiza, E. and Painter, M. (2006) 'Biodiversity conservation and the quality of life of Indigenous people in the Bolivian Chaco', *Human Organization* 65(1): 20–34.

Aridjis, H. (2000) 'Flight of kings', *Amicus Journal* (22)2: 26–29.

Arlen, C. (1995) 'Ecotour, hold the eco', *US News and World Report*, 29 May.

Armitage, D., Berkes, F. and Doubleday, N. (2007) 'Moving beyond the critiques of co-management: theory and practice of adaptive co-management', in D. Armitage, F. Berkes and N. Doubleday (eds), *Adaptive Co-management: Collaboration, Learning and Multi-Level Governance*, Vancouver, BC: UBC Press.

Armstrong, E.K. and Kern, C. (2011) 'Demarketing manages visitor demand in the Blue Mountains National Park', *Journal of Ecotourism* 10(1): 21–37.

Armstrong, E.K. and Weiler, B. (2002) 'Getting the message across: an analysis of messages delivered by tour operators in protected areas', *Journal of Ecotourism* 1(2/3) 104–121.

Association for Experiential Education (1993) *Manual of Accreditation Standards for Adventure Programs*, Boulder, CO: AEE.

Atwood, M. (1984) 'The Galápagos haven't changed much in a million years: now the tourists are coming', *Quest* (April): 37–45.

Axelrod, R. (1984) *The Evolution of Cooperation*, New York: Basic Books.

Ayala, H. (1996) 'Resort ecotourism: a paradigm for the 21st century', *Cornell Hotel and Restaurant Administration Quarterly* 37(5): 46–53.

Aylward, B., Allen, K., Echeverría, J. and Tosi, J. (1996) 'Sustainable ecotourism in Costa Rica: the Monteverde Cloud Forest Preserve', *Biodiversity and Conservation* 5(3): 315–343.

Azcárate, M.C. (2006) 'Between local and global, discourses and practices: rethinking ecotourism development in Celestún (Yucatán, México)', *Journal of Ecotourism* 5(1/2): 97–111.

Bachert, D.W. (1990) 'Wilderness education: a holistic model', in A.T. Easley, J.F. Passineau and B.L. Driver (eds), *The Use of Wilderness for Personal Growth, Therapy, and Education*, General Technical Report RM-193, Fort Collins, CO: USDA Forest Service.

Backman, S., Petrick, J. and Wright, B.A. (2001) 'Management tools and techniques: an integrated approach to planning', in D.B. Weaver (ed.), *The Encyclopedia of Ecotourism*, Wallingford, Oxon.: CAB International.

Badalamenti, F., Ramos, A.A., Voultsiadou, E., Sánchez Lizaso, J.L., D'Anna, G., Pipitone, C., Mas, J., Ruiz Fernandez, J.E., Whitmarsh, D. and Riggio, S. (2000) 'Cultural and socio-economic impacts of Mediterranean marine protected areas', *Environmental Conservation* 27(2): 110–125.

Baker, L., Woenne-Green, S. and the Mutitjulu Community (1992) 'The role of aboriginal ecological knowledge in ecosystem management', in J. Birckhead, I. De Lacy and L. Smith (eds), *Aboriginal Involvement in Packs and Protected Areas*, Canberra: Aboriginal Studies Press.

Balint, P.J. (2006) 'Improving community-based conservation near protected areas: the importance of development variables', *Environmental Management* 38(1): 137–148.

Ballantine, J. and Eagles, P.F.J. (1994) 'Defining the Canadian ecotourist', *Journal of Sustainable Tourism* 2(4): 210–214.

Barash, D.P. (1982) *Sociobiology and Behaviour*, 2nd edn, New York: Elsevier.

Barbier, E.B. (1987) 'The concept of sustainable economic development', *Environmental Conservation* 14(2): 101–109.

Barkin, D. (2000) 'The economic impacts of ecotourism: conflicts and solutions in Highland Mexico', in P.M. Goode, M.F. Price and F.M. Zimmermann (eds), *Tourism and Development in Mountain Regions*, London: CAB International.

Barnwell, K.E. and Thomas, M.P. (1995) 'Tourism and conservation in Dalyan, Turkey: with reference to the loggerhead turtle (*Caretta caretta*) and the Euphrates turtle (*Trionyx triunguis*)', *Environmental Education and Information* 14(1): 19–30.

Barrett, C. and Arcese, P. (1998) 'Wildlife harvest in integrated conservation and development projects: linking harvest to household demand, agricultural production, and environmental shocks in the Serengeti', *Land Economics* 74: 449–465.

Barrow, G. (1994) 'Interpretive planning: more to it than meets the eye', *Environmental Interpretation* 9(2): 5–7.

Barter, M., Newsome, D. and Calver, M. (2008) 'Preliminary quantitative data on behavioural responses of Australian pelican (*Pelecanus conspicillatus*) to human approach on Penguin Island, Western Australia', *Journal of Ecotourism* 7(2/3): 197–212.

Bartholomew, G.A. (1986) 'The role of natural history in contemporary biology', *Bioscience* 36: 324–329.

Bassin, Z., Breault, M., Flemming, J., Foell, S., Neufeld, J. and Priest, S. (1992) 'AEE organizational membership preference for program accreditation', *Journal of Experiential Education* 15(2): 21–27.

Bastmeijer, K. and Roura, R. (2004) 'Regulating Antarctic tourism and the precautionary principle', *American Journal of International Law* 98(4): 763–781.

Bauer, T. and Dowling, R. (2003) 'Ecotourism policies and issues in Antarctica', in D.A. Fennell and R.K. Dowling (eds), *Ecotourism: Policy and Strategy Issues*, Wallingford, Oxon.: CAB International.

Beadles Thurau, B., Carver, A.D., Mangun, J.C., Basman, C.M. and Bauer, G. (2007) 'A market segmentation analysis of cruise ship tourists visiting the Panama Canal watershed: opportunities for ecotourism development', *Journal of Ecotourism* 6(1): 1–18.

Beder, S. (1996) *The Nature of Sustainable Development*, 2nd edn, Newham, Australia: Scribe Publications.

Beaumont, N. (2011) 'The third criterion of ecotourism: are ecotourists more concerned about sustainability than other tourists?', *Journal of Ecotourism* 10(2): 135–148.

Belasco, J.A. and Stayer, R.C. (1993) *Flight of the Buffalo*, New York: Warner.

Beres, L. (1986) 'Contracting out! the pros and cons', paper presented at the CPRA National Conference, Edmonton, Alberta.

Berkes, F. (1984) 'Competition between commercial and sport fisherman: an ecological analysis', *Human Ecology* 12(4): 413–429.

—— (2004) 'Rethinking community-based conservation', *Conservation Biology* 18(3): 621–630.

Berridge, K.C. (2003) 'Pleasures of the brain', *Brain and Cognition* 52(1): 106–128.

Bhattacharya, A.K., Saksena, V. and Banerjee, S. (2006) 'Environmental auditing in ecotourism: a study on visitors' management in Van Vihar National park, Bhopal, M.P. (India)', *Indian Forester* February: 139–148.

Bjork, P. (2007) 'Definition paradoxes: from concept to definition', in James Higham (ed.), *Critical Issues in Ecotourism*, Burlington, MA: Elsevier, pp. 23–45.

Black, R. and Ham, S. (2005) 'Improving the quality of tour guiding: towards a model for tour guide certification', *Journal of Ecotourism* 4(3): 178–195.

Black, R. and King, B. (2002) 'Human resource development in remote island communities: an evaluation of tour-guide training in Vanuatu', *International Journal of Tourism Research* 4: 103–117.

Black, R., Ham, S. and Weiler, B. (2000) *Ecotour Guide Training in Less Developed Countries: Some Research Directions for the 21st Century*, Working Paper 07/00, Faculty of Business and Economics, Monash University, Caufield East, Victoria, Australia.

—— (2001) 'Ecotour guide training in less developed countries: some preliminary research findings', *Journal of Sustainable Tourism* 9(2): 147–156.

Blackstone Corporation (1996) *Developing an Urban Ecotourism Strategy for Metro Toronto: A Feasibility Assessment for the Green Tourism Partnership*, Toronto, Canada.

Blamey, R.K. (1995) *The Nature of Ecotourism*, Occasional Paper no. 21, Canberra, ACT: Bureau of Tourism Research.

—— (2001) 'Principles of ecotourism', in D.B. Weaver (ed.), *The Encyclopedia of Ecotourism*, New York: CAB International, pp. 5–22.

Blamey, R.K. and Braithwaite, V.A. (1997) 'A social values segmentation of the potential ecotourism market', *Journal of Sustainable Tourism* 5(1): 29–45.

Blane, J. and Jackson, R. (1994) 'The impact of ecotourism boats on the St Lawrence beluga whales', *Environmental Conservation* 21(3): 267–269.

Blangy, S. and Nielson, T. (1993) 'Ecotourism and minimum impact policy', *Annals of Tourism Research* 20(2): 357–360.

Blersch, D.M. and Kangas, P.C. (2013) 'A modelling analysis of the sustainability of ecotourism in Belize', *Environment, Development and Sustainability* 15: 67–80.

Bodansky, D. (1994) 'The precautionary principle in US environmental law', in T. O'Riordan and J. Cameron (eds), *Interpreting the Precautionary Principle*, London: Earthscan.

Bonham, C. and Mak, J. (1996) 'Private versus public financing of state destination promotion', *Journal of Travel Research* 35(2): 3–10.

Boo, E. (1990) *Ecotourism: The Potentials and Pitfalls*, Washington, DC: World Wildlife Fund.

—— (1992) *The Ecotourism Boom: Planning for Development and Management*, Washington, DC: Wildlands and Human Needs, World Wildlife Fund.

Boonzaier, E. (1996) 'Local responses to conservation in the Richtersveld National Park, South Africa', *Biodiversity and Conservation* 5(3): 307–314.

Borg, W.R. and Gall, M.D. (1989) *Educational Research: An Introduction*, 5th edn, New York: Longman.

Bottrill, C.G. and Pearce, D.G. (1995) 'Ecotourism: towards a key elements approach to operationalising the concept', *Journal of Sustainable Tourism* 3(1): 45–54.

Bourgouin, F. (2002) 'Information communication technologies and the potential for rural tourism SMME development: the case of the Wild Coast', *Development Southern Africa* 19(1): 191–212.

Bowie, N.E. (1986) 'Business ethics', in J.P. DeMarco and R.M. Fox (eds), *New Directions in Ethics: The Challenge of Applied Ethics*, New York: Routledge & Kegan Paul.

Bowler, P.J. (1993) *The Norton History of the Environmental Sciences*, New York: W.W. Norton.

Boyd, S.W. and Butler, R.W. (1996) 'Managing ecotourism: an opportunity spectrum approach', *Tourism Management* 17(8): 557–566.

Bramwell, B. and Lane, B. (2000) 'Collaboration and partnerships in tourism planning', in B. Bramwell and B. Lane (eds), *Tourism Collaboration and Partnerships: Politics, Practice and Sustainability*, Clevedon: Channel View Publications, pp. 1–19.

Brandon, K. (1996) *Ecotourism and Conservation: A Review of Key Issues*, Environment Department Paper no. 23, Washington, DC: World Bank.

Brechin, S.R., Wilshusen, P.R., Fortwangler, C.L. and West, P.C. (2002) 'Beyond the square wheel: toward a more comprehensive understanding of biodiversity conservation as social and political process', *Society and Natural Resources* 15: 41–64.

Bressan, R.A. and Crippa, J.A. (2005) 'The role of dopamine in reward and pleasure behavior – review of data from preclinical research', *Acta Psychiatrica Scandinavica* 111: 14–21.

Bright, A.D., Fishbein, M., Manfredo, M. and Bath, A. (1993) 'Application of the theory of reasoned action to the National Park Service's controlled burn', *Journal of Leisure Research* 25: 263–280.

British Columbia Ministry of Development, Industry and Trade (1991) *Developing a Code of Ethics: British Columbia's Tourism Industry*, Victoria, British Columbia: Ministry of Development, Trade, and Tourism.

Britton, R.A. (1977) 'Making tourism more supportive of small-state development: the case of St. Vincent', *Annals of Tourism Research* 4(5): 268–278.

Britton, S.G. (1982) 'The political economy of tourism in the Third World', *Annals of Tourism Research* 9(3): 331–358.

Brock, T.D. (1979) *Biology of Microorganisms*, 3rd edn, Englewood Cliffs, NJ: Prentice-Hall Inc.

Brodowsky, P.K., National Wildlife Federation (2010) *Ecotourists Save the World: The Environmental Volunteer's Guide to More Than 300 International Adventures to Conserve, Preserve, and Rehabilitate Wildlife and Habitats*. New York: Penguin.

Brody, B. (1983) *Ethics and Its Applications*, New York: Harcourt Brace Jovanovich.

Brookfield, H. (1975) *Interdependent Development*, London: Methuen.

Brown, K. (2002) 'Innovations for conservation and development', *Geographical Journal* 168(1): 6–17.

Brown, K., Turner, R.K., Hameed, H. and Bateman, I. (1997) 'Environmental carrying capacity and tourism development in the Maldives and Nepal', *Environmental Conservation* 24(4): 316–325.

Bryan, H. (1977) 'Leisure value systems and recreational specialization: the case of trout fishermen', *Journal of Leisure Research* 9: 174–187.

Buckley, R. (1994) 'A framework for ecotourism', *Annals of Tourism Research* 21(3): 661–665.

—— (2002) 'Tourism ecocertification in the international year of ecotourism', *Journal of Ecotourism* 1(2–3): 197–203.

—— (2003a) 'Partnerships in ecotourism: Australian political frameworks', *International Journal of Tourism Research* 6: 75–83.

—— (2003b) 'Ecological indicators of tourist impacts in parks', *Journal of Ecotourism* 2(1): 54–66.

—— (2004) 'Ecotourism land tenure and enterprise ownership: Australian case study', *Journal of Ecotourism* 3(3): 208–213.

—— (2005) 'In search of the narwhal: ethical dilemmas in ecotourism', *Journal of Ecotourism* 4(2): 129–134.

Buckley, R. and Pannell, J. (1990) 'Environmental impacts of tourism and recreation in national parks and conservation reserves', *Journal of Tourism Studies* 1(1): 24–32.

Buckley, R. and Sommer, M. (2000) *Tourism and Protected Areas: Partnerships in Principle and Practice*. Gold Coast, QLD: Cooperative Research Centre for Sustainable Tourism.

Budowski, G. (1976) 'Tourism and environmental conservation: conflict, coexistence, or symbiosis', *Environmental Conservation* 3(1): 27–31.

Bujold, P. (1995) 'Community development: making a better home', *Voluntary Action News*, 5–8.

Bull, A. (1991) *The Economics of Travel and Tourism*, Melbourne: Pitman.

Bullard, R.D. (1996) 'Environmental justice: it's more than waste facility siting', *Social Science Quarterly* 77: 493–499.

Burch, W.R. Jr (1988) 'Human ecology and environmental management', in J.K. Agee and D.R. Johnson (eds), *Ecosystem Management for Parks and Wilderness*, Seattle: University of Washington Press.

Burger, J., Gochfeld, M. and Niles, L.J. (1995) 'Ecotourism and birds in coastal New Jersey: contrasting responses to birds, tourists, and managers', *Environmental Conservation* 22(1): 56–65.

Burgoon, M., Hunsaker, F.G. and Dawson, E.J. (1994) *Human Communication*, 3rd edn, Thousand Oaks, CA: Sage.

Burns, G.L. MacBeth, J. and Moore, S. (2011) 'Should dingoes die? Principles for engaging ecocentric ethics in wildlife tourism management', *Journal of Ecotourism* 10(3): 179–196.

Burns, P. (1999) 'Dealing with dilemma', *In Focus* 33: 4–5.

Burr, S.W. (1995) 'Sustainable tourism development and use: follies, foibles, and practical approaches', in S.F. McCool and A.E. Watson (eds), *Linking Tourism, the Environment, and Sustainability*, USDA Technical Report INTGTR-323, Ogden, UT: US Department of Agriculture, Forest Service, Intermountain Research Station.

Burton, M. (2000) Minister of tourism – official opening address, in G. Wrath (ed.), Proceedings of the Antarctic tourism workshop. Christchurch, NZ: Antarctica New Zealand: 6–7.

Burton, R. and Wilson, J. (2001) 'Ecotourism resources on the Internet: a review of ecotourism websites', *International Journal of Tourism Research* 3: 72–75.

Busser, J.A. (1990) *Programming: For Employee Services and Recreation*, Champaign, IL: Sagamore.

Butcher, J. (2005) 'The moral authority of ecotourism: a critique', *Current Issues in Tourism* 8(2–3): 114–124.

—— (2006) 'Natural capital and the advocacy of ecotourism as sustainable development', *Journal of Sustainable Tourism* 14(6): 529–544.

Butler, J. (1992) 'Ecotourism: its changing face and evolving philosophy', International Union for Conservation of Nature and Natural Resources (IUCN), Fourth World Congress on National Parks and Protected Areas, Caracas, Venezuela, 10–12 February.

Butler, R.W. (1980) 'The concept of tourist area cycle of evolution: implications for management of resources', *Canadian Geographer* 24: 5–12.

—— (1985) 'Evolution of tourism in the Scottish Highlands', *Annals of Tourism Research* 12(3): 379–391.

—— (1990) 'Alternative tourism: pious hope or Trojan horse?', *Journal of Travel Research* 28(3): 40–45.

—— (1991) 'Tourism, environment, and sustainable development', *Environmental Conservation* 18(3): 201–209.

—— (1993) 'Integrating tourism and resource management: problems of complementarity', in M.E. Johnston and W. Haider (eds), *Communities, Resources and Tourism in the North*, Thunder Bay, Ontario: Centre for Northern Studies, Lakehead University.

—— (1999) 'Sustainable tourism: a state-of-the-art review', *Tourism Geographies* 1(1): 7–25.

Butler, R.W. and Fennell, D.A. (1994) 'The effects of North Sea oil development on the development of tourism', *Tourism Management* 15(5): 347–357.

Butler, R.W. and Waldbrook, L.A. (1991) 'A new planning tool: the tourism opportunity spectrum', *Journal of Tourism Studies* 2(1): 2–14.

Butler, R.W., Fennell, D.A. and Boyd, S.W. (1992) *The POLAR Model: A System for Managing the Recreational Capacity of Canadian Heritage Rivers*, Ottawa: Environment Canada.

Butynski, T. (1998) 'Is gorilla tourism sustainable', *Gorilla Journal* 16: 1–3.

Campbell, A., Devine, M. and Young, D. (1990) *A Sense of Mission*, London: Pitman Press.

Campbell, G.A., Straka, T.J., Franklin, R.M. and Wiggers, E.P. (2011) 'Ecotourism as a revenue-generating activity in South Carolina Lowcountry plantations', *Journal of Ecotourism* 10(2): 165–174.

Campbell, L.M. (1999) 'Ecotourism in rural developing communities', *Annals of Tourism Research* 26(3): 534–553.

—— (2002) 'Conservation narratives and the "received wisdom" of ecotourism: case studies from Costa Rica', *International Journal of Sustainable Development* 5(3): 300–325.

Canada, Government (1990) *Canada's Green Plan*, Ottawa: Supply and Services Canada.

Canada, Parliament (1993) *Statutes of Canada Revised Loose Leaf Edition*, Chapter N-14, Ottawa: Supply and Services Canada.

Canadian Environmental Advisory Council (CEAC) (1991) *A Protected Areas Vision for Canada*, Ottawa: Minister of Supply and Services.

—— (1992) *Ecotourism in Canada*, Ottawa: Minister of Supply and Services.

Canadian Tourism Commission (1995) *Adventure Travel in Canada: An Overview of Product, Market and Business Potential*, Ottawa: Tourism Canada.

Canadian Wildlife Servic e (1995) *Last Mountain Lake National Wildlife Area* (Catalogue no. CW66-86/1995E). Ottawa: Ministry of Supply and Services.

Canova, L. (1994) 'Tourism in the modern age', *All of Us* 15: (unknown).

Carpenter, G.M. and Howe, C.Z. (1985) *Programming Leisure Experiences: A Cyclical Approach*, Englewood Cliffs, NJ: Prentice-Hall.

Carr, L. and Mendelsohn, R. (2003) 'Valuing coral reefs: a travel cost analysis of the Great Barrier Reef', *AMBIO: A Journal of the Human Environment* 32(5): 353–357.

Carrier, J.G. and MacLeod, D.V.L. (2005) 'Bursting the bubble: the socio-cultural context of ecotourism', *Journal of the Royal Anthropological Institute* 11: 315–334.

Cater, E. (1994) Ecotourism in the third world: problems and prospects for sustainability, in E. Cater and G. Lowman (eds) *Ecotourism: A Sustainable Option?* Chichester: Wiley, pp. 69–86.

Cater, E. (2002) 'Ecotourism: the wheel keeps turning', paper presented at the Tourism and the Natural Environment Symposium, Eastbourne, UK, 23–25 October.

—— (2006) 'Ecotourism as a western construct', *Journal of Ecotourism* 5(1/2): 23–39.

Cater, E. and Lowman, G. (1994) *Ecotourism: A Sustainable Option?*, Chichester, UK: Wiley.

Catibog-Sinha, C. (2008) 'Zoo tourism: biodiversity conservation through tourism', *Journal of Ecotourism* 7 (2/3): 160–177.

Ceballos-Lascuráin, H. (1987) *Estudio de prefactibilidaad socioeconómica del turismo ecológico y anteproyecto asquitectónico y urbanístico del Centro de Turismo Ecológico de Sian Ka' an, Quintana Roo*, study completed for SEDUE, Mexico.

—— (1996a) *Tourism, Ecotourism, and Protected Areas*, Gland, Switzerland: International Union for the Conservation of Nature and Natural Resources.

—— (1996b) 'Ecotourism or a second Cancún in Quintana Roo?', *The Ecotourism Society Newsletter* third quarter: 1–3.

Center for Tourism Policy Studies (1994) *Ecotourism Opportunities for Hawaii's Visitor Industry*, Honolulu: Department of Business, Economic Development and Tourism.

Cerruti, J. (1964) 'The two Acapulcos', *National Geographic Magazine* 126(6): 848–878.

Chambers, E. (1999) *Native Tours: The Anthropology of Travel and Tourism*, Prospect Heights, IL: Waveland.

Chan, J.K.L. and Baum, T. (2007) 'Ecotourists' perception of ecotourism experience in Lower Kinabatangan, Sabah, Malaysia', *Journal of Sustainable Tourism* 15(5): 574–590.

Chapin, M. (1990) 'The silent jungle: ecotourism among the Kuna Indians of Panama', *Cultural Survival Quarterly* 14(1): 42–45.

Chase, A. (1987) 'How to save our national parks', *Atlantic Monthly* July: 35–44.

—— (1989) 'The Janzen heresy', *Condé Nast Traveler* November: 122–127.

Chase, L.C., Lee, D.R., Schulze, W.D. and Anderson, D.J. (1998) 'Ecotourism demand and differential pricing of national park access in Costa Rica', *Land Economics* 74(4): 466–482.

Chester, G. (1997) 'Australian ecotourism accreditation off and running', *Ecotourism Society Newsletter* second quarter: 9, 11.

Chipeniuk, R. (1988) 'The vacant niche: an argument for the re-creation of a hunter-gatherer component in the ecosystems of northern national parks', *Environments* 20(1): 50–59.

Chodos, R. (1977) *The Caribbean Connection*, Toronto: James Lorimer.

Christaller, W. (1963) 'Some considerations of tourism location in Europe: the peripheral regions – underdeveloped countries – recreation areas', *Regional Science Association Papers* 6: 95–105.

Christensen, N.A. (1995) 'Sustainable community-based tourism and host quality of life', in S.F. McCool and A. Watson (eds), *Linking Tourism, the Environment, and Sustainability*, USDA Forest Service General Technical Report INT-GTR-323, pp. 63–68.

Christiansen, D.R. (1990) 'Adventure tourism', in J.C. Miles and S. Priest (eds), *Adventure Education*, State College, PA: Venture Publishing.

Christie, M.F. and Mason, P.A. (2003) 'Transformative tour guiding: training tour guides to be critically reflective practitioners', *Journal of Ecotourism* 2(1): 1–16.

Chubb, M. and Chubb, H.R. (1981) *One-third of Our Time? An Introduction to Recreation Behavior and Resources*, New York: John Wiley.

Clarke, J. (1997) 'A framework of approaches to sustainable tourism', *Journal of Sustainable Tourism* 5(3): 224–233.

—— (2002) 'A synthesis of activity towards the implementation of sustainable tourism: ecotourism in a different context', *International Journal of Sustainable Development* 5(3): 232–249.

Clawson, M. and Knetsch, J.L. (1966) *Economics of Outdoor Recreation*, Baltimore, MD: Johns Hopkins University Press.

Clements, C.J., Schultz, J.H. and Lime, D.W. (1993) 'Recreation, tourism, and the local residents: partnership or coexistence?', *Journal of Park and Recreation Administration* 11(4): 78–91.

Cleverdon, R. and Kalisch, A. (2000) 'Fair trade in tourism', *International Journal of Tourism Research* 2: 171–187.

Coccossis, H. (1996) 'Tourism and sustainability: perspectives and implications', in G.K. Priestley, J.A. Edwards and H. Coccossis (eds), *Sustainable Tourism? European Experiences*, Wallingford, Oxon.: CAB International.

Cochran-Smith, M. (2003) 'Learning and unlearning: the education of teacher educators', *Teaching and Teacher Education* 19(1): 5–28.

Cockrell, D. and Detzel, D. (1985) 'Effects of outdoor leadership certification on safety, impacts, and program', *Trends* 22(3): 15–21.

Coghlan, A. and Fennell, D.A. (2009) 'Myth or substance: an examination of altruism as the basis of volunteer tourism', *Annals of Leisure Research* 12(3/4): 377–402.

Coghlan, A. and Kim, K.J. (2012) 'Interpretive layering in nature-based tourism: a simple approach for complex attractions', *Journal of Ecotourism* 11(3): 173–187.

Cohen, E. (1972) 'Toward a sociology of tourism', *Social Research* 39(1): 164–182.

—— (1978) 'The impact of tourism on the physical environment', *Annals of Tourism Research* 5(2): 215–237.

—— (1987) 'Alternative tourism: a critique', *Tourism Recreation Research* 12(2): 13–18.

—— (2002) 'Authenticity, equity and sustainability in tourism', *Journal of Sustainable Tourism* 10(4): 267–276.

Coleman, J.S. (1988) 'Social capital in the creation of human capital', *American Journal of Sociology* 84 (suppl.): S95S119.

Colvin, J.G. (1994) 'Capirona: a model of indigenous ecotourism', paper presented at the Second Global Conference: Building a Sustainable World through Tourism, Montreal, September.

Commonwealth of Australia (1997) *Ecotourism Snapshot: A Focus on Recent Market Research*, Canberra: Office on National Tourism.

Commonwealth Department of Tourism (1994) *National Ecotourism Strategy*, Canberra: AGPS.

Conservation International (1997) *Conservation International's Ecotourism Program Initiatives*, Washington, DC: CI.

Conservation Measures Partnership (2002) *Open Standards for the Practice of Conservation*. Available online at http://www.conservationmeasures.org

Conservation Measures Partnership (2004) *Open Standards for the Practice of Conservation*. Available online at http://www.conservationmeasures.org

Consulting and Audit Canada (1995) *What Tourism Managers Need to Know: A Practical Guide to the Development and Use of Indicators of Sustainable Tourism*, Ottawa: Consulting and Audit Canada, for the World Tourism Organization.

Cooper, C. (1995) 'Strategic planning for sustainable tourism: the case of the offshore islands of the UK', *Journal of Sustainable Tourism* 3(4): 191–209.

Cooper, C. and Jackson, S. (1989) 'Destination lifecycle: the Isle of Man case study', *Annals of Tourism Research* 16(3): 377–398.

Coppock, J.T. (1982) 'Tourism and conservation', *Tourism Management* 3: 270–276.

Corbett, R. and Findlay, H. (1998) *Your Risk Management Program: A Handbook for Sport Organization*, Ottawa: Centre for Sport and Law.

Cordell, J. (1993) 'Who owns the land? Indigenous involvement in Australian protected areas', in E. Kemf (ed.), *The Law of the Mother: Protecting Indigenous People in Protected Areas*, San Francisco, CA: Sierra Club.

Costanza, R., d'Arge, R., de Groot, R., Farber, S., Grasso, M., Hannon, B., Limburg, K., Naeem, S., O'Neill, R.V., Paruelo, J., Raskin, R.G., Sutton, P. and van den Belt, M. (1997) 'The value of the world's ecosystem services and natural capital', *Nature* 387: 253–260.

Countryside Commission (1990) *National Parks in Focus*, Cheltenham: Countryside Commission.

Crabtree, A.E. and Black, R.S. (n.d.) *Ecoguide Program: Guide Workbook*, Brisbane: Ecotourism Association of Australia.

Crittendon, A. (1975) 'Tourism's terrible toll', *International Wildlife* 5(2): 4–12.

Csikszentmihalyi, M. (1975) *Beyond Boredom and Anxiety*, San Francisco, CA: Jossey-Bass, Inc.

—— (1990) *Flow: The Psychology of Optimal Experience*, New York: HarperCollins.

Csikszentmihalyi, M. and Csikszentmihalyi, I.S. (1990) 'Adventure and the flow experience', in J.C. Miles and S. Priest (eds), *Adventure Education*, State College, PA: Venture Publishing.

Curtin, S. (2003) 'Whale-watching in Kaikoura: sustainable destination development?', *Journal of Ecotourism* 2(3): 173–195.

—— (2005) 'Nature, wild animals and tourism: an experiential view', *Journal of Ecotourism* 4(1): 1–15.

—— (2010) 'What makes for memorable wildlife encounters? Revelations from "serious" wildlife tourists', *Journal of Ecotourism* 9(2): 149–168.

Cutaş, C., Ştefan, P. and Gheorghe, S. (2011) 'The importance of ecomarketing in developing of ecotourism in Romania', *Scientific Papers Series Management, Economic Engineering in Agriculture and Rural Development* 11(3): 59–64.

Dahles, H. (2002) 'The politics of tour guiding: image management in Indonesia', *Annals of Tourism Research* 29(3): 783–800.

Daigle, J.J., Hrubes, D., and Ajzen, I. (2002) 'A comparative study of beliefs, attitudes, and values among hunters, wildlife viewers, and other outdoor recreationists', *Human Dimensions of Wildlife* 7: 1–19.

Daltabuit, M. and Pi-Sunyer, O. (1990) 'Tourism development in Quintana Roo, Mexico', *CS Quarterly* 14(1): 9–13.

D'Amore, L.J. (1993) 'A code of ethics and guidelines for socially and environmentally responsible tourism', *Journal of Travel Research* 31(3): 64–66.

Dankers, C. (2003) *Environmental and Social Standards, Certification and Labelling for Cash Crops.* Rome: FAO.

Davey, S. (1993) 'Creative communities: planning and comanaging protected areas', in E. Kemf (ed.), *The Law of the Mother: Protecting Indigenous Peoples in Protected Areas*, San Francisco, CA: Sierra Club Books.

Davidson, M. (1995) 'Community development', *Recreation Saskatchewan* 22: 5–6.

Davis, P.B. (1998) 'Beyond guidelines: a model for Antarctic tourism', *Annals of Tourism Research* 23(3): 546–553.

de Castro, F. and Bergossi, A. (1996) 'Fishing at Rio Grande (Brazil): ecological niche and competition', *Human Ecology* 24(3): 401–411.

De Graaf, D.G., Jordan, D.J. and De Graaf, K.H. (1999) *Programming for Parks, Recreation, and Leisure Services: A Servant Leadership Approach*, State College, PA: Venture Publishing.

De Groot, R.S. (1983) 'Tourism and conservation in the Galápagos Islands', *Biological Conservation* 26: 291–300.

de Haas, H.C. (2002) 'Sustainability of small-scale ecotourism: the case of Niue, South Pacific', *Current Issues in Tourism* 5(3–4): 319–337.

De la Barre, S. (2005) 'Not "ecotourism"?: wilderness tourism in Canada's Yukon Territory', *Journal of Ecotourism* 4(2): 92–107.

De Lacy, T. (1992) 'Towards an aboriginal land management curriculum', in J. Birckhead, T. De Lacy and L. Smith (eds), *Aboriginal Involvement in Parks and Protected Areas*, Canberra: Aboriginal Studies Press.

—— (1994) 'The Uluru/Kakadu model: Anangu Tjukurrpa. 50,000 years of Aboriginal law and land management changing the concept of national parks in Australia', *Society and Natural Resources* 7(5): 479–498.

De Lacy, T. and Lockwood, M. (1992) 'Estimating the non-market values of nature conservation resources in Australia', paper presented at the Fourth World Congress on Parks and Protected Areas, Caracas, 10–21 February.

Dearden, P. (1991) 'Parks and protected areas', in B. Mitchell (ed.), *Resource Management and Development*, Toronto: Oxford University Press.

Dearden, P. and Rollins, R. (1993) 'The times they are a-changin', in P. Dearden and R. Rollins (eds), *Parks and Protected Areas in Canada: Planning and Management*, Toronto: Oxford University Press.

Debbage, K.G. (1990) 'Oligopoly and the resort cycle in the Bahamas', *Annals of Tourism Research* 17(4): 513–527.

Deming, A. (1996) 'The edges of the civilized world: tourism and the hunger for wild places', *Orion* 15(2): 28–35.

Demoi, L.A. (1981) 'Alternative tourism: towards a new style in North-South relations', *Tourism Management* 2: 253–264.

Deruiter, D.S. and Donnelly, M.P. (2002) 'A qualitative approach to measuring determinants of wild-life value orientations', *Human Dimensions of Wildlife* 7: 251–271.

Devall, B. and Sessions, G. (1985) *Deep Ecology*, Salt Lake City, UT: Gibbs Smith.

Diamantis, D. (1999) 'The concept of ecotourism: evolution and trends', *Current Issues in Tourism* 2(2–3): 93–122.

—— (2000) 'Ecotourism and sustainability in Mediterranean islands', *Thunderbird International Business Review* 42(4): 427–443.

—— (2004) *Ecotourism*, London: Thomson.

Diamantis, D. and Ladkin, A. (1999) ' "Green" strategies in the tourism and hospitality industries', in F. Vellas and L. Becheral (eds), *The International Marketing of Travel and Tourism*, London: Macmillan.

Diamond, J. (1975) 'The island dilemma: lessons of modern biogeographic studies for the design of nature reserves', *Biological Conservation* 7: 3–15.

—— (2005) *Collapse: How Societies Choose to Fail or Succeed*, New York: Viking.

Dieke, P. (2001) 'Kenya and South Africa', in D.B. Weaver (ed.), *The Encyclopedia of Ecotourism*, London: CAB International.

Dodds, R. and Joppe, M. (2001) 'Promoting urban green tourism: the development of the other map of Toronto', *Journal of Vacation Marketing* 7(3): 261–267.

Dolnicar, S. (2002) 'A review of data-driven market segmentation in tourism', *Journal of Travel and Tourism Marketing* 12(1): 1–22.

—— (2010) 'Identifying tourists with smaller environmental footprints', *Journal of Sustainable Tourism* 18(6): 717–734.

Dolnicar, S., Crouch, G.I. and Long, P. (2008) 'Environmentaly friendly tourists: what do we really know about them?', *Journal of Sustainable Tourism* 16(2): 197–210. Also available at: http://ro.uow.edu.au/cgi/viewcontent.cgi?article=1478&context=commpapers

Donohoe, H.M. and Needham, R.D. (2006) 'Ecotourism: the evolving contemporary definition', *Journal of Ecotourism* 5(3): 192–210.

—— (2008) 'Internet-based ecotourism marketing: evaluating Canadian sensitivity to ecotourism tenets', *Journal of Ecotourism* 7(1): 15–43.

Dorsey, E.R., Steeves, H.L. and Porras, L.E. (2004) 'Advertising ecotourism on the internet: commodifying environment and culture', *New Media & Society* 6(6): 753–779.

Dowling, R.K. (1992) *The Ecoethics of Tourism: Guidelines for Developers, Operators and Tourists*, Canberra: Bureau of Tourism Research.

—— (1993) 'An environmentally based planning model for regional tourism development', *Journal of Sustainable Tourism* 1(1): 17–37.

Dowsley, M. (2009) 'Inuit-organised polar bear sport hunting in Nunavut territory, Canada', *Journal of Ecotourism* 8(2): 161–175.

Doxey, G.V. (1975) 'A causation theory of visitor-resident irritants: methodology and research inference', paper presented at the TTRA Conference, San Diego, California, pp. 195–198.

Drake, S.P. (1991) 'Local participation in ecotourism projects', in T. Whelan (ed.), *Nature Tourism: Managing for the Environment*, Washington, DC: Island Press.

Dredge, D. (2006) 'Networks, conflict and collaborative communities', *Journal of Sustainable Tourism* 14(6): 562–581.

D'Sa, E. (1999) 'Wanted: tourists with a social conscience', *International Journal of Contemporary Hospitality Management* 11(2–3): 64–68.

Duenkel, N. and Scott, H. (1994) 'Ecotourism's hidden potential: altering perceptions of reality', *Journal of Physical Education Recreation and Dance* October: 40–44.

Duffus, D.A. and Dearden, P. (1990) 'Non-consumptive wildlife-oriented recreation: a conceptual framework', *Biological Conservation* 53 (3): 213–231.

—— (2006a) 'Global environmental governance and the politics of ecotourism in Madagascar', *Journal of Ecotourism* 5(1/2): 128–144.

—— (2006b) 'The politics of ecotourism and the developing world', *Journal of Ecotourism* 5(1/2): 16.

Dufrene, M. (1973) *The Phenomenology of Aesthetic Experience*, Evanston, IL: Northwestern University Press.

Dunlap, R.E. and Heffernan, R.B. (1975) 'Outdoor recreation and environmental concern: an empirical examination', *Rural Sociology* 40: 18–30.

Durst, P.B. and Ingram, C.D. (1988) 'Nature-oriented tourism promotion by developing countries', *Tourism Management* 9(1): 39–43.

Dyess, R. (1997) 'Adventure travel or ecotourism?', *Adventure Travel Business* April: 2.

Eagles, P.F.J. (1992) 'The travel motivations of Canadian ecotourists', *Journal of Travel Research* 31(Fall): 3–7.

—— (1993) 'Parks legislation in Canada', in P. Dearden and R. Rollins (eds), *Parks and Protected Areas in Canada*, Toronto: Oxford University Press.

—— (1995) 'Tourism and Canadian parks: fiscal relationships', *Managing Leisure* 1: 16–27.

Eagles, P.F.J. and Wind, E. (1994) 'Canadian ecotours in 1992: a content analysis of advertising', *Journal of Applied Recreation Research* 19(1): 67–87.

Echeverría, J., Hanrahan, M. and Solorzano, R. (1995) 'Valuation of non-priced amenities provided by the biological resources within the Monteverde Cloud Forest Preserve, Costa Rica', *Ecological Economics* 13: 45–52.

Economic and Social Commission for Asia and the Pacific (ESCAP) (1978) 'The formulation of basic concepts and guidelines for preparation of tourism sub-regional master plans in the ESCAP region' (in Britton 1982).

Economist (1991) 'Travel and tourism: the pleasure principle', March: 3–22.

—— (2002) 'Poachers' return'. Available online at http://www.economist.com/node/1131330 (accessed 7 August 2014).

Ecotourism/Heritage Tourism Advisory Committee (1997) *Planning for the Florida of the Future*, Tallahassee, FL: Government Document.

Ecotourism Society (1993) *Ecotourism Guidelines for Nature Tour Operators*, North Bennington, VT: The Ecotourism Society.

Ecotourism Society of Saskatchewan (2000) *Saskatchewan Ecotourism Accreditation System*, Regina, Saskatchewan.

Edginton, C.R., Compton, D.M. and Hanson, C.J. (1980) *Recreation and Leisure Programming: A Guide for the Professional*, Philadelphia, PA: Saunders.

Edginton, C.R., Hanson, C.J., Edginton, S.R. and Hudson, S.D. (1998) *Leisure Programming: A Service Centred and Benefits Approach*, Boston, MA: WCB/McGraw-Hill.

Edwards, S., McLaughlin, W.J. and Ham, S. (1998) *Comparative Study of Ecotourism Policy in the Americas–1998: Volume II Latin America and the Caribbean*, Organization of America States, University of Idaho.

Ehrlich, P.R. (2000) *Human Natures: Genes, Cultures, and the Human Prospect*. New York: Penguin.

Ehrlich, P.R. and Ehrlich, A.H. (1992) 'The value of biodiversity', *Ambio* 21(3): 219–226.

Eidsvik, H. (1983) 'Biosphere reserves/opportunities for cooperation: a global perspective', in R.C. Scace and C.J. Martinka (eds), *Towards the Biosphere Reserve: Exploring Relationships between Parks and Adjacent Lands*, Kalispell, MT: Department of the Interior, National Park Service.

—— (1993) 'Canada, conservation, and protected areas', in P. Dearden and R. Rollins (eds), *Parks and Protected Areas in Canada*, Toronto: Oxford University Press.

Elliot, R. (1991) 'Environmental ethics', in P. Singer (ed.), *A Companion to Ethics*, Oxford: Blackwell.

Environment Canada (1990) *National Parks System Plan*, Ottawa: Supply and Services Canada.

Erisman, H.M. (1983) 'Tourism and cultural dependency in the West Indies', *Annals of Tourism Research* 10(3): 337–361.

Eubanks T.L. Jr, Stoll, J.R. and Ditton, R.B. (2004) 'Understanding the diversity of eight birder sub-populations: socio-demographic characteristics, motivations, expenditures and net benefits', *Journal of Ecotourism* 3(3): 151–172.

European Commission (2004) *Using natural and Cultural heritage to Develop Sustainable Tourism*,

1, 3. Available online at http://reports.europa.eu.int/comm/enterprise/services/tourism/studies/ecosystems/heritage.htm

Ewert, A. (1985) 'Why people climb: the relationship of participant motives and experience level to mountaineering', *Journal of Leisure Research* 17(3): 241–250.

Ewert, A. and Shultis, J. (1997) 'Resource-based tourism: an emerging trend in tourism experiences', *Parks and Recreation* September: 94–103.

Exmoor National Park Planning Committee (2013) Draft Exmoor National Park Local Plan. Available online at http://www.exmoor-nationalpark.gov.uk/about-us/meetings-agendas-reports/enpa-planning-committee/02-jul-2013/Draft-Exmoor-Local-Plan-for-consultation.pdf (accessed 9 July 2014).

Fairweather, J.R., Maslin, C. and Simmons, D.G. (2005) 'Environmental values and response to ecolabels among international visitors to New Zealand', *Journal of Sustainable Tourism* 13: 82–99.

Falk, J.H. and Dierking, L.D. (2000) *Learning from Museums: Visitor Experiences and the Making of Meaning,* Walnut Creek, CA: AltaMira Press.

Farquharson, M. (1992) 'Ecotourism: a dream diluted', *Business Mexico* 2(6): 8–11.

Farr, H. and Rogers, A. (1994) 'Tourism and the environment on the Isles of Scilly: conflict and complementarity', *Landscape and Urban Planning* 29: 1–17.

Farrell, B.H. and McLellan, R.W. (1987) 'Tourism and physical environment research', *Annals of Tourism Research* 14(1): 1–16.

Farrell, P. and Lundegren, H.M. (1993) *The Process of Recreation Programming*, State College, PA: Venture Publishing.

Farrell, T.A. and Marion, J.L. (2002) 'The protected area visitor impact management (PAVIM) framework: a simplified process for making management decisions', *Journal of Sustainable Tourism* 10(1): 31–51.

Fayos-Solá, E. (1996) 'Tourism policy: a midsummer night's dream?', *Tourism Management* 17(6): 405–412.

Federation of Nature and National Parks of Europe (1993) *Loving Them to Death? Sustainable Tourism in Europe's Nature and National Parks*, Eupen, Belgium: Kliemo.

Fennell, D.A. (1990) 'A profile of ecotourists and the benefits derived from their experience: a Costa Rican case study', unpublished master's thesis, University of Waterloo, Waterloo, Ontario.

—— (1998) 'Ecotourism in Canada', *Annals of Tourism Research* 25(1): 231–234.

—— (2000) 'Ecotourism on trial: the case of billfishing as ecotourism', *Journal of Sustainable Tourism* 8(4): 341–345.

—— (2001) 'Areas and needs in ecotourism research', in D.B. Weaver (ed.), *The Encyclopedia of Ecotourism*, Wallingford, Oxon: CABI, pp. 639–655.

—— (2002a) 'The Canadian ecotourist in Costa Rica: ten years down the road', *International Journal of Sustainable Development* 5(3): 282–299.

—— (2002b) *Ecotourism Programme Planning*, Wallingford, Oxon.: CAB International.

—— (2002c) 'Ecotourism: where we've been and where we're going', *Journal of Ecotourism* 1(1): 1–6.

—— (2003) 'Ecotourism in the South African context', *Africa Insight* 33(1): 1–8.

—— (2004) 'Towards interdisciplinarity in tourism: making a case through complexity and shared knowledge', *Recent Advances and Research Updates* 5(1): 99–110.

—— (2006a) 'Evolution in tourism: the theory of reciprocal altruism and tourist-host interactions', *Current Issues in Tourism* 9(2): 105–125.

—— (2006b) *Tourism Ethics*, Clevedon: Channel View.

—— (2008) 'Ecotourism and the myth of indigenous stewardship', *Journal of Sustainable Tourism*, 16(2): 129–149.

—— (2009) 'The nature of pleasure in pleasure travel', *Tourism Recreation Research*, 34(2): 123–134.

—— (2012a) 'Tourism, animals and ethics: utilitarianism', *Tourism Recreation Research* 37(3): 239–249.

—— (2012b) 'Tourism and animal rights', *Tourism Recreation Research* 37(2): 157–166.

—— (2012c) *Tourism and Animal Ethics*. London: Routledge.

—— (2012d) 'Ecotourism', in A. Holden and D. Fennell (eds), *The Routledge Handbook of Tourism and the Environment*, London: Routledge, pp. 323–333.

—— (2013a) 'The ethics of excellence in tourism research', *Journal of Travel Research* 52(4): 417–425.

—— (2013b) 'Contesting the zoo as a setting for ecotourism, and the design of a first principle', *Journal of Ecotourism* 12(1): 1–14.

—— (2013c) 'Tourism and animal welfare', *Tourism Recreation Research*, 38(3): 325–340.

—— (2013d) 'Tourism, animals and ecocentrism: a re-examination of the billfish debate', *Tourism Recreation Research* 38(2): 189–202.

Fennell, D.A. and Butler, R.W. (2003) 'A human ecological approach to tourism interactions', *Progress in Tourism and Hospitality Research* 5: 197–210.

Fennell, D.A. and Dowling, R.K. (2003) 'Ecotourism policy and planning: stakeholders, management and governance', in D.A. Fennell and R.K. Dowling (eds), *Ecotourism Policy and Planning*, Wallingford, Oxon.: CAB International.

Fennell, D.A. and Eagles, P.F.J. (1990) 'Ecotourism in Costa Rica: a conceptual framework', *Journal of Park and Recreation Administration* 8(1): 23–34.

Fennell, D. A. and Ebert, K. (2004) 'Tourism and the precautionary principle', *Journal of Sustainable Tourism* 12(6): 461–479.

Fennell, D.A. and Malloy, D.C. (1995) 'Ethics and ecotourism: a comprehensive ethical model', *Journal of Applied Recreation Research* 20(3): 163–184.

—— (1997) 'Measuring the ethical nature of tourist operators: a comparison', World Congress and Exhibition on Ecotourism, Rio de Janeiro, Brazil, 15–18 December, pp. 144–149.

—— (2007) *Codes of Ethics in Tourism: Practice, Theory, Synthesis*, Clevedon: Channel View.

Fennell, D.A. and Nowaczek, A. (2010) 'Moral and empirical dimensions of human-animal interactions in ecotourism: deepening an otherwise shallow pool of debate', *Journal of Ecotourism*, 9(3): 239–255.

Fennell, D.A. and Plummer, R. (2010) 'Sociobiology and adaptive capacity: evolving adaptive strategies to build environmental governance', in D. Armitage and R. Plummer (eds), *Adaptive Capacity and Environmental Governance*, Berlin: SpringerVerlag, pp. 243–261.

Fennell, D.A. and Przeclawski, K. (2003) 'Tourists and host communities', in S. Singh, D.J. Timothy, and R.K. Dowling (eds), *Tourism in Destination Communities*, Wallingford, Oxon.: CAB International.

Fennell, D.A. and Sheppard, V. (2011) 'Canada's 2010 Winter Olympic Legacy: applying an ethical lens to the post-games' sled dog cull', *Journal of Ecotourism* 10(3): 197–213.

Fennell, D.A. and Smale, B.J.A. (1992) 'Ecotourism and natural resource protection: implications of an alternative form of tourism for host nations', *Tourism Recreation Research* 17(1): 21–32.

Fennell, D.A. and Weaver, D.B. (1997) 'Vacation farms and ecotourism in Saskatchewan, Canada', *Journal of Rural Studies* 13(4): 467–475.

—— (2005) 'The ecotourium concept and tourism-conservation symbiosis', *Journal of Sustainable Tourism* 13(4): 373–390.

Fennel, D.A. and Weaver, D.B. (2005) 'The ecotourium concept and tourism symbiosis', *Journal of Sustainable Tourism*, 13(2): 373–390.

Fennell, D.A., Buckley, R. and Weaver, D.B. (2001) 'Policy and planning', in D.B. Weaver (ed.), *The Encyclopedia of Ecotourism*, Wallingford, Oxon.: CAB International.

Fennell, D.A., Plummer, R. and Marschke, M. (2008) 'Is adaptive co-management ethical?', *Journal of Environmental Management* 88(1): 62–75.

Fennell, J. (1989) 'Destination resorts', *Canadian Building* 39(5): 10–17.

Figgis, P. (2004) *Conservation on Private Lands: The Australian Experience*, IUCN Programme on protected Areas: Gland, Switzerland.

Finch, R. and Elder, J. (1990) *The Norton Book of Nature Writing*, New York: W.W. Norton and Co.

Fleckenstein, M.P. and Huebsch, P. (1999) 'Ethics in tourism: reality or hallucination', *Journal of Business Ethics* 19: 137–142.

Flinn, D. (1989) *Travellers in a Bygone Shetland*, Edinburgh: Scottish Academic Press.

Folke, C., Carpenter, S., Elmqvist, T., Gunderson, L., Holling, C.S., Walker, B., Bengtsson, J., Berkes, F., Colding, J., Danell, K., Falkenmark, M., Moberg, M., Gordon, L., Kaspersson, R., Kautsky, N., Kinzig, A., Levin, S.A., Mäler, K-G., Ohlsson, L., Olsson, P., Ostrom, E., Reid, W., Rockstöm, J., Savenije, S. and Svedin, U. (2002) *Resilience and Sustainable Development: Building Adaptive Capacity in a World of Transformations*, Environmental Advisory Council to the Swedish Government Scientific Background Paper.

Font, X. and Buckley, R. (2001) *Tourism Ecolabelling*, Wallingford, Oxon.: CAB International.

Font, X., Sanabria, R. and Skinner, E. (2003) 'Sustainable tourism and ecotourism certification: raising standards and benefits', *Journal of Ecotourism* 2(3): 213–218.

Font, X., Tapper, R. and Cochrane, J. (2006) 'Competitive strategy in a global industry: tourism', *Handbook of Business Strategy* 7(1): 51–55.

Ford, P. and Blanchard, J. (1993) *Leadership and Administration of Outdoor Pursuits*, State College, PA: Venture.

Forestell, P.H. (1993) 'If Leviathan has a face, does Gaia have a soul?: incorporating environmental education in marine eco-tourism programs', *Ocean and Coastal Management* 20(3): 267–282.

Forman, R.T.T. (1990) 'Ecologically sustainable landscapes: the role of spatial configuration', in I.S. Zonneveld and R.T.T. Forman (eds), *Changing Landscapes: An Ecological Perspective*, New York: Springer Verlag.

Forstner, K. (2004) 'Community ventures and access to markets: the role of intermediaries in marketing rural tourism products', *Development Policy Review* 22(5): 497–514.

Forsyth, T. (1995) 'Business attitudes to sustainable tourism: self-regulation in the UK outgoing tourism industry', *Journal of Sustainable Tourism* 3(4): 210–231.

Fowkes, J. and Fowkes, S. (1991) 'Roles of private sector ecotourism in protected areas', *Parks* 2(3): 26–30.

Fox, A. (1996) 'Kakadu, tourism and the future', *Australian Natural History* 21(7): 266–271.

Fox, R.M. and DeMarco, J.P. (1986) 'The challenge of applied ethics', in R.M. Fox and J.P. DeMarco (eds), *New Directions in Ethics: The Challenge of Applied Ethics*, New York: Routledge & Kegan Paul.

Francis, G. (n.d.) 'Ecosystem management', paper presented at the Tri-national Conference on the North American Experience Managing Transboundary Resources: United States and the Boundary Commissions.

—— (1985) 'Biosphere reserves: innovations for cooperation in the search for sustainable development', *Environments* 17(3): 21–38.

Frangialli, F. (1997) 'Keynote address to the World Ecotour '97 Conference', Rio de Janeiro, 15–18 December.

Frost, J.E. and McCool, S.F. (1988) 'Can visitor regulations enhance recreational experiences?', *Environmental Management* 12(1): 5–9.

Fulgosi, D. (2006) 'Man v. monkey', *National Post*, Thursday, 9 November, section A22.

Funnell, D.C. and Bynoe, P.E. (2007) 'Ecotourism and institutional structures: the case of North Rupununi, Guyana', *Journal of Ecotourism* 6(3): 163–183.

Galley, G. and Clifton, J. (2004) 'The motivational and demographic characteristics of research ecotourists: Operation Wallacea volunteers in south-east Sulawesi, Indonesia', *Journal of Ecotourism* 3(10): 69–82.

Garavan, T.N. (1997) 'Training, development, education and learning: different or the same?', *Journal of European Industrial Training* 21(2): 39–50

Gardner, J.E. (1993) 'Environmental non-government organizations (ENGOs) and sustainable development', in S. Lerner (ed.), *Environmental Stewardship: Studies in Active Earthkeeping*, Department of Geography Publication Series, no. 39, University of Waterloo, Waterloo, Ontario.

Garrod, B. (2003) 'Local participation in the planning and management of ecotourism: a revised model approach', *Journal of Ecotourism* 2(1): 33–53.

Garrod, B. and Fennell, D.A. (2004) 'An analysis of whalewatching codes of ethics', *Annals of Tourism Research* 31(2): 334–352.

Gass, M. and Williamson, J. (1995) 'Accreditation for adventure programs', *Journal of Physical Education, Recreation and Dance* 22–27 January.

Geist, V. (1994) 'Wildlife conservation as wealth', *Nature* 368: 45–46.

Genot, H. (1995) 'Voluntary environmental codes of conduct in the tourism sector', *Journal of Sustainable Tourism* 3(3): 166–172.

Getz, D. (1986) 'Models in tourism planning: towards integration of theory and practice', *Tourism Management* 7(1): 21–32.

—— (1990) *Festivals, Special Events, and Tourism*, New York: Van Nostrand Reinhold.

Getz, D. and Jamal, D. (1994) 'The environment–community symbiosis: a case for collaborative tourism planning', *Journal of Sustainable Tourism* 2(3): 152–173.

Giannecchini, J. (1993) 'Ecotourism: new partners, new relationships', *Conservation Biology* 7(2): 429–432.

Gilchrist, N.L. (1998) 'Winter hiking and Camping', in N.J. Dougherty IV (ed.), *Outdoor Recreation Safety*, Champaign, IL: Human Kinetics.

Gilmore, G.D. and Campbell, M.D. (1996) *Needs Assessment Strategies for Health Education and Health Promotion*, Madison, WI: WCB Brown and Benchmark.

Gjerdalen, G. and Williams, P.W. (2000) 'An evaluation of the utility of a whale watching code of conduct', *Tourism Recreation Research* 25(2): 27–37.

Glasbergen, P. (1998) 'The question of environmental governance', in P. Glasbergen (ed.), *Co-operative Environmental Governance*, Dordrecht, Netherlands: Kluwer Academic Press, pp. 1–18.

Globe '90 (1990) *An Action Strategy for Sustainable Tourism Development*, Ottawa: Tourism Canada.

Gnoth, J. (1997) 'Tourism motivation and expectation formation', *Annals of Tourism Research* 24: 283–304.

Goeldner, C.R., Ritchie, J.R.B. and McIntosh, R.W. (2000) *Tourism: Principles, Practices, Philosophies*, New York: John Wiley & Sons, Inc.

Goodall, B. (1994) 'Environmental auditing: current best practice (with special reference to British tourism firms)', in A.V. Seaton (ed.), *Tourism: The State of the Art*, Chichester: John Wiley.

Goodall, B. and Cater, E. (1996) 'Self-regulation for sustainable tourism?', *Ecodecision* 20(Spring): 43–45.

Goodwin, H. (1995) 'Tourism and the environment', *Biologist* 42(3): 129–133.

—— (1996) 'In pursuit of ecotourism', *Biodiversity and Conservation* 5(3): 277–291.

Gössling, S. (2000) 'Sustainable tourism development in developing countries: some aspects of energy use', *Journal of Sustainable Tourism* 8(5): 410–425.

—— (2002) 'Human–environmental relations within tourism', *Annals of Tourism Research* 29(2): 539–556.

—— (2006) 'Ecotourism as experience tourism', in S. Gössling and J. Hultman (eds), *Ecotourism in Scandinavia: Lessons in Theory and Practice*, Wallingford, Oxon: CAB International, pp. 89–97.

Gössling, S., Borgstrom Hansson, H., Horstmeier, O. and Saggel, S. (2002) 'Ecological footprint analysis as a tool to assess tourism sustainability', *Ecological Economics* 43: 199–211

Gouldner, A. (1960) 'The norm of reciprocity: a preliminary statement', *American Sociological Review* 25: 161–178.

Government of British Columbia (2011a) *Premier Appoints Task Force to Review Dog Killings*. Accessed 28 February 2011 at http://www2.news.gov.bc.ca/news_releases_2009-2013/2011PREM0006-000094.htm

—— (2011b) *Premier Announces Canada's Toughest Animal Cruelty Laws*. Accessed 28 February 2012 at http://www2.news.gov.bc.ca/news_releases_2009-2013/2011PREM0030-000340.htm

—— (2012) *Sled Dog Code of Practice*. Victoria, BC: Government of British Columbia and Ministry of Agriculture.

Graburn, N. (1989) 'Tourism: the sacred journey', in V.L. Smith (ed.), *Hosts and Guests: The Anthropology of Tourism* (2nd edn), Philadelphia, PA: University of Pennsylvania Press.

Graefe, A.R., Vaske, J.J. and Kuss, F.R. (1984a) 'Resolved issues and remaining questions about social carrying capacity', *Leisure Sciences* 6(4): 497–507.

—— (1984b) 'Social carrying capacity: an integration and synthesis of twenty years of research', *Leisure Sciences* 6: 395–431.

Gratzer, M., Winiwarter, W. and Werthner, H. (2006) *State of the Art in eTourism*. Available online at http://homepage. univie.ac.at/werner.winiwarter/secec.pdf (accessed 22 November).

Gray, B. (1989) *Collaborating: Finding Common Ground for Multiparty Problems*, San Francisco, CA: Jossey-Bass.

Gray, L.C. and Moseley, W.G. (2005) 'A geographical perspective on poverty–environment interaction', *Geographical Journal* 171(1): 9–23.

Gray, N. and Campbell, L.M. (2007) 'A decommodified experience? Exploring aesthetic, economic and ethical values for volunteer ecotourism in Costa Rica', *Journal of Sustainable Tourism* 15(5): 463–482.

Gray, P.A., Duwors, E., Villeneuve, M. and Boyd, S. (2003) 'The socioeconomic significance of nature-based recreation in Canada', *Environmental Monitoring and Assessment* 86: 129–147.

Green, H. and Hunter, C. (1992) 'The environmental impact assessment of tourism development', in P. Johnson and B. Thomas (eds), *Perspectives on Tourism Policy*, London: Mansell.

Green Tourism Association (1999) *Map Project Evaluation Report*. Available online at www.green-tourism.on.ca (accessed October 2001).

Grenier, A.A. (1998) *Ship-based Polar Tourism in the Northwest Passage: A Case Study*, Rovaniemi: University of Lapland.

Grenier, D., Kaae, B.C., Miller, M.L. and Mobley, R.W. (1993) 'Ecotourism, landscape architecture and urban planning', *Landscape and Urban Planning* 25(1/2): 1–16.

Grumbine, E. (1996) 'Beyond conservation and preservation in American environmental values', in B.L. Driver, D. Dustin, T. Baltic, G. Elsner and G. Peterson (eds), *Nature and the Human Spirit: Toward an Expanded Land Management Ethic*, State College, PA: Venture.

Grupo de Trabalho Interministerial MICT/MMA (1994) *Diretrizes para uma Política Nacional de ecotourism/Coordenaçao de Silvio Magalhaes Barros II e Denis Hamú m de La Penha*, Brasilia: EMBRATUR.

Guignon, C. (1986) 'Existential ethics', in R.M. Fox and J.P. DeMarco (eds), *New Directions in Ethics: The Challenge of Applied Ethics*, New York: Routledge & Kegan Paul.

Gullison, R.E., Rice, R.E. and Blundell, A.G. (2000) 'Marketing species conservation', *Nature* 404: 923–924.

Gunn, C.A. (1972) *Vacationscape: Designing Tourist Regions*, University of Texas: Bureau of Business Research.

—— (1988) *Tourism Planning*, New York: Taylor & Francis.

—— (1994) *Tourism Planning: Basics, Concepts, Cases*, 3rd edn, Washington, DC: Taylor & Francis.

Gunnarsdotter, Y. (2006) 'Hunting tourism as ecotourism: conflicts and opportunities', in S. Gössling and J. Hultman (eds), *Ecotourism in Scandinavia: Lessons in Theory and Practice*, Wallingford, Oxon, CAB International, pp. 178–192.

Gurung, C. (1995) 'Ecolodges and their role in integrated conservation and development', paper presented at the Second International Ecolodge Forum and Field Seminar, San José, Costa Rica, 22–29 October.

Hadley, J. and Crow, P. (1995) 'Some guidelines for the architecture of ecotourism facilities', in *The Ecolodge Sourcebook for Planners and Managers*, North Bennington, VT: The Ecotourism Society.

Hadwen, W.L., Hill, W. and Pickering, C.M. (2008) 'Linking visitor impact research to visitor impact monitoring in protected areas', *Journal of Ecotourism* 7(1): 87–93.

Halbertsma, N.F. (1988) 'Proper management is a must', *Naturopa* 59: 23–24.

Hall, C.M. (1992) 'Adventure, sport and health tourism', in B. Weiler and C.M. Hall (eds), *Special Interest Tourism*, London: Belhaven Press.

—— (1999) 'Rethinking collaboration and partnership: a public policy perspective', *Journal of Sustainable Tourism* 7(3–4): 274–289.

Hall, C.M. and Boyd, S. (2005) 'Nature-based tourism in peripheral areas: introduction', in C.M. Hall and S. Boyd (eds), *Nature-based Tourism in Peripheral Areas: Development or Disaster?*, Toronto: Channel View Publications.

Hall, C.M. and Jenkins, J. (1995) *Tourism and Public Policy*, London: Routledge.

Hall, C.M. and McArthur, S. (1998) *Integrated Heritage Management: Principles and Practice*, London: Stationery Office.

Hall, C.M. and Page, S.J. (1999) *The Geography of Tourism and Recreation: Environment, Place and Space*, London: Routledge.

Hall, C.M. and Wouters, M. (1995) 'Issues in Antarctic tourism', in C.M. Hall and M.E. Johnston (eds), *Polar Tourism: Tourism in the Arctic and Antarctic Regions*, London: John Wiley.

Hall, S. (ed.) (1993) *Ethics in Hospitality Management: A Book of Readings*, East Lansing, MI: Educational Institute of the American Hotel and Motel Association.

Halpern, S. (1998) 'A fragile kingdom', *Audubon* 100(2): 37–45, 99–101.

Ham, S. and Weiler, B. (2000) *Six Principles for Tour Guide Training and Sustainable Development in Developing Countries*, Working Paper 102/00, Faculty of Business and Economics, Monash University, Caufield East, Victoria, Australia.

Ham, S.H. and Weiler, B. (2002) 'Interpretation as the centrepiece of sustainable wildlife tourism', in R. Harris, T. Griffin and P. Williams (eds) *Sustainable Tourism: A Global Perspective* (pp. 35–44). Oxford: Butterworth Heinemann.

Hammitt, W.E. and Cole, D.C. (1987) *Wildland Recreation: Ecology and Management*, New York: John Wiley.

Harrington, I. (1971) 'The trouble with tourism unlimited', *New Statesman* 82: 176.

Harroun, L.A. (1994) *Potential Frameworks for Analysis of Ecological Impacts of Tourism in Developing Countries*, Washington, DC: World Wide Fund for Nature.

Harroy, J.P. (1974) 'A century in the growth in the "national park" concept throughout the world', in H. Elliot (ed.), *Second World Conference on National Parks*, Gland, Switzerland: International Union for the Conservation of Nature and Natural Resources.

Hart, J. and Hart, T. (2003) 'Rules of engagement for conservation', *Conservation in Practice* 4(1): 14–22.

Hasse, J.C. and Milne, S. (2005) 'Participatory approaches and geographical information systems (PAGIS) in tourism planning', *Tourism Geographies* 7(3): 272–289.

Hawkes, S. and Williams, P. (1993) *The Greening of Tourism: From Principles to Practice*, Burnaby, British Columbia: Centre for Tourism Policy and Research, Simon Fraser University.

Hawkins, D. (1994) 'Ecotourism: opportunities for developing countries', in W. Theobald (ed.), *Global Tourism: The Next Decade*, Oxford: Butterworth.

Hawkins, J.P., Roberts, C.M., Van't Hof, T., De Meyer, K., Tratalos, J. and Aldam, C. (1999) 'Effects of recreational scuba diving on Caribbean coral and fish communities', *Conservation Biology* 13: 888–897.

Hayek, F.A. ([1944]/1994) *The Road to Serfdom*, Chicago: University of Chicago Press.

Hays, S.P. (1959) *Conservation and the Gospel of Efficiency*, Cambridge, MA: Harvard University Press.

Hayward, S.J., Gomez, V.H. and Sterrer, W. (1981) *Bermuda's Delicate Balance*, Hamilton: Bermuda National Trust.

Haywood, M. (1986) 'Can the tourist area lifecycle be made operational?', *Tourism Management* 7(3): 154–167.

Heald, D. (1984) 'Privatization: analysing its appeal and limitations', *Fiscal Studies* 5(1): 9–15.

Heidegger, M. (1966) *Discourse on Thinking*, New York: Harper Torchbooks.

Heinen, J.T. and Shrestha, S.K. (2006) 'Evolving policies for conservation: an historical profile of the protected area system of Nepal', *Journal of Environmental Planning and Management* 49(1): 41–58.

Hendee, J.C., Stankey, G.H. and Lucas, R.C. (1990) *Wilderness Management*, Golden, CO: International Wilderness Leadership Foundation.

Henderson, N. (1992) 'Wilderness and the nature conservation ideal: Britain, Canada, and the United States contrasted', *Ambio* 21(6): 394–399.

Henry, I.P. and Jackson, G.A.M. (1996) 'Sustainability of management processes and tourism products and contexts', *Journal of Sustainable Tourism* 4(1): 17–28.

Herfindahl, O. (1961) 'What is conservation', in O.C. Herfindahl (ed.), *Three Studies in Mineral Economics*, Washington, DC: Resources for the Future.

Hetzer, N.D. (1965) 'Environment, tourism, culture', *LINKS* (July); reprinted in *Ecosphere* (1970) 1(2): 1–3.

Hezri, A.A. and Hasan, M.N. (2006) 'Towards sustainable development? The evolution of environmental policy in Malaysia', *Natural Resources Forum* 30: 37–50.

Higginbottom, K. (2004) 'Wildlife tourism: an introduction', in K. Higginbottom (ed.), *Wildlife Tourism: Impacts, Management and Planning*, Altona, VIC: Common Ground.

Higgins, B.R. (1996) 'The global structure of the nature tourism industry: ecotourists, tour operators, and local businesses', *Journal of Travel Research* 35(2): 11–18.

Higgins-Desbiolles, F. (2009) 'Indigenous ecotourism's role in transforming ecological consciousness', *Journal of Ecotourism* 8(2): 144–160.

Higham, J. and Carr, A. (2002) 'Ecotourism visitor experiences in Aotearoa/New Zealand: challenging the environmental values of visitors in pursuit of pro-environmental behaviour', *Journal of Sustainable Tourism* 10(4): 277–294.

Higham, J. and Lück, M. (2002) 'Urban ecotourism: a contradiction in terms?', *Journal of Ecotourism* 1(1): 36–51.

Higham, J.E.S. and Bejder, L. (2008) 'Managing wildlife-based tourism: edging slowly towards sustainability?', *Current Issues in Tourism* 11(1): 75–83.

Higham, J.E.S. and Dickey, A. (2007) 'Benchmarking ecotourism in New Zealand: A *c.* 1999 analysis of activities offered and resources utilised by ecotourism businesses', *Journal of Ecotourism* 6(1): 67–74.

Higham, J.E.S., Lusseau, D. and Hendry, W. (2008) 'Wildlife viewing: the significance of the viewing platforms', *Journal of Ecotourism* 7(2/3): 137–146.

Hill, J., Woodland, W. and Gough, G. (2007) 'Can visitor satisfaction and knowledge about tropical rainforests be enhanced through biodiversity interpretation, and does this promote a positive attitude towards ecosystem conservation?', *Journal of Ecotourism* 6(1): 75–85.

Hill, R. (2006) 'The effectiveness of agreements and protocols to bridge between Indigenous and non-Indigenous toolboxes for protected area management: a case study from the wet tropics of Queensland', *Society and Natural Resources* 19: 577–590.

Hill, T., Nel, E. and Trotter, D. (2006) 'Small-scale, nature-based tourism as pro-poor development intervention: two examples in Kwazulu-Natal, South Africa', *Singapore Journal of Tropical Geography* 27: 163–175.

Hills, T. and Lundgren, J. (1977) 'The impact of tourism in the Caribbean: a methodological study', *Annals of Tourism Research* 4(5): 248–267.

Hinch, T. (1998) 'Ecotourists and indigenous hosts: diverging views on their relationship with nature', *Current Issues in Tourism* 1(1): 120–124.

Hitchcock, R.K. (1993) 'Toward self-sufficiency', *Cultural Survival Quarterly* 17(2): 51–53.

Hitchner, S.L., Lapu Apu, F., Tarawe, L., Galih, S., Nabun Aran, S. and Yesaya, E. (2009) 'Community-based transboundary ecotourism in the Heart of Borneo: a case study of the Kelabit Highlands of Malaysia and the Kerayan Highlands of Indonesia', *Journal of Ecotourism* 8(2): 193–213.

Hiwasaki, L. (2006) 'Community-based tourism: a pathway to sustainability for Japan's protected areas', *Society and Natural Resources* 19: 675–692.

Hjalager, A.-M. (1996) 'Tourism and the environment: the innovation connection', *Journal of Sustainable Tourism* 4(4): 201–218.

Hockings, M. (1994) 'A survey of the tour operator's role in marine park interpretation', *Journal of Tourism Studies* 5(1): 16–28.

Hoffman, W.M., Frederisk, R.E. and Schwartz, M.S. (2001) *Business Ethics: Readings and Cases in Corporate Morality*, Boston, MA: McGraw-Hill.

Holden, A. (2003) 'In need of a new environmental ethics for tourism?', *Annals of Tourism Research* 30(1): 95–108.

Holland, S.M., Ditton, R.B. and Graefe, A.R. (1998) 'An ecotourism perspective on billfish industries', *Journal of Sustainable Tourism* 6(2): 97–116.

Holloway, J.C. (1994) *The Business of Tourism*, 4th edn, London: Pitman.

Honderich, T. (1995) *The Oxford Companion to Philosophy*, New York: Oxford University Press.

Honey, M. (2003) 'Protecting Eden: setting green standards for the tourism industry', *Equipment* 45(6): 8–14.

Honey, M. (2008) *Ecotourism and Sustainable Development: Who Owns Paradise?* 2nd edn, Washington, DC: Island Press.

Honey, M. and Rome, A. (2001) *Protecting Paradise: Certification Programs for Sustainable Tourism and Ecotourism*, Washington, DC: Institute for Policy Studies.

Hope, K.R. (1980) 'The Caribbean tourism sector: recent performance and trends', *Tourism Management* 1(3): 175–183.

Hough, J.L. (1988) 'Obstacles to effective management of conflicts between national parks and surrounding human communities in developing countries', *Environmental Conservation* 15(2): 129–136.

Hovinen, G.R. (1981) 'The tourist cycle in Lancaster County, Pennsylvania', *Canadian Geographer* 25(3): 286–289.

Howell, B.J. (1994) 'Weighing the risks and rewards of involvement in cultural conservation and heritage tourism', *Human Organization* 53: 150–159.

Howes, L., Scarpaci, C. and Parsons, E.C.M. (2012) 'Ineffectiveness of a marine sanctuary zones to protect burrunan dolphins (*Tursiops australis* sp.nov.) from commercial tourism in Port Phillip Bay, Australia', *Journal of Ecotourism* 11(3): 188–201.

Hoyman, M.M. and McCall, J.R. (2013) 'Is there trouble in paradise? The perspectives of Galapagos community leaders on managing economic development and environmental conservation through ecotourism policies and the Special Law of 1998', *Journal of Ecotourism* 12(1): 33–48.

Hoyt, E. (2000) *Whale Watching 2000: Worldwide Tourism Numbers, Expenditures and Expanding Socioeconomic Benefits*, Yarmouth Port: International Fund for Animal Welfare.

—— (2005) 'Sustainable ecotourism on the Atlantic Islands, with special reference to whale watching, marine protected areas and sanctuaries for cetaceans', *Biology and Environment: Proceedings of the Irish Academy* 105B(3): 141–154.

Hudson, S. and Miller, G. (2005) 'The responsible marketing of tourism: the case of Canadian mountain holidays', *Tourism Management* 26: 133–142.

Hughes, G. (1995) 'The cultural construction of sustainable tourism', *Tourism Management* 16(1): 49–59.

Hughes, K., J. Packer and R. Ballantyne (2011) 'Using post-visit action resources to support family conservation learning following a wildlife tourism experience', *Environmental Education Research* 17(3): 307–328.

Hughes, M. and Morrison-Sanders, A. (2002) 'Impact of trail-side interpretive signs on visitor knowledge', *Journal of Ecotourism* 1(2/3): 122–132.

—— (2005) 'Influence of on-site interpretation intensity on visitors to natural areas', *Journal of Ecotourism* 4(3): 161–177.

Hughes, M., Newsome, D. and Macbeth, J. (2005) 'Visitor perceptions of captive wildlife tourism in a Western Australian natural setting', *Journal of Ecotourism* 4(2): 73–91.

Hultman, J. and Andersson Cederholm, E. (2006) 'The role of nature in Swedish ecotourism', in S. Gössling and J. Hultman (eds), *Ecotourism in Scandinavia: Lessons in Theory and Practice*, Wallingford, UK: CAB International, pp. 76–88.

Hultsman, J. (1995) 'Just tourism: an ethical framework', *Annals of Tourism Research* 22(3): 553–567.

Hunter, C. and Shaw, J. (2006) 'Applying the ecological footprint to ecotourism scenarios', *Environment Conservation* 32(4): 294–304.

Hunter, C.J. (1995) 'On the need to re-conceptualise sustainable tourism development', *Journal of Sustainable Tourism* 3(3): 155–165.

Husbands, W. (1981) 'Centres, peripheries, tourism and socio-spatial development', *Ontario Geography* 17: 37–59.

Hvenegaard, G.T. (1994) 'Ecotourism: a status report and conceptual framework', *Journal of Tourism Studies* 5(2): 24–35.

—— (2002) 'Using tourist typologies for ecotourism research', *Journal of Ecotourism* 1(1): 7–18.

Inbakaran, R.J., Jackson, M.S. and Chhetri, P. (2004) 'Spatial representation of resident attitudes in tourism product regions of Victoria, Australia', in K.A. Smith and C. Schott (eds), *Proceedings of the New Zealand Tourism and Hospitality Research Conference*, Wellington, 8–10 December, pp. 146–158.

Ingram, C.D. and Durst, P.B. (1987) 'Marketing nature-oriented tourism for rural development and wildlands management in developing countries: a bibliography', General Technical Report SE-44, Asheville, NC: US Dept of Agriculture, Forest Service, Southeastern Forest Experiment Station.

—— (1989) 'Nature-oriented tour operators: travel to developing countries', *Journal of Travel Research* 28(2): 11–15.

Inskeep, E. (1987) 'Environmental planning for tourism', *Annals of Tourism Research* 14(1): 118–135.

Isaacs, J.C. (2000) 'The limited potential of ecotourism to contribute to wildlife conservation', *Wildlife Society Bulletin* 28(1): 61–69.

Infield, M. (2002) 'The culture of conservation: exclusive landscapes, beautiful cows and conflict over Lake Mburo National Park, Uganda', unpublished PhD thesis, School of Development Studies, University of East Anglia, Norwich.

Iso-Ahola, S. (1982) 'Toward a social psychological theory of tourism motivation: a rejoinder', *Annals of Tourism Research* 9(2): 256–262.

Issaverdis, J.-P. (2001) 'The pursuit of excellence: benchmarking, accreditation, best practice and auditing', in D.B. Weaver (ed.), *The Encyclopedia of Ecotourism*, Wallingford, Oxon.: CAB International.

Iturregui, P. and Dutschke, M. (2005) 'Liberalisation of environmental goods and services and climate change', discussion paper no. 335 of the Hamburg Institute of International Economics, August.

Iverson, T. (1997) 'Ecolabelling and tourism', TRINET communication, 22 June.

Jackson, S. (2007) 'Attitudes towards the environment and ecotourism of stakeholders in the UK tourism industry with particular reference to ornithological tour operators', *Journal of Ecotourism* 6(1): 34–62.

Jamal, T. (2004) 'Virtue ethics and sustainable tourism pedagogy: phronesis, principles and practice', *Journal of Sustainable Tourism* 12(6): 530–545.

Jamal, T. and Getz, D. (1995) 'Collaboration theory and community tourism planning', *Annals of Tourism Research* 22(1): 186–204.

Jamal, T., Borges, M. and Stronza, A. (2006) 'The institutionalization of ecotourism: certification, cultural equity and praxis', *Journal of Ecotourism* 5(3): 145–175.

Jansen-Verbeke, M. and Dietvorst, A. (1987) 'Leisure, recreation, tourism: a geographic view on integration', *Annals of Tourism Research* 14(3): 361–375.

Jenkins, C.L. (1991) 'Tourism development strategies', in L.J. Lickorish (ed.), *Developing Tourism Destinations*, Harlow: Longman, pp. 61–78.

—— (1994) 'Tourism in developing countries: the privatisation issue', in A.V. Seaton (ed.), *Tourism: The State of the Art*, Chichester: John Wiley.

Jenkins, J. and Wearing, S. (2003) 'Ecotourism and protected areas in Australia', in D.A. Fennell and R.K. Dowling (eds), *Ecotourism Policy and Planning*, Wallingford, Oxon.: CAB International.

Johnson, D. (2006) 'Providing ecotourism excursions for cruise passengers', *Journal of Sustainable Tourism* 14(1): 43–54.

Johnson, D.R. and Agee, J.K. (1988) 'Introduction to ecosystem management', in J.K. Agee and D.R. Johnson (eds), *Ecosystem Management for Parks and Wilderness*, Seattle: University of Washington Press.

Johst, D. (1982) 'Does wilderness designation increase recreation use?', unpublished report of the Bureau of Land Management, Washington, DC.

Jones, H. (1972) 'Gozo: the living showpiece', *Geographical Magazine* 45(1): 53–57.

Jones, S. (2005) 'Community-based ecotourism: the significance of social capital', *Annals of Tourism Research* 32(2): 303–324.

Jones, T., Wood, D., Catlin, J. and Norman, B. (2009) 'Expenditure and ecotourism: predictors of expenditure for whale shark tour participants', *Journal of Ecotourism* 8(1): 32–50.

Joppe, M. (1996) 'Sustainable community tourism development revisited', *Tourism Management* 17(7): 475–479.

Jordan, M. (2001) 'Brazil lays claims to the rain forest', *Wall Street Journal*, Eastern edition, 31 August: A6, A11.

Juric, B., Cornwell, T.B. and Mather, D. (2002) 'Exploring the usefulness of an ecotourism interest scale', *Journal of Travel Research* 40: 259–269.

Jurowski, C. (1996) 'Tourism means more than money to the host community', *Parks and Recreation* 31(9): 110–118.

Jurowski, C., Muzaffer, U., Williams, D.R. and Noe, F.P. (1995) 'An examination of preferences and evaluations of visitors based on environmental attitudes: Biscayne Bay National Park', *Journal of Sustainable Tourism* 3(2): 73–85.

Kaae, B.C. (2006) 'Perceptions of tourism by national park residents in Thailand', *Tourism and Hospitality Planning and Development* 3(1): 19–33.

Kahle, L.R., Beatty, S.E. and Homer, P. (1986) 'Alternative measurement approaches to consumer values: the list of values (LOV) and values and life style (VALS)', *Journal of Consumer Research* 13(3): 404–409.

Kaosa-ard, M. (2002) 'Development and management of tourism products: the Thai experience', *Chiang Mai University Journal* 1(3): 289–301.

Kaosa-ard, M. (with 16 others) (1993) *A Review Report on Tourism for the Master Plan Research Report*, prepared for the Tourism Authority of Thailand (TAT), Thailand Development Research Institute, Bangkok, Thailand.

Karlsson, S.E. (2005) 'The social and cultural capital of a place and their influence on the production of tourism: a theoretical reflection based on an illustrative case study', *Scandinavian Journal of Hospitality and Tourism* 5(2): 102–115.

Karwacki, J. and Boyd, C. (1995) 'Ethics and ecotourism', *Business Ethics* 4(4): 225–232.

Kavallinis, I. and Pizam, A. (1994) 'The environmental impacts of tourism: whose responsibility is it anyway? The case study of Mykonos', *Journal of Travel Research* 33(2): 26–32.

Kearsley, G., Hall, C.M. and Jenkins, J. (1997) 'Tourism planning in natural areas: introductory comments', in C.M. Hall, J. Jenkins and G. Kearsley (eds), *Tourism Planning and Policy in Australia and New Zealand: Cases, Issues and Practice*, Sydney: Irwin, pp. 66–74.

Kelkit, A., Ozel, A.E. and Demirel, O. (2005) 'A study of the Kazdagi (Mt. Ida) National Park: an ecological approach to the management of tourism', *International Journal of Sustainable Development and World Ecology* 12: 141–148.

Kellert, S.R. (1985) 'Birdwatching in American society', *Leisure Sciences* 7(3): 343–360.

—— (1987) 'The contributions of wildlife to human quality of life', in D.J. Decker and G.R. Goff (eds), *Valuing Wildlife: Economic and Social Perspectives*, Boulder, CO: Westview.

Kelly, J.R. and Godbey, G. (1992) *The Sociology of Leisure*, State College, PA: Venture Publishing.

Kenchington, R.A. (1989) 'Tourism in the Galapágos Islands: the dilemma of conservation', *Environmental Conservation* 16(3): 227–236.

Kerstetter, D., Hou, J.-S. and Lin, C.-H. (2004) 'Profiling Taiwanese ecotourists using a behavioural approach', *Tourism Management* 25: 491–498.

Khan, M.M. (1997) 'Tourism development and dependency theory: mass tourism v. ecotourism', *Annals of Tourism Research* 24(4): 988–991.

Kibicho, W. (2006) 'Tourists to Amboseli National Park: a factor-cluster segmentation analysis', *Journal of Vacation Marketing* 12(3): 218–231.

King, D.A. and Stewart, W.P. (1996) 'Ecotourism and commodification: protecting people and places', *Biodiversity and Conservation* 5(3): 293–305.

Kiss, A. (2004) 'Is community-based ecotourism a good use of biodiversity conservation funds?', *TRENDS in Ecology and Evolution* 19(5): 232–237.

Knowles, M. (1980) *The Modern Practice of Adult Education*, New York: Cambridge University Press.

Kohl, J. (2005) 'Putting environmental interpretation to work for conservation in a park setting: conceptualizing principal conservation strategies', *Applied Environmental Education and Communication* 4: 31–42.

Kohlberg, L. (1981) *Essays on Moral Development*, Volume I: *The Philosophy of Moral Development*, New York: Harper & Row.

—— (1984) *Essays on Moral Development*, Volume II: *The Psychology of Moral Development*, New York: Harper & Row.

Kontogeorgopoulos, N. (2004a) 'Conventional tourism and ecotourism in Phuket, Thailand: conflicting paradigms or symbiotic partners?', *Journal of Ecotourism* 3(2): 87–108.

—— (2004b) 'Ecotourism and mass tourism in southern Thailand: spatial interdependence, structural connections, and staged authenticity', *GeoJournal* 61: 1–11.

—— (2005) 'Community-based ecotourism in Phuket and Ao Phangnga, Thailand: partial victories and bittersweet remedies', *Journal of Sustainable Tourism* 13(1): 4–23.

Kraus, R. (1997) *Recreation Programming: A Benefits-driven Approach*, Toronto: Allyn & Bacon.

Kraus, R. and Allen, L. (1997) *Research and Evaluation in Recreation, Parks and Leisure Studies*, 2nd edn, Scottsdale, AZ: Gorsuch Scarisbrick.

Kraus, R. and Curtis, J. (1990) *Creative Management in Recreation, Parks, and Leisure Services*, 5th edn, St Louis, MO: Times Mirror/Mosby.

Krech, S. (1999) *The Ecological Indian: Myth and History*, New York: W.W. Norton & Co.

Kretchman, J.A. and Eagles, P.F.J. (1990) 'An analysis of the motives of ecotourists in comparison to the general Canadian population', *Society and Leisure* 13(2): 499–507.

Krippendorf, J. (1977) *Les dévoreurs des paysages*, Lausanne: 24 Heures.

—— (1982) 'Towards new tourism policies', *Tourism Management* 3: 135–148.

—— (1987) 'Ecological approach to tourism marketing', *Tourism Management* 8(2): 174–176.

Krishna, A. (2001) 'Moving from the stock of social capital to the flow of benefits: the role of agency', *World Development* 29: 925–943.

Kroshus Medina, L. (2005) 'Ecotourism and certification: confronting the principles and pragmatics of socially responsible tourism', *Journal of Sustainable Tourism* 13(3): 281–295.

Krug, W. (2002) *Private Supply of Protected Land in Southern Africa: A Review of Markets, Approaches, Barriers and Issues*, Report for the Working Party on Global and Structural Policies, Paris, Organisation for Economic Co-operation and Development, JT00124118.

Kuo, I.-L. (2002) 'The effectiveness of environmental interpretation at resource-sensitive destinations', *International Journal of Tourism Research* 4: 87–101.

Kur, N.T. and Hvenegaard, G.T. (2012) 'Promotion of ecotourism principles by whale-watching companies' marketing efforts', *Tourism in Marine Environments* 8(3): 145–151.

Kusler, J.A. (1991) 'Ecotourism and resource conservation: introduction to issues', in J.A. Kusler (ed.), *Ecotourism and Resource Conservation: A Collection of Papers*, Volume 1, Madison, WI: Omnipress.

Kutay, K. (1989) 'The new ethic in adventure travel', *Buzzworm: The Environmental Journal* 1(4): 31–34.

Kwan, P., Eagles, P.F.J. and Gebhardt, A. (2010) 'Ecolodge patrons' characteristics and motivations: a study of Belize', *Journal of Ecotourism* 9(1): 1–20.

LaFranchi, H. (1998) 'Bye-bye to Brazil's bio-paradise?', *Christian Science Monitor* 91(8): 12–14.

Laarman, J.G. and Durst, P.B. (1987) 'Nature travel and tropical forests', FPEI Working Paper Series, Southeastern Center for Forest Economics Research, North Carolina State University, Raleigh.
—— (1993) 'Nature tourism as a tool for economic development and conservation of natural resources', in J. Nenon and P.B. Durst (eds), *Nature Tourism in Asia: Opportunities and Constraints for Conservation and Economic Development*, Washington, DC: US Forest Service.
Laarman, J.G. and Gregersen, H. (1994) 'Making nature-based tourism contribute to sustainable development', *EPAT/MUCIA Policy Brief* 5: 1–6.
—— (1996) 'Pricing policy in nature-based tourism', *Tourism Management* 17(4): 247–254.
Ladkin, A. and Martinez Bertramini, A. (2002) 'Collaborative tourism planning: a case study of Cusco, Peru', *Current Issues in Tourism* 5(2): 71–93.
Lai, P.-H. and Shafer, S. (2005) 'Marketing ecotourism through the Internet: an evaluation of selected ecolodges in Latin America and the Caribbean', *Journal of Ecotourism* 4(3): 143–160.
Lane, M.B. (2001) 'Affirming new directions in planning theory: comanagement of protected areas', *Society and Natural Resources* 14: 657–671.
Langholz, J. and Brandon, K. (2001) 'Privately owned protected areas', in D.B. Weaver (ed.), *The Encyclopedia of Ecotourism*, Wallingford, Oxon.: CAB International.
Langholz, J.A. and Lassoie, J.P. (2001) 'Perils and promise of privately owned protected areas', *BioScience* 51(12): 1079–1085.
—— (2002) 'Combining conservation and development on private lands: lessons from Costa Rica', *Environment, Development and Sustainability* 3: 309–322.
Langholz, J.A., Lassoie, J.P. and Schelhas, J. (2000) 'Incentives for biological conservation: Costa Rica's private wildlife refuge program', *Conservation Biology* 14(6): 1735–1743.
Lau, C.K.H. and Johnston, C.S. (2002) *Eco-gateway! Auckland's Emerging Role as an Ecotourism Destination*. Available online at www.gdrc.org/uem/eco-tour/etour-define.html (accessed 20 September 2006).
Law, R. and Wong, J. (2003) 'Successful factors for a travel web site: perceptions of on-line purchasers in Hong Kong', *Journal of Hospitality and Tourism Research* 27(1): 118–124.
Lawton, L. and Weaver, D.B. (2001) 'Ecotourism in modified spaces', in D.B. Weaver (ed.), *Encyclopedia of Ecotourism*, Wallingford, Oxon.: CAB International.
Lea, J.P. (1993) 'Tourism development ethics in the Third World', *Annals of Tourism Research* 20: 701–715.
Leader-Williams, N. (2002) 'Animal conservation, carbon and sustainability', *Philosophical Transactions of the Royal Society of London* A 360: 1787–1806.
Lebel, L., Anderies, J., Campbell, B., Folke, C., Hatfield-Dodds, S., Hughes, T.P. and Wilson, J. (2006) 'Governance and the capacity to manage resilience in regional socio-ecological systems' 11(1): 19. Available online at http://www.ecologyandsociety.org/vol11/iss1/art9/.
Lee, K.N. (1993) *Compass and Gyroscope: Integrating Science and Politics for the Environment*, Washington, DC: Island Press.
Lee, S. and Jamal, T. (2008) 'Environmental justice and environmental equity in tourism: missing links to sustainability', *Journal of Ecotourism*, 7(1): 44–67.
Leiper, N. (1981) 'Towards a cohesive curriculum in tourism: the case for a distinct discipline', *Annals of Tourism Research* 8(1): 69–84.
—— (1990) 'Tourist attraction systems', *Annals of Tourism Research* 17(3): 367–384.
Lemelin, H., Fennell, D.A. and Smale, B.J.A. (2008) 'Polar bear viewers as deep ecotourists: how specialized are they?', *Journal of Sustainable Tourism*, 16(1): 42–62.
Lemelin, H., McCarville, R. and Smale, B. (2002) *The Effects of Context on Price Expectations for Wildlife Viewing Opportunities in Churchill, Manitoba*, Report for the Manitoba Department of Conservation, and Parks Canada, Waterloo, Ontario.
Lemelin, R.H. (2007) 'Finding beauty in the dragon: the role of dragonflies in recreation and tourism', *Journal of Ecotourism* 6(2): 139–145.
Lemelin, R.H., Smale, B. and Fennell, D. (2008) 'Polar bear viewers as deep ecotourists: how specialized are they?', *Journal of Sustainable Tourism* 16 (1): 42–62.

León, C. and González, M. (1995) 'Managing the environment in tourism regions: the case of the Canary Islands', *European Environment* 5(6): 171–177.

Leslie, D. (1994) 'Sustainable tourism or developing sustainable approaches to lifestyle', *World Leisure and Recreation* 36(3): 30–36.

Lew, A. (1987) 'A framework of tourist attraction research', *Annals of Tourism Research* 14(4): 553–575.

Li, Y. (2004) 'Exploring community tourism in China: the case of Nanshan cultural tourism zone', *Journal of Sustainable Tourism* 12(3): 175–193.

Lickorish, L.J. (1991) 'Roles of government and the private sector', in L.J. Lickorish (ed.), *Developing Tourism Destinations*, Harlow: Longman.

Liddle, M. (1997) *Recreation Ecology: The Ecological Impact of Outdoor Recreation and Ecotourism*, London: Chapman and Hall.

Lindberg, K. (1991) *Policies for Maximising Nature Tourism's Ecological and Economic Benefits*, Washington, DC: World Resources Institute.

Lindsey, P.A., Alexander, R., Mills, M.G.L., Romanach, S. and Woodroffe, R. (2007) 'Wildlife viewing preferences of visitors to protected areas in South Africa: implications for the role of ecotourism in conservation', *Journal of Ecotourism* 6(1): 19–33.

Lipske, K. (1992) 'How a monkey saved the jungle', *International Wildlife* 22(1): 38–43.

Liu, F. (2001) *Environmental Justice Analysis: Theories, Methods, and Practice*, New York: Lewis Publishers.

Liu, J.C. (1994) *Pacific Island Ecotourism: A Public Policy and Planning Guide*, Honolulu: Office of Territorial and International Affairs.

Liu, Z. (2003) 'Sustainable tourism development: a critique', *Journal of Sustainable Tourism* 11(6): 459–475.

Loomis, L. and Graefe, A.R. (1992) 'Overview of NPCA's Visitor Impact Management Process', paper presented at the Fourth World Congress on Parks and Protected Areas, Caracas, 11–21 February.

López-Espinosa, R. (2002) 'Evaluating ecotourism in natural protected areas of La Paz Bay, Baja California Sur, Mexico: ecotourism or nature-based tourism?', *Biodiversity and Conservation* 11: 1539–1550.

Lothian, W.F. (1987) *A Brief History of Canada's National Parks*, Ottawa: Supply and Services Canada.

Lovejoy, T. (1992) 'Looking to the next millennium', *National Parks* January–February: 41–44.

Lovelock, B. and Robinson, K. (2005) 'Maximizing economic returns from consumptive wildlife tourism in peripheral areas: white-tailed deer hunting on Stewart Island/Rakiura, New Zealand', in C.M. Hall and S. Boyd (eds), *Nature-based Tourism in Peripheral Areas: Development or Disaster?*, Toronto: Channel View Publications.

Lovelock, C.H. and Weinberg, C.B. (1984) *Marketing for Public and Nonprofit Managers*, Toronto: Wiley.

Low, B.S. (1996) 'Behavioral ecology of conservation in traditional societies', *Human Nature* 7(4): 353–379.

Lucas, R.C. (1964) 'Wilderness perception and use: the example of the Boundary Waters Canoe Area', *Natural Resources Journal* 3(3): 394–411.

Lucas, R.C. and Stankey, G.H. (1974) 'Social carrying capacity for backcountry recreation', in *Outdoor Recreation Research: Applying the Results*, USDA Forest Service General Technical Report NC-9, North Central Forest Experiment Station, St Paul, Minnesota, pp. 14–23.

Lück, M. (2002a) 'Large-scale ecotourism: a contradiction in itself?', *Current Issues in Tourism* 5(3–4): 361–370.

—— (2002b) 'Looking into the future of ecotourism and sustainable tourism', *Current Issues in Tourism* 5(3–4): 371–374.

—— (2003) 'Education on marine mammal tours as agent for conservation: but do tourists want to be educated?', *Ocean and Coastal Management* 46: 943–956.

Luhrman, D. (1997) 'WTO Manila meeting', internet communication, 23 May, WTO Press and Communications.

Lumsdon, L.M. and Swift, J.S. (1998) 'Ecotourism at a crossroads: the case of Costa Rica', *Journal of Sustainable Tourism* 6(2): 155–172.

Lynes, J.K. and Dredge, D. (2006) 'Going green: motivations for environmental commitment in the airline industry: a case study of Scandinavian Airlines', *Journal of Sustainable Tourism* 14(2): 116–138.

Mabey, C. (1994) 'Youth leadership: commitment for what?', in S. York and D. Jordan (eds), *Bold Ideas: Creative Approaches to the Challenge of Youth Programming*, Institute for Youth Leaders: University of Northern Iowa.

Machura, L. (1954) 'Nature protection and tourism: with particular reference to Austria', *Oryx* 2(5): 307–311.

Mackie, J.L. (1977) *Ethics: Inventing Right and Wrong*, Harmondsworth: Penguin Books.

Madrigal, R. (1995) 'Personal values, traveler personality type, and leisure travel style', *Journal of Leisure Research* 27(2): 125–142.

Madrigal, R. and Kahle, L.R. (1994) 'Predicting vacation activity preferences on the basis of value-system segmentation', *Journal of Travel Research* 32(3): 22–28.

Maher, P.T., Steel, G. and McIntosh, A. (2003) 'Antarctica: tourism, wilderness, and "ambassador-ship" ', *USDA Forest Service Proceedings RMRS-P-27*, pp. 204–210.

Mahoney, E.M. (1988) 'Recreation and tourism marketing', unpublished paper, Michigan State University, Ann Arbor, Michigan.

Maille, P. and Mendelsohn, R. (1993) 'Valuing ecotourism in Madagascar', *Journal of Environmental Management* 38: 213–218.

Malloy, D.C. and Fennell, D.A. (1998a) 'Ecotourism and ethics: moral development and organiza-tional cultures', *Journal of Travel Research* 36(4): 47–56.

—— (1998b) 'Codes of ethics and tourism: an explanatory content analysis', *Tourism Management* 19(5): 453–461.

Manfredo, M., Teel, T.L. and Bright, A.D. (2003) 'Why are public values toward wildlife changing?', *Human Dimensions of Wildlife* 8: 287–306.

Mangott, A.H., Birtles, R.A. and Marsh, H. (2011) 'Attraction of dwarf minke whales *Balaenoptera acutorostrata* to vessels and swimmers in the Great Barrier Reef World Heritage Area: the management challenges of an inquisitive whale', *Journal of Ecotourism* 10(1): 64–76.

Mannell, R. and Kleiber, D. (1997) *A Social Psychology of Leisure*, State College, NY: Venture.

Manning, T. (1996) 'Tourism: where are the limits?', *Ecodecision* 20(Spring): 35–39.

Maple, L.C., Eagles, P.F.J. and Rolfe, H. (2010) 'Birdwatchers' specialisation characteristics and national park tourism planning', *Journal of Ecotourism* 9(3): 219–238.

Marinelli, L. (1997) 'Ecotourists take to the Hawaiian hills', *Toronto Star*, 25 October.

Markwell, K. (1998) 'Taming the chaos of nature: cultural construction and lived experience in nature-based tourism', PhD dissertation, University of Newcastle, Newcastle, NSW, Australia.

Maroney, R.L. (2006) 'Community based wildlife management planning in protected areas: the case of Altai argali in Mongolia', *USDA Forest Proceedings RMRS-P-39*, Proceedings of the Conference on Transformation Issues and Future Challenges, 27 January Salt Lake City, UT.

Marsh, G.P. (1864) *Man and Nature; Or, Physical Geography as Modified by Human Action*, New York: Scribner.

Maslow, A. (1954) *Motivation and Personality*, New York: Harper & Row.

Mason, J.G. (1960) *How to Be a More Effective Executive*, New York: McGraw-Hill.

Mason, P. (1994) 'A visitor code for the Arctic', *Tourism Management* 15(2): 93–97.

—— (1997a) 'Ecolabelling and tourism', TRINET communication, 22 June.

—— (1997b) 'Tourism codes of conduct in the Arctic and Sub-Arctic Region', *Journal of Sustainable Tourism* 5(2): 151–165.

Mason, P. and Christie, M. (2003) 'Tour guides as critically reflective practitioners: a proposed training model', *Tourism Recreation Research* 28(1): 23–33.

Mason, P. and Mowforth, M. (1995) 'Codes of conduct in tourism', *Occasional Papers in Geography*, no. 1, Department of Geographical Sciences, University of Plymouth.

—— (1996) 'Codes of conduct in tourism', *Progress in Tourism and Hospitality Research* 2(2): 151–167.

Mathieson, A. and Wall, G. (1982) *Tourism: Economic, Physical, and Social Impacts*, London: Longman.

Mau, R. (2008) 'Managing for conservation and recreation: the Ningaloo whale shark experience', *Journal of Ecotourism* 7(2/3): 213–225.

May, R.M. (1992) 'How many species inhabit the earth?', *Scientific American* 267(4): 42–48.

Mayo, E.F. (1975) 'Tourism and the national parks: a psychographic and attitudinal study', *Journal of Leisure Research* 14(1): 14–18.

Mayr, E. (1988) *Toward a New Philosophy of Biology: Observations of an Evolutionist*, Cambridge: Belknap.

Mbaiwa, J.E. (2003) 'The socio-economic and environmental impacts of tourism development on the Okavango Delta, north-western Botswana', *Journal of Arid Environments* 54: 447–467.

—— (2005) 'The problems and prospects of sustainable tourism development in the Okavango Delta, Botswana', *Journal of Sustainable Tourism* 13(3): 203–227.

MacArthur, R.H. and Wilson, E.O. (1963) 'An equilibrium theory of insular zoogeography', *Evolution* 17: 373–387.

McArthur, S. (1997) 'Introducing the National Ecotourism Accreditation Program', *Australian Parks and Recreation* 33(2): 30–34.

Macbeth, J. (1994) 'To sustain is to nurture, to nourish, to tolerate and to carry on: can tourism? Trends in sustainable rural tourism development', *Parks and Recreation Magazine* 31(1): 42–45.

MacCannell, D. (1989) *The Tourist: A New Theory of the Leisure Class*, New York: Schocken Books.

McCarville, R. (1993) 'Keys to quality leisure programming', *Journal of Physical Education, Recreation and Dance* October: 32–39.

McCool, S.F. (1985) 'Does wilderness designation lead to increased recreational use?', *Journal of Forestry* January: 39–41.

—— (1995) 'Linking tourism, the environment, and concepts of sustainability: setting the stage', in S.F. McCool and A.E. Watson (eds), *Linking Tourism, the Environment, and Sustainability*, USDA Technical Report INT-GTR-323, Ogden, UT: US Department of Agriculture, Forest Service, Intermountain Research Station.

McFarlane, B.L. and Boxall, P.C. (1996) 'Participation in wildlife conservation by birdwatchers', *Human Dimensions of Wildlife* 1(3): 1–14.

McIntosh, C. (1992) 'Eco-tourism shows promise for the north', *Northern Ontario Business* 12(1): 9.

MacKay, K. and McIlraith, A. (1997) *Churchill Visitor Study: Seasonal Overview*, Winnipeg: Health, Leisure and Human Performance Research Institute, University of Manitoba.

McKenzie, J.F. and Smeltzer, J.L. (1997) *Planning, Implementing, and Evaluating Health Promotion Pograms: A Primer*, 2nd edn, Toronto: Allyn & Bacon.

McKercher, B. (1993a) 'The unrecognized threat to tourism: can tourism survive "sustainability"?', *Tourism Management* 14(2): 131–136.

—— (1993b) 'Some fundamental truths about tourism: understanding tourism's social and environ-mental impacts', *Journal of Sustainable Tourism* 1(1): 6–16.

—— (2010) 'Academia and the evolution of ecotourism', *Tourism Recreation Research* 35(1): 15–26.

McKercher, B. and du Cros, H. (2003) 'Testing a cultural tourism typology', *International Journal of Tourism Research* 5: 45–58.

MacKinnon, B. (1995) 'Beauty and the beasts of ecotourism', *Business Mexico* 5(4): 44–47.

McNeely, J.A. (1988) *Economics and Biological Diversity*, Gland, Switzerland: International Union for the Conservation of Nature and Natural Resources.

—— (1993) 'People and protected areas: partners in prosperity', in E. Kemf (ed.), *The Law of the Mother: Protecting Indigenous People in Protected Areas*, San Francisco, CA: Sierra Club.

Medeiros de Araujo, L. and Bramwell, B. (1999) 'Stakeholder assessment and collaborative tourism planning: the case of Brazil's Costa Dourada project', *Journal of Sustainable Tourism* 7(3–4): 356–378.

Mehmetoglu, M. (2007) 'Nature-based tourism: a contrast to everyday life', *Journal of Ecotourism* 6(2): 111–126.

Meletis, Z.A. and Campbell, L.M. (2007) 'Call it consumption! Re-conceptualising ecotourism as consumption and consumptive', *Geography Compass* 1(4): 850–870.

Menkhaus, S. and Lober, D.J. (1996) 'International ecotourism and the valuation of tropical rainforests in Costa Rica', *Journal of Environmental Management* 47: 1–10.

Merry, W. (1994) *St. John Ambulance Official Wilderness First-Aid Guide*, Toronto: McClelland & Stewart Inc.

Metelka, C.J. (1981) *The Dictionary of Tourism*, Urbana-Champaign, IL: Merton House Co.

—— (1990) *The Dictionary of Hospitality, Travel and Tourism*, Albany, NY: Delmar.

Meyer-Arendt, K.J. (1985) 'The Grand Isle, Louisiana resort cycle', *Annals of Tourism Research* 12: 449–465.

Midgley, M. (1994) *The Ethical Primate: Humans, Freedom and Morality*, London: Routledge.

Mieczkowski, Z.T. (1981) 'Some notes on the geography of tourism: a comment', *Canadian Geographer* 25: 186–191.

Milgrath, L. (1989) 'An inquiry into values for a sustainable society: a personal statement', in L. Milgrath (ed.), *Envisioning a Sustainable Society*, Albany, NY: SUNY Press.

Mill, R.C. and Morrison, A.M. (1985) *The Tourism System*, Englewood Cliffs, NJ: Prentice-Hall.

Miller, B.R. (1996) 'The global structure of the nature tourism industry: ecotourists, tour operators, and local businesses', *Journal of Travel Research* 35(2): 11–18.

Miller, G. (2001) 'Corporate responsibility in the UK tourism industry', *Tourism Management* 22: 589–598.

Miller, K.R. (1976) 'Global dimensions of wildlife management in relation to development and environmental conservation in Latin America', Proceedings of Regional Expert Consultation on Environment and Development, Bogota, 5–10 July, Santiago, Chile: Food and Agriculture Organization.

—— (1989) *Planning National Parks for Ecodevelopment: Methods and Cases from Latin America*, Washington, DC: Peace Corps.

Miller, M.L. and Kaye, B.C. (1993) 'Coastal and marine ecotourism: a formula for sustainable development?', *Trends* 30(2): 35–41.

Mimi (2011) 'Consumptive use', *Environment*. Available online at <http://en.mimi.hu/environment/index_environment.html> (accessed 10 January 2011).

Mintzberg, H. (1996) 'The myth of "Society Inc."', *The Globe and Mail Report on Business Magazine* October: 113 117.

Mirabella, M. (1998) 'Brazil hopes new push for ecotourism can help save Amazon rain forest'. Available online at www.cnn.com/travel/news/9801/28/amazon.ecotourism (accessed 15 May 2002).

Mitchell, B. (1989) *Geography and Resource Analysis*, New York: Longman.

—— (1994) 'Sustainable development at the village level in Bali, Indonesia', *Human Ecology* 22(2): 189–211.

Mitchell, G.E. (1992) *Ecotourism Guiding: How to Start your Career as an Ecotourism Guide*, n.p.: G.E. Mitchell.

Mitchell, L.S. (1984) 'Tourism research in the United States: a geographical perspective', *Geojournal* 9(1): 5–15.

Mlinarić, I.B. (1985) 'Tourism and the environment: a case for Mediterranean cooperation', *International Journal of Environmental Studies* 25: 239–245.

Mohd Shahwahid, H.O., Mohd Iqbal, M.N., Amira Mas, A.Y.U. and Farah, M.S. (2013) 'Assessing service quality of community-based ecotourism: a case study from Kampung Kuantan Firefly Park', *Journal of Tropical Forest Science* 25(1): 22–33.

Moore, J. (1984) 'The evolution of reciprocal sharing', *Ethological Sociobiology* 5: 4–14.

Moore, S. and Carter, B. (1993) 'Ecotourism in the 21st century', *Tourism Management* 14(2): 123–130.

Moran, D. (1994) 'Contingent valuation and biodiversity: measuring the user surplus of Kenyan protected areas', *Biodiversity and Conservation* 3: 663–684.

Moreno, P.S. (2005) 'Ecotourism along the Meso-American Caribbean reef: the impacts of foreign investment', *Human Ecology* 33(2): 217–244.

Morrison, A.M., Hsieh, S. and Wang, C.-Y. (1992) 'Certification in the travel and tourism industry', *Journal of Tourism Studies* 3(2): 32–40.

Moscardo, G., Morrison, A.M. and Pearce, P.L. (1996) 'Specialist accommodation and ecologically sustainable tourism', *Journal of Sustainable Tourism* 4(1): 29–54.

Moskwa, E. (2010) 'Ecotourism in the rangelands: landholder perspectives on conservation', *Journal of Ecotourism* 9(3): 175–186.

Mountjoy, A.B. (1971) *Developing the Underdeveloped Countries*, New York: John Wiley.

Mowforth, M. (1993) *Eco-tourism: Terminology and Definitions*, Research Report Series, no. 1, Department of Geographical Sciences, University of Plymouth.

Müller, D.K. and Huuva, S.K. (2009) 'Limits to Sami tourism development: the case of Jokkmokk, Sweden', *Journal of Ecotourism* 8(2): 115–127.

Muloin, S. (1998) 'Wildlife tourism: the psychological benefits of whale watching', *Pacific Tourism Review* 2: 199–213.

Munasinghe, M. (1994) 'Economic and policy issues in natural habitats and protected areas', in L. Munasinghe and J. McNeely (eds), *Protected Area Economics and Policy: Linking Conservation and Sustainable Development*, Washington, DC: World Bank.

Munro, J.K., Morrison-Saunders, A. and Hughes, M. (2008) 'Environmental interpretation evaluation in natural areas', *Journal of Ecotourism* 7(1): 1–12.

Munt, I. and Higinio, E. (1993) 'Belize: eco-tourism gone awry', *In Focus* 9: 18–19.

Murphy, P.E. (1983) 'Tourism as a community industry: an ecological model of tourism development', *Tourism Management* 4(3): 180–193.

—— (1985) *Tourism: A Community Approach*, London: Routledge.

Musa, G., Hall, C.M. and Higham, J.E.S. (2004) 'Tourism sustainability and health impacts in high altitude adventure, cultural and ecotourism destinations: a case study of Nepal's Sagarmatha National Park', *Journal of Sustainable Tourism* 12(4): 306–331.

Myers, N. (1980) *The Sinking Ark*, Oxford: Pergamon Press.

Myles, P.B. (2003) 'Contribution of wilderness to survival of the adventure and ecotourism markets', *USDA Forest Service Proceedings RMRS-P-27*, pp. 185–187.

Nagendra, H., Karmacharya, M. and Marna, B. (2005) 'Evaluating forest management in Nepal: views from across space and time', *Ecology and Society* 10(1). Available online at http://www.ecologyandsociety.org/vol110/iss1/art24 (accessed 8 November 2006).

Naidoo, R. and Adamowicz, W.L. (2005) 'Biodiversity and nature-based tourism at forest reserves in Uganda', *Environment and Development Economics* 10: 159–178.

Nash, R. (1982) *Wilderness and the American Mind*, New Haven: Yale University Press.

Naylon, J. (1967) 'Tourism: Spain's most important industry', *Geography* 52: 23–40.

Nelson, J. and Gami, N. (2003) 'Enhancing equity in the relationship between protected areas and indigenous and local communities in Central Africa, in the context of global change', *CEESP-WCPA-IUCN Theme on Indigenous and Local Communities, Equity and Protected Areas*. Available online at http://www.iucn.org/themes/ceesp/Publications/TILCEPA/CCA-Nelson-Gami.pdf (accessed 8 November 2006).

Nelson, J.G. (1991) 'Sustainable development, conservation strategies, and heritage', in B. Mitchell (ed.), *Resource Management and Development*, Oxford: Oxford University Press.

—— (1993) 'Planning and managing national parks and protected areas: a human ecological approach', paper presented at the Ecosystem Management for Managers Workshop, University of Waterloo, Waterloo, Ontario, 18–21 January.

—— (1994) 'The spread of ecotourism: some planning implications', *Environmental Conservation* 21(1): 248–255.

Nelson, J.G., Butler, R. and Wall, G. (1993) *Tourism and Sustainable Development: Monitoring, Planning, Managing*, Waterloo, Ontario: Department of Geography Publication Series, no. 37, University of Waterloo.

Nelson, V. (2010) 'Promoting energy strategies on Eco-Certified accommodation websites', *Journal of Ecotourism* 9(3): 187–200.

Nepal, S. (2004) 'Indigenous ecotourism in central British Columbia: the potential for building capacity in the Tl'azt'en nations territories', *Journal of Ecotourism* 3(3): 173–194.

Neto, F. (2003) 'A new approach to sustainable tourism development: moving beyond environmental protection', *Natural Resources Forum* 27: 212–222.

Newsome, D., Moore, S.A. and Dowling, R.K. (2002) *Natural Area Tourism: Ecology, Impacts, and Management*, Clevedon: Channel View Publications.

Niezgoda, A. (2004) *Problems of Implementing Sustainable Tourism in Poland*. Available online at http://www.puereview.ac.poznan.pl/2004v4n1/3-niezgoda.pdf (accessed 28 November).

Nolan, R. and Rotherham, I. (2012) 'Volunteer perceptions of an ecotourism experience: a case study of ecotourism to the coral reefs of Southern Negros in the Philippines', *Journal of Ecotourism* 11(3): 153–172.

Norris, R. (1992) 'Can ecotourism save natural areas?', *Parks*, January–February: 31–34.

Notzke, C. (1994) 'Aboriginal people and protected areas', paper presented at Saskatchewan's Protected Areas Conference, University of Regina, Regina, Saskatchewan, June.

Novelli, M., Barnes, J.I. and Humavinan, M. (2006) 'The other side of the ecotourism coin: consumptive tourism in Southern Africa', *Journal of Ecotourism* 5(1/2): 62–79.

Nowaczek, A. and Smale, B. (2005) 'Traveller evaluations of ecotour operator ethics at a community-owned and operated site in Tambopata Reserve, Peru', paper presented at the 11th Canadian Congress on Leisure Research, 17–20 May, Nanaimo, BC.

—— (2010) 'Exploring the predisposition of travellers to qualify as ecotourists: the Ecotourist Predisposition Scale', *Journal of Ecotourism* 9(1): 45–61.

Nyaupane, G.P. and Thapa, B. (2004) 'Evaluation of ecotourism: a comparative assessment in the Annapurna Conservation Area Project, Nepal', *Journal of Ecotourism* 3(1): 20–45.

O'Connor, S., Campbell, R., Cortez H. and Knowles T. (2009) *Whale Watching Worldwide: Tourism Numbers, Expenditures and Expanding Economic Benefits*, Yarmouth, MA: International Fund for Animal Welfare.

O'Gara, G. (1996) 'A natural history of the Yellowstone tourist', *Sierra* 81(2): 54–59, 83–85.

Ogutu, Z.A. (2002) 'The impact of ecotourism on livelihood and natural resource management in Eselenkei, Amboseli ecosystem, Kenya', *Land Degeneration and Development* 13: 251–256.

OK, K. (2006) 'Multiple criteria activity selection for ecotourism planning in Iğneada', *Turkish Journal of Agriculture and Forestry* 30: 153–164.

Olesen, R.M. and Schettini, P. (1994) 'From classroom to cornice: training the adventure tourism professional', in A.V. Seaton (ed.), *Tourism: The State of the Art*, Chichester: John Wiley.

Olsen, M.E., Lodwick, D.G. and Dunlop, R.E. (1992) *Viewing the World Ecologically*. Boulder, CO: Westview.

Olsson, P., Folke, C. and Berkes, F. (2004) 'Adaptive comanagement for building resilience in social-ecological systems', *Environmental Management* 34(1): 75–90.

Ontario Ministry of Tourism and Recreation (1985a) *Planning Recreation: A Manual of Principles and Practices*, Toronto: Government of Ontario.

—— (1985b) *Better Planning for Better Recreation*, Toronto: Government of Ontario.

Oppermann, M. (1997) 'Ecolabelling and tourism', TRINET communication, 22 June.

Oprah.com (2013) *3 Ways To Do Good (and Feel Good) on Your Next Vacation*. Available online at http://www.oprah.com/world/Eco-Tourism-Volunteering-Eco-Vacations/2 (accessed 15 September).

Orams, M. (1996) 'Using interpretation to manage nature-based tourism', *Journal of Sustainable Tourism* 4(2): 81–94.

Orams, M.B. (1995) 'Towards a more desirable form of ecotourism', *Tourism Management* 16(1): 3–8.

—— (2002) 'Humpback whales in Tonga: an economic resource for tourism', *Coastal Management* 30: 361–380.

O'Reilly, A.M. (1986) 'Tourism carrying capacity: concepts and issues', *Tourism Management* 7(4): 254–258.

Ormsby, A. and Mannle, K. (2006) 'Ecotourism benefits and the role of local guides at Masoala National Park, Madagascar', *Journal of Sustainable Tourism* 14(3): 271–287.

Ortolano, L. (1984) *Environmental Planning and Decision Making*, New York: John Wiley.

Osland, G.E. and Mackoy, R. (2004) 'Ecolodge performance goals and evaluations', *Journal of Ecotourism* 3(2): 109–126.

Ostrom, E. (1999) 'Coping with tragedies of the commons', *Annual Review Political Science* 2: 493–535.

Page, S. and Dowling, R. (2002) *Ecotourism*, Essex, UK: Pearson Education.

Page, S. and Thorn, K. (2002) 'Towards sustainable tourism development and planning in New Zealand: the public sector response revisited', *Journal of Sustainable Tourism* 10(3): 222–238.

Palacio, V. and McCool, S.F. (1997) 'Identifying ecotourists in Belize through benefit segmentation: a preliminary analysis', *Journal of Sustainable Tourism* 5(3): 234–243.

Palmer, N.J. (2006) 'Economic transition and the struggle for local control in ecotourism development: the case of Kyrgyzstan', *Journal of Ecotourism* 5(1/2): 40–61.

Parker, S. and Khare, A. (2005) 'Understanding success factors for ensuring sustainability in ecotourism development in southern Africa', *Journal of Ecotourism* 4(1): 32–46.

Parsons, E.C.M., Fortuna, C.M., Ritter, F., Rose, N.A., Simmonds, M.P., Weinrich, M., Williams, R. and Panigada, S. (2006) 'Glossary of whale watching terms', *Journal of Cetacean Research and Management* 8 (suppl.): 249–251.

Passmore, J. (1974) *Man's Responsibility for Nature*, New York: Scribner.

Patterson, T., Gulden, T., Cousins, K. and Kraev, E. (2004) 'Integrating environmental, social and economic systems: a dynamic model of tourism in Dominica', *Ecological Modelling* 175: 121–136.

Payne, D. and Dimanche, F. (1996) 'Towards a code of conduct for the tourism industry: an ethics model', *Journal of Business Ethics* 15: 997–1007.

Payne, R.J. and Graham, R. (1993) 'Visitor planning and management in parks and protected areas', in P. Dearden and R. Rollins (eds), *Parks and Protected Areas in Canada*, Toronto: Oxford University Press.

Pearce, D.G. (1985) 'Tourism and environmental research: a review', *International Journal of Environmental Studies* 25: 247–255.

—— (1991) *Tourist Development*, London: Longman.

Pearce, P.L. (1982) *The Social Psychology of Tourist Behaviour*, Oxford: Pergamon Press.

Peattie, K. (1995) *Environmental Marketing Management: Meeting the Green Challenge*, London: Pitman.

Pegas, F., Coghlan, A. and Rocha, V. (2012) 'An exploration of a mini-guide programme: training local children in sea turtle conservation and ecotourism in Brazil', *Journal of Ecotourism* 11(1): 48–55.

Pennisi, L.A., Holland, S.M. and Stein, T.V. (2004) 'Achieving bat conservation through tourism', *Journal of Ecotourism* 3(3): 195–207.

Perkins, H. and Grace, D.A. (2009) 'Ecotourism: supply of nature or tourist demand?', *Journal of Ecotourism* 8(3): 223–236.

Peterson, G. (1996) 'Four corners of human ecology: different paradigms of human relationships with the earth', in B.L. Driver, D. Dustin, T. Baltic, G. Elsner and G. Peterson (eds), *Nature and the Human Spirit: Toward an Expanded Land Management Ethic*, State College, PA: Venture.

Pettersson, R. (2006) 'Ecotourism and indigenous people: positive and negative impacts of Sami tourism', in S. Gössling and J. Hultman (eds), *Ecotourism in Scandinavia: Lessons in Theory and Practice*, Wallingford, UK: CAB International, pp. 166–177.

Pfafflin, G.F. (1987) 'Concern for tourism: European perspective and response', *Annals of Tourism Research* 14(4): 576–579.

Philipsen, J. (1995) 'Nature-based tourism and recreation: environmental change, perception, ideology and practice', in G.J. Ashworth and A.G.J. Dietvorst (eds), *Tourism and Spatial Transformations: Implications for Policy and Planning*, Wallingford, Oxon.: CAB International.

Phillips, A. (1985) 'Socio-economic development in the "national parks" of England and Wales', *Parks* 10(1): 1–5.

Phillips, H. (2003) 'The pleasure seekers', *New Scientist* 11 October. Available online at http://wire-heading.com/pleasure.html (accessed 24 October 2008).

Pinchot, G. (1910) *The Fight for Conservation*, New York: Harcourt Brace.

—— (1947) *Breaking New Ground*, Seattle: University of Washington Press.

Pinhey, T.K. and Grimes, M.D. (1979) 'Outdoor recreation and environmental concern: a re-examination of the Dunlap-Heffernan thesis', *Leisure Sciences* 2: 1–11.

Piper, L.A. and Yeo, M. (2011) 'Ecolabels, ecocertification and ecotourism', *Sustainable Tourism: Socio-Cultural, Environmental and Economics Impact*, 279–294.

Pipitone, C., Mas, J., Ruiz Fernandez, J.A., Whitmarsh, D. and Riggio, S. (2000) 'Cultural and socio-economic impacts of Mediterranean marine protected areas', *Environmental Conservation* 27(2): 110–125.

Pitt, D.G. and Zube, E.H. (1987) 'Management of natural environments', *Handbook of Environmental Psychology* 1: 1009–1041.

Plog, S. (1973) 'Why destination areas rise and fall in popularity', *Cornell Hotel and restaurant Administration Quarterly*, November: 13–16.

Plummer, R., Kulczycki, C. and Stacey, C. (2006) 'How are we working together? A framework to assess collaborative arrangements in nature-based tourism', *Current Issues in Tourism* 9(6): 499–515.

Pollock, N.C. (1971) 'Serengeti', *Geography* 56(2): 145–147.

Poon, A. (1993) *Tourism, Technology and Competitive Strategies*, Wallingford, Oxon.: CAB International.

Pouta, E., Neuvonen, M. and Sievänen, T. (2006) 'Determinants of nature trip expenditures in southern Finland: implications for nature tourism development', *Scandinavian Journal of Hospitality and Tourism* 6(2): 118–135.

Powell, R.B. and Ham, S.H. (2008) 'Can ecotourism interpretation really lead to pro-conservation knowledge, attitudes and behaviour? Evidence from the Galapagos Islands', *Journal of Sustainable Tourism*, 16(4): 467–489.

PPT (2002) *What Is Pro-poor Tourism*. Available online at www.propoortourism.org.uk (accessed 6 June).

Preece, N., van Oosterzee, P. and James, D. (1995) *Biodiversity Conservation and Ecotourism: An Investigation of Linkages, Mutual Benefits and Future Opportunities*, Canberra: Department of the Environment, Sport and Territories.

Pretty, J. and Pimbert, M. (1995) 'Trouble in the Garden of Eden', *Guardian* (London), 13 May, section D.

Price, A.R.G. and Firaq, I. (1996) 'The environmental status of reefs on Maldivian resort islands: a preliminary assessment for tourism planning', *Aquatic Conservation: Marine and Freshwater Ecosystems* 6(2): 93–106.

Priest, S. (1990) 'The semantics of adventure education', in J.C. Miles and S. Priest (eds), *Adventure Education*, State College, PA: Venture Publishing.

Priskin, J. (2003) 'Tourist perceptions of degradation caused by coastal nature-based recreation', *Environmental Management* 32(2): 189–204.

Przeclawski, K. (1996) 'Deontology of tourism', *Progress in Tourism and Hospitality Research* 2: 239–245.

Puhakka, R. and Siikamäki, P. (2012) 'Nature tourists' response to ecolabels in Oulanka PAN Park, Finland', *Journal of Ecotourism* 11(1): 56–73.

Puppim de Oliveira, J.A. (2003) 'Governmental responses to tourism development: three Brazilian case studies', *Tourism Management* 24: 97–110.

—— (2005) 'Tourism as a force for establishing protected areas: the case of Bahia, Brazil', *Journal of Sustainable Tourism* 13(1): 24–49.

Putnam, R.D. (2000) *Bowling Alone: The Collapse and Revival of American Community*, New York: Simon & Schuster.

Putz, F.E. and Pinard, M.A. (1993) 'Reduced-impact logging as a carbon off-set method', *Conservation Biology* 7(4): 755–757.

Puustinen, J., Pouta, E., Neuvonen, M. and Sievänen, T. (2009) 'Visits to national parks and the provision of natural and man-made recreation and tourism resources', *Journal of Ecotourism* 8(1): 18–31.

Quinn, B. (1990) 'The essence of adventure', in J.C. Miles and S. Priest (eds), *Adventure Education*, State College, PA: Venture Publishing.

Rainforest Alliance (2008) 'Rainforest Alliance Welcomes the United Nations Commitment to the establishment of the Sustainable Tourism Stewardship Council'. Available online at http://www.rainforest-alliance.org/newsroom/news/un-stsc (accessed 4 April 2013).

Rajotte, F. (1980) 'Tourism in the Pacific', in F. Rajotte and R. Crocombe (eds), *Pacific Tourism: As Islanders See It*, Suva, Fiji: Institute of Pacific Studies, University of the South Pacific, pp. 1–14.

Ramírez, E. (2005) 'Domestic ecotourism opportunities in Barra de Santiago Estuary, El Salvador', unpublished masters thesis, Lund University, Helsingborg, Sweden.

Rattan, J.K., Eagles, P.F.J. and Mair, H.L. (2012) 'Volunteer tourism: its role in creating conservation awareness', *Journal of Ecotourism* 11(1): 1–15.

Ray, R. (2000) *Management Strategies in Athletic Training*, 2nd edn, Champaign, IL: Human Kinetics.

Redclift, M. (1987) *Sustainable Development: Exploring the Contradictions*, London: Methuen.

Rees, W. and Wackernagel, M. (1996) *Our Ecological Footprint: Reducing Human Impact on the Earth*, Gabriola Island, BC: New Society Publishers.

Reid, H. (2001) 'Contractual national parks and the Makuleke community', *Human Ecology* 29(2): 135–155.

Reid, R., Stone, M. and Whiteley, T. (1995) *Economic Value of Wilderness Protection and Recreation in British Columbia*, British Columbia Ministries of Forests and Environment, and Land and Parks. WP-6-012, December.

Reidenbach, R.E. and Robin, D.P. (1988) 'Some initial steps toward improving the measurement of ethical evaluations of marketing activities', *Journal of Business Ethics* 7: 871–879.

—— (1990) 'Toward the development of a multidimensional scale for improving evaluations of business ethics', *Journal of Business Ethics* 9: 639–653.

Reimer, J.K. and Walter, P. (2013) 'How do you know it when you see it? Community-based ecotourism in the Cardamom Mountains of southwestern Cambodia', *Tourism Mangement* 34: 122–132.

Reingold, L. (1993) 'Identifying the elusive ecotourist', in *Going Green*, a supplement to *Tour and Travel News*, 25 October, pp. 36–37.

Relph, E. (1976) *Place and Placelessness*, New York: Methuen.

REST (2003) *Community-based Tourism: The Sustainability Challenge*. Available online at http://www.iadb.org/int/jpn/ English/support_files/REST-ENG.pdf (accessed 2 March 2004).

Reynolds, L. (1992) 'Montserrat to target "ecotourists"', *Globe and Mail* 22 February, travel section.

Reynolds, P.C. and Braithwaite, D. (2001) 'Towards a conceptual framework for wildlife tourism', *Tourism Management* 22: 31–42.

Rhodes, R.A.W. (1997) *Understanding Governance: Policy Networks, Governance, Reflexivity and Accountability*, Milton Keynes: Open University Press.

Richards, G. and Hall, D. (2000) 'The community: A sustainable concept in tourism development?', in G. Richards and D. Hall (eds), *Tourism and Sustainable Community Development*, New York: Routledge, pp. 1–13.

Richter, L.K. (1991) 'Political issues in tourism policy: a forecast', in D.E. Hawkins and J.R.B. Ritchie (eds), *World Travel and Tourism Review*, Volume 1, London: CAB International, pp. 189–193.

Riedmiller, S. (2001) 'Private sector investment in marine protected areas: experiences of the Chumbe Island coral park in Zanzibar/Tanzania', paper presented at the ICRI–UNEP– CORDIO Regional Workshop for the Indian Ocean, November 26–28, 2001, Maputo, Mozambique.

Rinne, P. and Saastamoinen, O. (2005) 'Local economic role of nature-based tourism in Kuhmo municipality, eastern Finland', *Scandinavian Journal of Hospitality and Tourism* 5(2): 89–101.

Rivera, J. (2004) 'Institutional pressures and voluntary environmental behaviour in developing countries: evidence from the Costa Rican hotel industry', *Society and Natural Resources* 17: 779–797.

Rivera, M.A. and Croes, R. (2010) 'Ecotourists' loyalty: will they tell about the destination or will they return?', *Journal of Ecotourism* 9(2): 85–103.

Rivers, P. (1973) 'Tourist troubles', *New Society* 23: 250.

Robinson, M. (1999) 'Collaboration and cultural consent: refocusing sustainable tourism', *Journal of Sustainable Tourism* 7(3–4): 379–397.

Rocharungsat, P. (2004) 'Community-based ecotourism: the perspectives of three stakeholder groups', in K.A. Smith and C. Schott (eds), *Proceedings of the New Zealand Tourism and Hospitality Research Conference*, Wellington, 8–10 December, pp. 335–347.

Rodger, K., Smith, A., Newsome, D. and Moore, S.A. (2011) 'Developing and testing an assessment framework to guide the sustainability of the marine wildlife tourism industry', *Journal of Ecotourism* 10(2): 149–164.

Roe, D. and Urquhart, P. (2002) *Pro-poor Tourism: Harnessing the World's Largest Industry for the World's Poor*, London: International Institute for Environment and Development. Available online at www.propoortourism.org.uk/Dilys%20IIED%20paper.pdf (accessed 27 June).

Rogerson, C.M. (2006) 'Pro-poor local economic development in South Africa: the role of pro-poor tourism', *Local Environment* 11(1): 37–60.

Rollins, R. (1993) 'Managing the national parks', in P. Dearden and R. Rollins (eds), *Parks and Protected Areas in Canada*, Toronto: Oxford University Press.

Rollins, R. and Dearden, P. (1993) 'Challenges for the future', in P. Dearden and R. Rollins (eds), *Parks and Protected Areas in Canada*, Toronto: Oxford University Press.

Romeril, M. (1985) 'Tourism and the environment: towards a symbiotic relationship', *International Journal of Environmental Studies* 25: 215–218.

Romero, S. (2000) 'Amazon seeks to green economy', *Toronto Star*, 19 June, section E6.

Ross, G.F. (2003) 'Workstress response perceptions among potential employees: the influence of ethics and trust', *Tourism Review* 58(1): 25–33.

Rozzi, R., Massardo, F., Anderson, C.B., Heidinger, K. and Silander, Jr, J.A. (2006) 'Ten principles for biocultural conservation at the southern tip of the Americas: the approach of the Omora Ethnobotanical Park', *Ecology and Society* 11(1): 43. Available online at http://www.ecology and society.org/vol11/iss1/art43.

Ruitenbeek, J. and Cartier, C. (2001) *The Invisible Wand: Adaptive Co-management as an Emergent Strategy in Complex Bio-economic Systems*, Center for International Forestry Research Occasional Paper, no. 34, (online), http://www.cifor.cigar.org.

Russell, D., Bottrill, C. and Meredith, G. (1995) 'International ecolodge survey', in *The Ecolodge Sourcebook for Planners and Managers*, North Bennington, VT: The Ecotourism Society.

Russell, R.V. (1982) *Planning Programs in Recreation*, St Louis, MO: Mosby.

Ryan, C. (1991) *Recreational Tourism: A Social Science Perspective*, New York: Routledge.

—— (1997) 'Ecolabelling and tourism', TRINET communication, 22 June.

—— (2002) 'Tourism and cultural proximity: examples from New Zealand', *Annals of Tourism Research* 29(4): 952–971.

Ryan, C., Hughes, K. and Chirgwin, S. (2000) 'The gaze, spectacle and ecotourism', *Annals of Tourism Research* 27(1): 148–163.

Ryel, R. and Grasse, T. (1991) 'Marketing ecotourism: attracting the elusive ecotourist', in T. Whelan (ed.), *Nature Tourism: Managing for the Environment*, Washington, DC: Island Press.

Saarinen, J. (2005) 'Tourism in the northern wildernesses: Wilderness discourses and the development of nature-based tourism in northern Finland', in C.M. Hall and S. Boyd (eds), *Nature-based Tourism in Peripheral Areas: Development or Disaster?*, Clevedon: Channel View Publications, pp. 36–49.

Saayman, M. and Saayman, A. (2006) 'Estimating the economic contribution of visitor spending in the Kruger National Park to the regional economy', *Journal of Sustainable Tourism* 14(1): 67–81.

Sabatier, P. (1988) 'An advocacy coalition framework for policy change and the role of policy-oriented learning therein', *Policy Sciences* 21: 129–168.

Sadler, B. (1989) 'National parks, wilderness preservation, and native peoples in northern Canada', *Natural Resources Journal* 29: 185–204.

—— (1992) 'Introduction', in S. Hawkes and P. Williams (eds), *The Greening of Tourism: From Principles to Practice*, Centre for Tourism Policy and Research, Simon Fraser University, British Columbia.

Sander, B. (2012) 'The importance of education in ecotourism ventures: lessons from Rara Avis ecolodge, Costa Rica', *International Journal of Sustainable Society* 4(4): 389–404.

Sanderson, H.T. and Koester, S. (2000) 'Co-management of tropical coastal zones: the case of the Soufriere Marine Management Area, St Lucia, WI', *Coastal Management* 28: 87–97.

Sarrasin, B. (2013) 'Ecotourism, poverty and resources management in Ranomafana, Madagascar', *Tourism Geographies* 15(1): 3–24.

Sasidharan, V., Sirakaya, E. and Kerstetter, D. (2002) 'Developing countries and tourism ecolabels', *Tourism Management* 23: 161–174.

Satria, A., Matsuda, Y. and Sano, M. (2006) 'Questioning community based coral reef management systems: case study of *awig-awig* in Gili Indah, Indonesia', *Environment, Development and Sustainability* 8: 99–118.

Saul, J.R. (2001) *On Equilibrium*, Toronto: Penguin/Viking.

Scace, R.C., Grifone, E. and Usher, R. (1992) *Ecotourism in Canada*, Consulting report prepared for the Canadian Environmental Advisory Council, Hull, Quebec: Minister of Supply and Services.

Scalet, C.G. and Adelman, I.R. (1995) 'Accreditation of fisheries and wildlife programs', *Fisheries* 20(2): 8–13.

Schein, E.H. (1985) *Organizational Culture and Leadership*, San Francisco: Jossey-Bass.

Schermerhorn, J.R. (1996) *Management*, 5th edn, New York: John Wiley & Sons.

Schmidley, D.J. (2005) 'What it means to be a naturalist and the future of natural history at American universities', *Journal of Mammology* 86(3):449–456.

Seale, R.G. (1992) 'Aboriginal societies, tourism and conservation: the case of Canada's Northwest Territories', paper presented at the Fourth World Congress on Parks and protected Areas, Caracas, 10–21 February.

Searle, M.S. and Brayley, R.E. (1993) *Leisure Services in Canada*, State College, PA: Venture Publishing, Inc.

Seidl, A., Guiliano, F. and Pratt, L. (2006) 'Cruise tourism and community economic development in Central America and the Caribbean: the case of Costa Rica', *PASOS. Revista de Turismo y Patrimonio Cultural* 4(2): 213–224.

Sekercioglu, C.H. (2002) 'Impacts of birdwatching on human and avian communities', *Environmental Conservation* 29: 282–289.

Serio-Silva, J.C. (2006) 'Las Islas de los Changos (the Monkey Island): the economic impact of ecotourism in the region of Los Tuxtlas, Veracruz, Mexico', *American Journal of Primatology* 68: 499–506.

Sessoms, H.D. (1991) 'Certifying park and recreation: the American experience', *Recreation Canada*, 21–23 July.

Shackleford, P. (1985) 'The World Tourism Organisation: 30 years of commitment to environmental protection', *International Journal of Environmental Studies* 25: 257–263.

Sharpley, R. (2006) 'Ecotourism: a consumption perspective', *Journal of Ecotourism* 5(1/2): 7–22.

Shelby, B. and Heberlein, T. (1986) *Carrying Capacity in Recreational Settings*, Oregon: Oregon State University Press.

Shelby, B. and Vaske, J.J. (1991) 'Using normative data to develop evaluative standards for resource management: a comment on three recent papers', *Journal of Leisure Research* 23(2): 173 187.

Shepherd, N. (2002) 'How ecotourism can go wrong: the cases of SeaCanoe and Siam Safari, Thailand', *Current Issues in Tourism* 5(3–4): 309–318.

Sherman, P.B. and Dixon, J.A. (1991) 'The economics of nature tourism: determining if it pays', in T. Whelan (ed.), *Nature Tourism: Managing for the Environment*, Washington, DC: Island Press.

Shores, J.N. (1992) 'The challenge of ecotourism: a call for higher standards', paper presented at the Fourth World Congress on Parks and Protected Areas, Caracas, 10–21 February.

Short, J.R. (1991) *Imagined Country*, New York: Routledge, Chapman & Hall.

Shundich, S. (1996) 'Ecotourists: dollars, sense and the environment', *Hotels* March: 34–40.

Silberberg, T. (1995) 'Cultural tourism and business opportunities for museums and heritage sites', *Tourism Management* 16(5): 361–365.

Silva, G. and McDill, M.E. (2004) 'Barriers to ecotourism supplier success: a comparison of agency and business perspectives', *Journal of Sustainable Tourism* 12(4): 289–305.

Silverberg, K.E., Backman, S.J. and Backman, K.F. (1996) 'A preliminary investigation into the psychographics of nature-based travelers to the Southeastern United States', *Journal of Travel Research* 35(2): 19–28.

Silverman, G.S. (1992) 'Accrediting undergraduate programs in environmental health science and protection', *Environmental Professional* 14(4): 319–24.

Simpson, J. (1983) 'The discovery of Shetland from *The Pirate* to the tourist board', in D. Withrington (ed.), *Shetland and the Outside World 1469–1969*, New York: Oxford University Press.

Simpson, K. (2001) 'Strategic planning and community involvement as contributors to sustainable tourism development', *Current Issues in Tourism* 4(1): 3–41.

Sims, R.R. (1991) 'The institutionalization of organizational ethics', *Journal of Business Ethics* 10: 493–506.

Singer, M.G. (1986) 'Ethics, science and moral philosophy', in J.P. DeMarco and R.M. Fox (eds), *New Directions in Ethics: The Challenge of Applied Ethics*, New York: Routledge & Kegan Paul.

Singer, P. (2009) *Animal Liberation*, Toronto: Harper Perennial.

Singh, R.B. and Mishra, D.K. (2004) 'Green tourism in mountain regions: reducing vulnerability and promoting people and place centric development in the Himalayas', *Journal of Mountain Science* 1(1): 57–64.

Sinha, B.C., Qureshi, Q., Uniyal, V.K. and Sen, S. (2012) 'Economics of wildlife tourism: contribution to livelihoods of communities around Kanha tiger reserve, India', *Journal of Ecotourism* 11(3): 207–218.

Sirakaya, E. and Uysal, M. (1997) 'Can sanctions and rewards explain conformance behaviour of tour operators with ecotourism guidelines?', *Journal of Sustainable Tourism* 5(4): 322–332.

Siurua, H. (2006) 'Nature above people: Rolston and "fortress" conservation in the south', *Ethics & the Environment* 11(1): 71–96.

Skadberg, X., Jamal, T.B. and Skadberg, A.N. (2005) 'An IT and GIS exploration of web-based nature tourism enterprises in the rural agriculture sector in Texas', *International Journal of Services Technology and Management* 6(2): 120–134.

Slocombe, D.S. and Nelson, J.G. (1992) 'Management issues in hinterland national parks: a human ecological approach', *Natural Areas Journal* 12(4): 206–215.

Smale, B.J.A. and Reid, D.G. (1995) 'Public policy on recreation and leisure in urban Canada', in R. Loreto and T. Price (eds), *Urban Policy Issues: Canadian Perspectives*, 2nd edn, Toronto: Oxford University Press.

Smelser, N.J. (1963) *Theory of Collective Behavior*, New York: Free Press.

Smith, A.J., Scherrer, P. and Dowling, R. (2009) 'Impacts on aboriginal spirituality and culture from tourism in the coastal waterways of the Kimberley region, North West Australia', *Journal of Ecotourism* 8(2): 82–98.

Smith, S.L.J. (1990a) *Dictionary of Concepts in Recreation and Leisure Studies*, New York: Greenwood Press.

—— (1990b) *Tourism Analysis*, New York: Longman.

Smith, V.L. (1989) 'Introduction', in V.L. Smith (ed.), *Hosts and Guests: The Anthropology of Tourism*, Philadelphia: University of Pennsylvania Press.

Snell, M.B. (2001) 'Gorillas in the crossfire', *Sierra* 86(6): 30–34.

Sofield, T. and Li, F.M.S. (2003) 'Processes in formulating an ecotourism policy for nature reserves in Yunnan Province, China', in D.A. Fennell and R.K. Dowling (eds), *Ecotourism Policy and Planning*, Wallingford, Oxon.: CAB International.

Solomon, B., Corey-Luse, C. and Halvorsen, K. (2004) 'The Florida manatee and eco-tourism: toward a safe minimum standard', *Ecological Economics* 50: 101–115.

Sorice, M.G., Shafer, C.S. and Ditton, R.B. (2006) 'Managing endangered species within the use-preservation paradox: the Florida manatee (*Trichechus manatus latirostris*) as a tourism attraction', *Environmental Management* 37(1): 69–83.

Southgate, C.R.J. (2006) 'Ecotourism in Kenya: the vulnerability of communities', *Journal of Ecotourism* 5(1/2): 80–96.

Spinage, C. (1998) 'Social change and conservation misrepresentation in Africa', *Oryx* 32(4): 265–276.

Sproule, K.W. (1996) 'Community-based ecotourism development: identifying partners in the process', paper presented at the Ecotourism Equation: Measuring the Impacts (ISTF) Conference, Yale School of Forestry and Environmental Studies, 12–14 April.

Staiff, R., Bushell, R. and Kennedy, P. (2002) 'Interpretation in national parks: some critical questions', *Journal of Sustainable Tourism* 10(2): 97–113.

Stanford, C.B. (1999) 'Gorilla warfare', *The Sciences* (July–August): 18–23.

Stankey, G.H. and McCool, S.F. (1984) 'Carrying capacity in recreational settings: evolution, appraisal, and application', *Leisure Sciences* 6(4): 453–473.

Stark, J.C. (2002) 'Ethics and ecotourism: connections and conflicts', *Philosophy and Geography* 5(1): 101–113.

Starmer-Smith, C. (2004, 6 November) 'Eco-friendly tourism on the rise', *Daily Telegraph Travel*, 4.

Stebbins, R.A. (1996) 'Cultural tourism as serious leisure', *Annals of Tourism Research* 23(4): 948–950.

Steele, P. (1993) 'The economics of eco-tourism', *In Focus* 9: 7–9.

—— (1995) 'Ecotourism: an economic analysis', *Journal of Sustainable Tourism* 3(1): 29–44.

Stein, T.V., Clark, J.K. and Rickards, J.L. (2003) 'Assessing nature's role in ecotourism development in Florida: perspectives of tourism professionals and government decision-makers', *Journal of Ecotourism* 2(3): 155–172.

Stern, P.C., Dietz, T., Dolsak, N., Ostrom, E. and Stonich, S. (2002) 'Knowledge and questions after 15 years of research', in E. Ostrom, T. Dietz, N. Dolsak, P.C. Stern, S. Stonich and E.U. Weber (eds), *The Drama of the Commons*. Washington, DC: National Academy Press, pp. 445–486.

Stevens, B. (1994) 'An analysis of corporate ethical code studies: "Where do we go from here?"', *Journal of Business Ethics* 13: 63–69.

Stoll, J.R., Ditton, R.B. and Stokes, M.E. (2009) 'Sturgeon viewing as nature tourism: to what extent do participants value their viewing experiences and the resources upon which they depend?', *Journal of Ecotourism* 8(3): 254–268.

Stonehouse, B. (2001) 'Polar environments', in D.B. Weaver (ed.), *The Encyclopedia of Ecotourism*, Wallingford, Oxon.: CAB International.

Strong, P. and Morris, S.R. (2010). 'Grey seal (*Halichoerus grypus*) disturbance, ecotourism and the Pembrokeshire marine code around Ramsey Island', *Journal of Ecotourism* 9(2): 117–132.

Stronza, A. (2001) 'Anthropology of tourism: forging new ground for ecotourism and other alternatives', *Annual Review of Anthropology* 30: 261–283.

—— (2007) 'The economic promise of ecotourism conservation', *Journal of Ecotourism* 6(3): 210–230.

Stronza, A. and Gordillo, J. (2008) 'Community views on ecotourism', *Annals of Tourism Research* 35(2): 448–468.

Sulaiman, Y. (2006) 'Enter lottery for right to visit ecotourism destination?', *ETurboNews*, Wednesday 27 September, p. 1.

Swarbrooke, J. and Horner, S. (1999) *Consumer Behaviour in Tourism*, Oxford: Butterworth–Heinemann.

Swart, S. and Saayman, M. (1997) 'Legislative restrictions on the tourism industry: a South African perspective', *World Leisure and Recreation* 39(1): 24–30.

Sydee, J. and Beder, S. (2006) 'The right way to go? Earth Sanctuaries and market-based conservation', *Capitalism Nature Socialism* 17(1): 83–98.

Tang, T., Zhong, L. and Cheng, S. (2012) 'Tibetan attitudes towards community participation and ecotourism', *Journal of Resources and Ecology* 3(1): 8–15.

Tangley, L. (1988) 'Who's polluting Antarctica?', *BioScience* 38(9): 590–594.

Tao, C.-H., Eagles, P.F.J. and Smith, S.L.J. (2004) 'Profiling Taiwanese ecotourists using a self-definition approach', *Journal of Sustainable Tourism* 12(2): 149–168.

Taylor, J.E., Dyer, G.A., Stewart, M., Yunez-Naude, A. and Ardila, S. (2003) 'The economics of ecotourism: a Galápagos Islands economy-wide perspective', *Economic Development and Cultural Change* 51: 977–997.

Taylor, P.W. (1989) *Respect for Nature: A Theory of Environmental Ethics*, Princeton, NJ: Princeton University Press.

Tepelus, C. (2008) 'Reviewing the IYE and WSSD processes and impacts on the tourism sustainability agenda', *Journal of Ecotourism* 7(1): 77–86.

Texas Parks and Wildlife (1996) *Nature Tourism in the Lone Star State*, Austin, TX: Texas Parks and Wildlife Development.

Thaman, K.H. (2002) 'Shifting sights: the cultural challenge of sustainability', *International Journal of Sustainability in Higher Education* 3(3): 233–242.

The International Ecotourism Society (TIES) (2006) *Fact Sheet: 'Global ecotourism'*, Bennington, VT: Author.

Theerapappisit, P. (2003) 'Mekong tourism development: capital or social mobilization?', *Tourism Recreation Research* 28(1): 47–56.

Theophile, K. (1995) 'The forest as a business: is ecotourism the answer?', *Journal of Forestry* 93(3): 25–27.

Thomlinson, E. and Getz, D. (1996) 'The question of scale in ecotourism: case study of two small ecotour operators in the Mundo Maya region of Central America', *Journal of Sustainable Tourism* 4(4): 183–200.

Thompson, P. (1995) 'The errant e-word: putting ecotourism back on track', *Explore*, 73: 67–72.

Tibbetts, J. (1995–6) 'A walk on the wild side', *Coastal Heritage* 10(3): 3–9.

TIES (2012) 'SEE turtles offers an expenses paid week of volunteering with sea turtle conservation project in Coats Rica'. Available online at http://www.ecotourism.org/news/volunteer (accessed 15 September 2013).

Tilley, F. (1999) 'The gap between the environmental attitudes and the environmental behaviour of small firms', *Business Strategy and the Environment* 8: 238–248.

Timothy, D. (1998) 'Cooperative tourism planning in a developing destination', *Journal of Sustainable Tourism* 6(1): 52–68.

Tims, D. (1996) 'The perspective of outfitters and guides', in B.L. Driver, D. Dustin, T. Baltic, G. Elsner and G. Peterson (eds), *Nature and the Human Spirit: Toward an Expanded Land Management Ethic*, State College, PA: Venture.

Tipa, G. and Welch, R. (2006) 'Comanagement of natural resources: issues of definition from an indigenous community perspective', *Journal of Applied Behavioral Science* 42(3): 373–391.

Tisdell, C. (1995) 'Investment in ecotourism: assessing its economics', *Tourism Economics* 1(4): 375–387.

Tisdell, C. and Wilson, C. (2004) 'Economics of wildlife tourism', in K. Higginbottom (ed.), *Wildlife Tourism: Impacts, Management and Planning*, Altona, VIC: Common Ground.

Tompkins, L. (1996) *A Description of Wilderness Tourism and Outfitting in the Yukon*, Whitehorse: Department of Tourism.

Tongson, E. and Dygico, M. (2004) 'User fee system for marine ecotourism: the Tubbataha reef experience', *Coastal Management* 32: 17–23.

Topelko, K.N. and Dearden, P. (2005) 'The shark watching industry and its potential contribution to shark conservation', *Journal of Ecotourism* 4(2): 108–128.

Tourism Concern (1992) *Beyond the Green Horizon: Principles for Sustainable Tourism*, United Kingdom: World Wildlife Fund.

Tourism Industry Association of Canada (1995) *Code of Ethics and Guidelines for Sustainable Tourism*, Ottawa: Tourism Industry Association of Canada.

Travis, A.S. (1982) 'Physical impacts: trends affecting tourism', *Tourism Management* 3: 256–262.

Tremblay, P. (2001) 'Wildlife tourism consumption: consumptive or non-consumptive?', *International Journal of Tourism Research* 3: 81–86.

—— (2008) 'Wildlife in the landscape: a top end perspective on destination-level wildlife and tourism management', *Journal of Ecotourism* 7(2/3): 179–196.

Tribe, J. (2002) 'Education for ethical tourism action', *Journal of Sustainable Tourism* 10: 309–324.

Tribe, J., Font, X., Griffiths, N., Vickery, R. and Yale, K. (2000) *Environmental Management for Rural Tourism and Recreation*, London: Cassell.

Trivers, R. (1971) 'The evolution of reciprocal altruism', *Quarterly Review of Biology* 46: 35–57.

Tuan, Y.-F. (1971) 'Geography, phenomenology, and the study of human nature', *The Canadian Geographer* 14: 193–201.

Tubb, K.N. (2003) 'An evaluation of the effectiveness of interpretation within Dartmoor National Park in reaching the goals of sustainable tourism development', *Journal of Sustainable Tourism* 11(6), 476–498.

Turnbull, C. (1981) 'East African safari', *Natural History* 90(5): 26–34.

Uhlik, K.S. (1995) 'Partnership step by step: a practical model of partnership formation', *Journal of Park and Recreation Administration* 13(4): 13–24.

UNEP/WTO (2002) *Quebec Declaration on Ecotourism*. Available online at www.uneptie.org/pc/tourism/documents/ecotourism/WESoutcomes/Quebec-Declar-eng.pdf (accessed 31 August).

United Nations Environment Programme Industry and Environment (1995) *Environmental Codes of Conduct for Tourism*, Technical Report no. 29, Paris: UNEP.

UNWTO (2013) 'UNWTO Highlights: 2013 Edition'. Available online at http://dtxtq4w60xqpw.cloudfront.net/sites/all/files/pdf/unwto_highlights13_en_hr_0.pdf (accessed 15 July 2013).

Upchurch, R.S. and Ruhland, S.K. (1995) 'An analysis of ethical work climate and leadership relationship in lodging operations', *Journal of Travel Research* 34(2): 36–42.

Upreti, B.N. (1985) 'The park-people interface in Nepal: problems and new directions', in J.A. McNeely, J.W. Thorsell and S. R. Chalise (eds), *People and Protected Areas in the Hindu Kush–Himalaya*, Kathmandu, Nepal: King Mahendra Trust for Nature Conservation and the International Centre for Integrated Mountain Development.

Urich, P.B., Day, M.J. and Lynagh, F. (2001) 'Policy and practice in karst landscape protection: Bohol, the Philippines', *Geographical Journal* 167(4): 305–323.

Urry, J. (1992) 'The tourist gaze and the "environment" ', *Theory, Culture and Society* 9: 1–26.

U.S. Department of the Interior (1993) *Guiding Principles of Sustainable Design*, Denver, CO: Denver Service Center.

US Forest Service (1994) *Recreation Executive Report*. Washington, DC: Department of Interior, US Forest Service.

Valentine, P.S. (1993) 'Ecotourism and nature conservation: a definition with some recent developments in Micronesia', *Tourism Management* 14(2): 107–115.

Van Amerom, M. (2006) 'African foreign relations as a factor in ecotourism development: the case of South Africa', *Journal of Ecotourism* 5(1/2): 112–127.

van der Merwe, C. (1996) 'How it all began: the man who "coined" ecotourism tells us what it means', *African Wildlife* 50(3): 7–8.

van Liere, K.D. and Noe, F.P. (1981) 'Outdoor recreation and environmental attitudes: further examination of the Dunlap–Heffernan thesis', *Rural Sociology* 46: 501–513.

Veal, A.J. (1992) *Research Methods for Leisure and Tourism: A Practical Guide*, Harlow: Longman Group.

Vespestad, M. and Lindberg, F. (2010) 'Understanding nature-based tourist experiences: An ontological analysis', *Current Issues in Tourism* 14(6): 563–580.

Veverka, J.A. (1994) 'Interpretation as a management tool', *Environmental Interpretation* 9(2): 18–19.

Vincent, V.C. and Thompson, W. (2002) 'Assessing community support and sustainability for ecotourism development', *Journal of Travel Research* 41: 153–160.

Vogeler, I. and DeSouza, A. (1980) *Dialectics of Third World Development*, New Jersey: Allanheld, Osmun.

Vovelli, M., Barnes, J.I. and Humavindu, M. (2006) 'The other side of the ecotourism coin: consumptive tourism in southern Africa', *Journal of Ecotourism* 5(1/2): 62–79.

Waayers, D., Newsome, D. and Lee, D. (2006) 'Observations of non-compliance behaviour by tourists to a voluntary code of conduct: a pilot study of turtle tourism in the Exmouth region, Western Australia', *Journal of Ecotourism* 5(3): 211–222.

Wagar, J.A. (1964) 'The carrying capacity of wildlands for recreation', Society of American Foresters, *Forest Service Monograph* 7: 23.

Waite, G. (1999) 'Naturalizing the "primitive": a critique of marketing Australia's indigenous peoples as "hunter-gatherers" ', *Tourism Geographies* 1(2): 142–163.

Waitt, G., Lane, R. and Head, L. (2003) 'The boundaries of nature tourism', *Annals of Tourism Research* 30(3): 523–545.

Wall, G. (1982) 'Cycles and capacity: incipient theory or conceptual contradiction', *Tourism Management* 3(3): 188–192.

—— (1993) 'International collaboration in the search for sustainable tourism in Bali, Indonesia', *Journal of Sustainable Tourism* 1(1): 38–47.

—— (1994) 'Ecotourism: old wine in new bottles?', *Trends* 31(2): 4–9.

Wall, G. and Wright, C. (1977) *The Environmental Impact of Outdoor Recreation*, Publication Series, no. 11, Department of Geography, University of Waterloo, Ontario.

Wallace, G.N. (1993) 'Wildlands and ecotourism in Latin America', *Journal of Forestry* 91(2): 37–40.

Wallace, G.N. and Pierce, S.M. (1996) 'An evaluation of ecotourism in Amazonas, Brazil', *Annals of Tourism Research* 23(4): 843–873.

Walle, A.H. (1995) 'Business ethics and tourism: from micro to macro perspectives', *Tourism Management* 16(4): 263–268.

Walpole, M.J., Goodwin, J.J. and Ward, K.G.R. (2001) 'Pricing policy for tourism in protected areas: lessons from Komodo National Park, Indonesia', *Conservation Biology* 15: 218–227.

Walter, P.G. (2013) 'Theorising visitor learning in ecotourism', *Journal of Ecotourism* 12(1): 15–32.

Walters, R.D.M. and Samways, M.J. (2001) 'Sustainable dive ecotourism on a South African coral reef', *Biodiversity and Conservation* 10: 2167–2179.

Watkinson, R. (2002) 'Frogs or cassowaries: cooperative marketing with the tourism industry', *Journal of Ecotourism* 1(2/3): 181–188.

Watson, R. (1996) 'Risk management: a plan for safer activities', *Canadian Association for Health, Physical Education, Recreation and Dance Journal*, Spring: 13–17.

Wearing, S. (1994) 'Social and cultural perspectives in training for indigenous ecotourism development', unpublished paper.

—— (1995) 'Professionalisation and accreditation of ecotourism', *Leisure and Recreation* 37(4): 31–36.

Wearing, S. and Neil, J. (1999) *Ecotourism: Impacts, Potential and Possibilities*, Oxford: Butterworth–Heinemann.

Weaver, D.B. (1990) 'Grand Cayman Island and the resort cycle concept', *Journal of Travel Research* 29(2): 9–15.

—— (1991) 'Alternative to mass tourism in Dominica', *Annals of Tourism Research* 18: 414–432.

—— (1993) 'Ecotourism in the small island Caribbean', *GeoJournal* 31: 457–465.

—— (1995) 'Alternative tourism in Monserrat', *Tourism Management* 16(8): 593–604.

—— (1998) *Ecotourism in the Less Developed World*, London: CAB International.

—— (1999) 'Magnitude of ecotourism in Costa Rica and Kenya', *Annals of Tourism Research* 26(4): 792–816.

—— (2001a) 'Ecotourism as mass tourism? Contradiction or reality?', *Cornell Hotel and Restaurant Administration Quarterly* April: 104–112.

—— (2001b) *Ecotourism*, Milton, Queensland: John Wiley & Sons.

—— (2001c) *The Encyclopedia of Ecotourism*, Wallingford, Oxon.: CAB International.

—— (2002a) 'Asian ecotourism: patterns and themes', *Tourism Geographies* 4(2): 153–172.

—— (2002b) 'The evolving concept of ecotourism and its potential impacts', *International Journal of Sustainable Development* 5(3): 251–264.

Weaver, D.B. and Fennell, D.A. (1997) 'The vacation farm industry of Saskatchewan: a profile of operators', *Tourism Management* 18(6): 357–365.

Weaver, D.B. and Lawton, L. (2002) 'Overnight ecotourist market segmentation in the Gold Coast Hinterland of Australia', *Journal of Travel Research*, 40: 270–280.

—— (2007) 'Progress in tourism management twenty years on: the state of contemporary ecotourism research', *Tourism Management* 28: 1168–1179.

Weaver, D.B. and Schluter, R. (2001) 'Latin America and the Caribbean', in D.B. Weaver (ed.), *The Encyclopedia of Ecotourism*, Wallingford, Oxon.: CAB International.

Weaver, G. and Wishard-Lambert, V. (1996) 'Community tourism development: an opportunity for park and recreation departments', *Parks and Recreation* 31(9): 78–83.

Weeden, C. (2001) 'Ethical tourism: an opportunity for competitive advantage?', *Journal of Vacation Marketing* 8(2): 141–153.

Weiler, B. (1993) 'Nature-based tour operators: are they environmentally friendly or are they faking it?', *Tourism Recreation Research* 18(1): 55–60.

Weiler, B. and Davis, D. (1993) 'An exploratory investigation into the roles of the nature-based tour leader', *Tourism Management* 14(2): 91–98.

Weiler, B. and Ham, S. (2001) 'Tour guides and interpretation', in D.B. Weaver (ed.), *The Encyclopedia of Ecotourism*, Wallingford, Oxon.: CAB International.

—— (2002) 'Tour guide training: a model for sustainable capacity building in developing countries', *Journal of Sustainable Tourism* 10(1): 52–69.

Weiler, B. and Richins, H. (1995) 'Extreme, extravagant and elite: a profile of ecotourists on Earthwatch expeditions', *Tourism Recreation Research* 20(1): 29–36.

Weinberg, A., Bellows, S. and Ekster, D. (2002) 'Sustaining ecotourism: insights and implications from two successful case studies', *Society and Natural Resources* 15: 371–380.

Welford, R. and Ytterhus, B. (1998) 'Conditions for the transformation of eco-tourism into sustainable tourism', *European Environment* 8: 193–201.

Welford, R., Ytterhus, B. and Eligh, J. (1999) 'Tourism and sustainable development: an analysis of policy and guidelines for managing provision and consumption', *Sustainable Development* 7(4) 165–177.

Wells, M., Brandon, K. and Hannah, L. (1992) *People and Parks: Linking Protected Area Management with Local Communities*, Washington, DC: World Bank.

Weschler, I.R. (1962) *Issues in Human Relations Training*, Washington, DC: National Training Laboratories.

Western, D. (1993) 'Defining ecotourism', in K. Lindberg and D.E. Hawkins (eds), *Ecotourism: A Guide for Planners and Managers*, North Bennington, VT: The Ecotourism Society.

Western, D. and Thresher, P. (1973) *Development Plans for Amboseli*, Nairobi: World Bank Report.

Weston, S.A. (1996) *Commercial Recreation and Tourism: An Introduction to Business Oriented Recreation*, Toronto: Brown & Benchmark.

Wheeler, M. (1994) 'The emergence of ethics in tourism and hospitality', *Progress in Tourism, Recreation, and Hospitality Management* 6: 46–56.

Wheeller, B. (1994) 'Egotourism, sustainable tourism and the environment: a symbiotic, symbolic or shambolic relationship', in A.V. Seaton (ed.), *Tourism: The State of the Art*, Chichester: John Wiley.

—— (2004) 'The truth? The hole truth. Everything but the truth. Tourism and knowledge: a septic sceptic's perspective', *Current Issues in Tourism* 7(6): 467–477.

White, D. (1993) 'Tourism as economic development for native people living in the shadow of a protected area: a North American case study', *Society and Natural Resources* 6: 339–345.

White, L. Jr (1971) 'The historic roots of our ecologic crisis', in R.M. Irving and G.B. Priddle (eds), *Crisis*, London: Macmillan.

Wickens, E. (2002) 'The sacred and the profane: a tourist typology', *Annals of Tourism Research* 29(3): 834–851.

Wight, P.A. (1993a) 'Sustainable ecotourism: balancing economic, environmental and social goals within an ethical framework', *Journal of Tourism Studies* 4(2): 54–66.

—— (1993b) 'Ecotourism: ethics or eco-sell?', *Journal of Travel Research* 21(3): 3–9.

—— (1995) 'Greening of remote tourism lodges', paper presented at Shaping Tomorrow's North: The Role of Tourism and Recreation, Lakehead University, Thunder Bay, Ontario, 12–15 October.

—— (1996) 'North American ecotourists: market profile and trip characteristics', *Journal of Travel Research* 34(4): 2–10.

Wilcove, D S. and Eisner, T. (2000) 'The impending extinction of natural history', *Chronicle Review, Chronicle of Higher Education* 47(3): B24.

Wilensky, H.L. (1964) 'The professionalization of everyone?', *American Journal of Sociology* 70(2): 137–158.

Wiles, R. and Hall, T. (2005) 'Can interpretive messages change park visitors' views on wildland fire?', *Journal of Interpretation Research* 10 (2): 18–35.

Williacy, S. and Eagles, P.F.J. (1990) *An Analysis of the Federation of Ontario Naturalists' Canadian Nature Tours Programme*, Department of Recreation and Leisure Studies, University of Waterloo, Waterloo, Ontario.

Williams, P.W. (1992) 'A local framework for ecotourism development', *Western Wildlands* 18(3): 14–19.

Wilson, A. (1992) *The Culture of Nature*, Cambridge, MA: Blackwell.

Wilson, C. and Tisdell, C. (2003) 'Conservation and economic benefits of wildlife-based marine tourism: sea turtles and whales as case studies', *Human Dimensions of Wildlife* 8: 49–58.

Wilson, E.O. (1984) *Biophilia*, Cambridge, MA: Harvard University Press.

—— (2002) *The Future of Life*, New York: Vintage Books.

Wilson, M. (1987) 'Nature oriented tourism in Ecuador: assessment of industry structure and development needs', Forestry Private Enterprise Initiative, North Carolina State University, Raleigh, North Carolina, no. 20.

Windsor, R., Baranowski, T., Clark, N. and Cutter, G. (1994) *Evaluation of Health Promotion, Health Education, and Disease Prevention Programs*, Mountain View, CA: Mayfeld.

Winkler, R. (2006) 'Subsistence farming, poaching and ecotourism: social versus communal welfare maximization in wildlife conservation'. Available online at http://www.uni-kiel.de/ifw/konfer/wsumwelt/winkler.pdf (accessed 28 September).

Winpenny, J.T. (1982) 'Issues in the identification and appraisals of tourism projects in developing countries', *Tourism Management* 3(4): 218–221.

Winson, A. (2006) 'Ecotourism and sustainability in Cuba: does socialism make a difference?', *Journal of Sustainable Tourism* 14(1): 6–23.

Wise, R.A. (2004) 'Dopamine, learning and motivation', *Nature Reviews Neuroscience* 5(6): 483–495.

Wollenberg, K.C., Jenkins, R.K.B., Randrianavelona, R., Rampilamanana, R., Ralisata, M., Ramanandraibe, A., Ravoahangimalala, O.R. and Vences, M. (2011) 'On the shoulders of lemurs: pinpointing the ecotouristic potential of Madagascar's unique herpetofauna', *Journal of Ecotourism* 10(2): 101–117.

Wood, M.E. (1991) 'Formulating the Ecotourism Society's Regional Action Plan', in J.A. Kusler (ed.), *Ecotourism and Resource Conservation*. Madison, WI: Madison Publishers, pp. 80–89.

World Commission on Environment and Development (1987) *Our Common Future*, Oxford: Oxford University Press.

Worthen, B.R., Sanders, J.R. and Fitzpatrick, J.L. (1997) *Program Evaluation: Alternative Approaches and Practical Guidelines*, New York: Longman.

Wright, J.R. (1983) *Urban Parks in Ontario Part I: Origins to 1860*, Toronto: Ministry of Tourism and Recreation.

—— (1987) 'The university and the recreation profession', *Recreation Canada* 45(3): 14–18.

Wunder, S. (2000) 'Ecotourism and economic incentives: an empirical approach', *Ecological Economics* 32: 465–479.

Wyatt, S. (1997) 'Dialogue, reflection, and community', *Journal of Experiential Education* 20(2): 80–85.

Yaman, A.R. and Mohd, A. (2004) 'Community-based ecotourism: a new proposition for sustainable development and environmental conservation in Malaysia', *Journal of Applied Sciences* 4(4): 583–589.

Yaman, H.R. (2003) 'Skinner's naturalism as a paradigm for teaching business ethics: a discussion from tourism', *Teaching Business Ethics* 7: 107–122.

Yee, J.G. (1992) *Ecotourism Market Survey: A Survey of North American Ecotourism Tour Operators*, San Francisco: PATA.

Yeoman, I., Munro, C. and McMahon-Beattie, U. (2006) 'Tomorrow's: world, consumer and tourist', *Journal of Vacation Marketing* 12(2): 174–190.

Yi-fong, C. (2012) 'The indigenous ecotourism and social development in Taroko National Park area and San-Chan tribe, Taiwan', *GeoJournal* 77: 805–815.

Young, B. (1983) 'Touristization of a traditional Maltese fishing-farming village: a general model', *Tourism Management* 4(1): 35–41.

Young, E.H. (1999) 'Balancing conservation with development in small-scale fisheries: is ecotourism an empty promise?', *Human Ecology* 27(4): 581–620.

Young, F.J.L. (1964) *The Contracting Out of Work: Canadian and USA Industrial Relations Experience*, Industrial Relations Centre, Queen's University, Kingston, Ontario.

Yudina, O. and Fennell, D.A. (2013) 'Ecofeminism in the tourism context: a discussion of the use of other-than-human animals as food in tourism', *Tourism Recreation Research* 38(1): 55–69.

Zeppel, H. (2006) *Indigenous Ecotourism: Sustainable Development and Management*, Wallingford, Oxon.: CABI.

Zeppel, H. and Muloin, S. (2008) 'Aboriginal interpretation in Australian wildlife tourism', *Journal of Ecotourism* 7(2/3): 116–136.

Zhuang, H., Lassoie, J.P. and Wolf, S.A. (2011) 'Ecotourism development in China: prospects for expanded roles for non-governmental organisations', *Journal of Ecotourism* 10(1): 46–63.

Ziffer, K. (1989) *Ecotourism: The Uneasy Alliance*, Working Paper, no. 1, Washington, DC: Conservation International.

Zimmerman, E.W. (1951) *World Resources and Industries*, New York: Harper.

Zwirn, M., Pinsky, M. and Rahr, G. (2005) 'Angling ecotourism: issues, guidelines, and experience from Kamchatka', *Journal of Ecotourism* 4(1): 16–31.

Index

Note: page numbers in *italic* refer to Figures; page numbers in **bold** refer to Tables.